A GUIDE TO FIELD IDENTIFICATION

SEASHELLS

OF NORTH AMERICA

by R. TUCKER ABBOTT

Illustrated by GEORGE F. SANDSTRÖM

Under the Editorship of
HERBERT S. ZIM

GOLDEN PRESS • NEW YORK
Western Publishing Company, Inc.
Racine, Wisconsin

FOREWORD

The Golden Field Guides are designed to aid in the identification of species in the field. Like its companion volumes, BIRDS OF NORTH AMERICA and TREES OF NORTH AMERICA, this guide introduces the reader to a large and diverse group of animals. The author, Dr. R. Tucker Abbott, who holds the du Pont Chair of Malacology at the Delaware Museum of Natural History, has done much to popularize malacology, the study of mollusks. In conjunction with an increasing interest in life found in the seas, this book should serve constantly as a stimulus and as a stepping-stone to more advanced studies.

H.S.Z.

PREFACE

SEASHELLS OF NORTH AMERICA introduces the world of marine mollusks and provides identification of the common shells of the Atlantic and Pacific coasts of North America. Some 850 species, most of them illustrated in color, are described or mentioned, with emphasis on habitats, habits, and identifying characters.

Although the identification of our common mollusks in the field or laboratory is emphasized, an attempt is also made in this book to summarize the salient biological characteristics of many of the better known species. There is a great difference between the collecting of dead shells and the study of the living mollusks. To understand how snails and clams live and fight for survival is to know many of the principles of marine biology.

The author would like to thank many colleagues and shell enthusiasts for supplying information and specimens necessary for the completion of this book. Special thanks are tendered to Mrs. Sally D. Kaicher, who wrote first-draft descriptions of many species and helped in outlining various family introductions. Dr. William K. Emerson and Mr. William E. Old, Jr., both of the American Museum of Natural History in New York, very kindly read the manuscript and made many valuable suggestions. The artist is indebted to the Academy of Natural Sciences of Philadelphia for making available its magnificent collections of mollusks.

R.T.A.

 Produced in the U.S.A. by Western Publishing Company, Inc. Published by Golden Press, New York, N.Y. Library of Congress Catalog Card Number: 68-10083. ISBN 0-307-13657-4

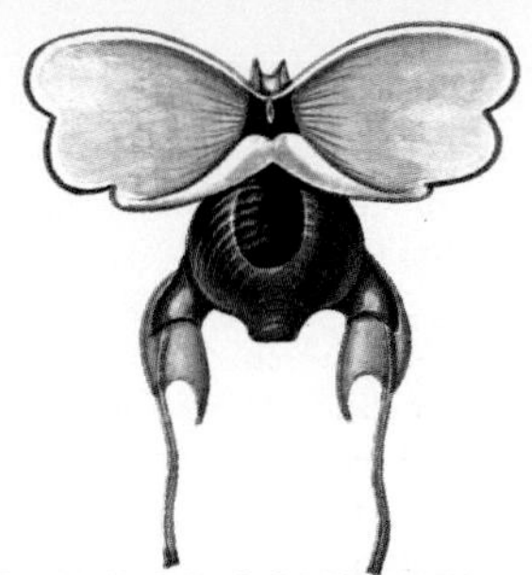

TABLE OF CONTENTS

HOW TO USE THIS BOOK

The basic method of using this guide is simple. You look at a shell and then at the pictures in the book until you are satisfied as to the shell's identity. Reading the text gives further confirmation and additional information.

The guide becomes even more valuable, however, if you peruse its pages in spare moments, not so much to memorize the names or identities of particular shells but to become increasingly conscious of the characteristics of families and genera. Gradually, and with a reasonable chance of success, you can quickly pigeonhole a specimen as a cowrie, a cone, a scallop, or a whelk. The families are arranged in a natural sequence, from the most primitive to the most advanced. Museum research collections and nearly all reference books use this order. Familiarization with this system will send you most quickly to the section of the book in which detailed scrutiny of the specimen, the illustration, and the text are likely to yield the name of the species.

Note that all species have their own geographical distribution patterns. This will assist in narrowing down the identity. A species may have a very spotty distribution within a given range, of course, and its abundance may vary greatly, depending upon local conditions. The depth ranges given in this book are approximations in many cases, and, again, local conditions or accidental migration can place a shell far below or above its stated depth limits.

A number of species have been introduced from Europe, or from one coast of our continent to the other, usually during experimental attempts to naturalize foreign edible oysters or clams. Not infrequently Atlantic species, such as the Channeled Whelk *(Busycon)* and the False Angel Wing *(Petricola)* turn up along the California coast. The identifier must not be unduly influenced by the supposed natural range of such roaming species.

In the identification of mollusks you should learn to expect some degree of natural variation in color, pattern, size, and form. Some species are more variable than others. This is often indicated in the description. The illustrations also may show several color forms. In groups that show more than the usual variability, identification should be made with caution. The section on biology will give you a general understanding of the growth and development of shells so that you will be aware of these factors in checking a specimen you have in hand. One of the chief sources of confusion is the differences in shape between young and adult snail shells, since the latter may sometimes have thickened or flaring edges. Where possible we have illustrated unusual cases.

Problems begin when, after reasonable effort, you cannot identify a specimen. First of all, bear in mind the limitations of this book. Some 6,000 species of marine mollusks belonging to about 170 families are found in American waters, and since a field guide cannot be a definitive reference, a selection had to be made. Priority was given to those species found in the intertidal zone and in shallow waters of the United States and Canada. Most such species are included in this guide; the exceptions are those that are exceedingly rare and those that are very minute.

Deeper waters are less well explored, and material is usually not available to the amateur. Here, the larger, more attractive, and more interesting species have been selected and illustrated. The text often refers to characteristics of similar species, and introductions give the number of species in the genus or family so that you will have an idea about the percentage covered in this guide.

As a further step, other guides and pertinent references are listed in the bibliography, on pages 270–271. These, in turn, have extensive bibliographies. Some of these books cover a more limited area and hence treat the species in greater detail. In using other references do not be surprised at some discrepancies in scientific names used from book to book. Despite two centuries of research, there are great gaps in the biological knowledge of mollusks. The field is encumbered with outdated descriptions of invalid species. The reevaluation of American species still continues, and this means that even the current names and classifications, used in this book, will undoubtedly change. Studies of the relationships among various orders, families, and genera suggest the possibility of name changes on a larger scale as the framework of mollusk evolution becomes clarified. The amateur must recognize that changes of this sort are taking place. Thus, there may be many good reasons why he fails to identify a specimen.

When self-help fails, the amateur may sometimes find it possible to obtain professional assistance. Experts at universities and the curators at museums are often willing to aid a serious amateur with his problems. Remember that these scientists are fully occupied with their professional duties and other tasks; it is wise to inquire by telephone or by letter if the expert can help you. Make your request a modest one, involving not more than a few species. It is a good idea to have each species accompanied by a small temporary label bearing the locality and any other data. If material is shipped, pack it securely in crumpled newspaper and in a cardboard carton. In a letter attached outside you can enclose the return postage and a shipping label.

In many larger cities, especially along the coasts, there are shell clubs whose members are glad to help and at whose meetings there are opportunities to exchange specimens and information.

In using your own copy of this book, don't hesitate to add marginal notes if your study includes new information about the living animals. You will probably write more field notes than the margins provide space for, but these extended notes will give you more satisfaction than anything in print.

WHAT IS A MOLLUSK?

Over a million species of invertebrate animals have been classified into ten or more major groups, or phyla. One of the oldest and most successful groups, with about 100,000 living species, is the phylum Mollusca. The mollusks include such familiar forms as clams, oysters, snails, and conchs. Although they may be superficially classified as shelled animals, the mollusks as a group are so diverse that a simple definition is difficult. Mollusks were originally marine animals. Present-day species have become adapted to life in a great number of different ecological niches. They live in mud, sand, or rocks and on land as well as in fresh and salt waters. A few are parasitic. Other invertebrates, such as crabs and sand dollars, produce shells of a different origin.

The mollusk body is divided into three regions: the head, the foot, and the visceral mass. The head is well-developed in the snails and squids, which have eyes, tentacles, and a well-formed mouth. In the clams, or bivalves, the head does not appear as a distinct structure. Associated with the body, though sometimes classified as a separate basic structure in the mollusks, is a thin sheet of tissue—the mantle—which secretes the limy shell characteristic of most mollusks. The mantle cavity between the mantle and the visceral mass contains the gills and the waste ducts from the intestines and the kidneys.

Inside the mouth cavity, mollusks usually have a ribbonlike set of teeth known as the radula. These rasping teeth vary in number and size. The radula is absent in the pelecypods, or bivalves.

The visceral mass contains the internal organs of circulation, respiration, excretion, digestion, and reproduction. The heart has two chambers and circulates colorless or blue blood. The paired kidneys, with tubes, open into the mantle cavity. The digestive tube begins at the mouth, continues through a stomach, which often has dead-end sacs, then through a long intestine that also terminates in the mantle cavity. In mollusks the sexes are usually separate, but there are many hermaphroditic species with both male and female sex organs. In most mollusks the fertilized egg develops into a trochophore larval stage that is free-swimming and bears bands of minute cilia. This is followed by a more advanced larval stage, the veliger in most groups (p. 16).

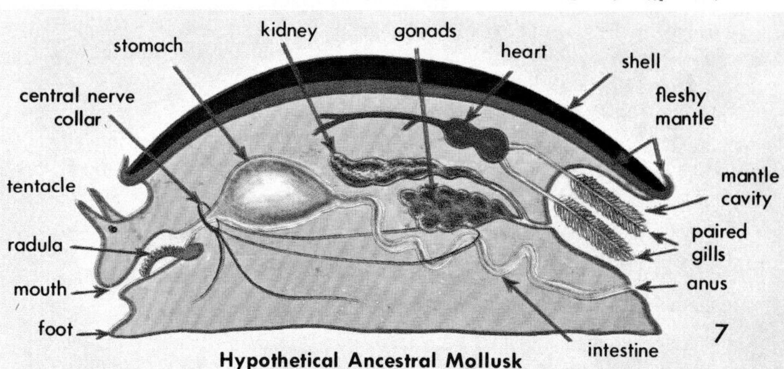

Hypothetical Ancestral Mollusk

MAJOR GROUPS OF MOLLUSKS

The phylum Mollusca is divided into six classes on the basis of major differences in such anatomical features as the foot, shell, mouth parts, and breathing organs. The various classes represent widely differing degrees of evolution and specialization, as well as numbers of species. There are representatives of all the classes in the oceans, but only two have invaded fresh waters, namely the bivalves and univalves. The latter are also found on land.

Class MONOPLACOPHORA, containing the primitive gastroverms, is represented by only five living deep-sea species, although many fossil species are known. The shells are spoon-shaped, resembling limpets, but on the inside there are five or six pairs of muscle scars. The internal organs and gills are paired and each arranged in a separate segment. There are no eyes or tentacles. A row of radular teeth is in the pharynx. The anus is at the posterior end.

Class AMPHINEURA, or chitons, are rock-dwelling marine mollusks having eight shelly plates that arise embryologically from a single shell gland. The rather primitive animal is encircled by a leathery border, the girdle, which holds the shelly plates in place. Underneath is a large, broad foot and a head that lacks eyes and tentacles. There is a well-developed radula. Sexes are separate. Some species brood their young. Most of the 500 or so species are herbivorous, but some are carnivorous.

Class GASTROPODA is the largest class of mollusks, with about 80,000 living species. Univalves usually have a single, coiled or cap-shaped shell. A few lack shells. Most have a radula, well-developed tentacles, and a pair of eyes. Many produce an operculum on the back of the foot; this serves to seal the opening of the shell. In all gastropods, the late embryonic stage undergoes a peculiar twisting of 180°, called "torsion," so that many of the posterior internal organs face forward into the mantle cavity.

The univalves are commonly divided into three subclasses. The **Prosobranchia** snails are the most numerous and are mainly aquatic. They have gills in the mantle cavity. There are two orders: the herbivorous, marine Archeogastropoda, such as the top and turban snails, which have numerous radular teeth, and the omnivorous fresh- and salt-water Caenogastropoda, such as the conchs and mud snails, which have relatively few radular teeth. Snails of the subclass **Opisthobranchia** are marine snails, such as the sea hares and bubble snails, which commonly have a much-reduced shell. The gills are generally posterior to the heart. Male and female reproductive organs are found in each individual. Most are herbivorous, but many are parasitic on clams and other sea creatures. The subclass **Pulmonata,** not treated in this book, includes the land snails, garden slugs, and pond snails, all of which have a modified "lung" and usually lack an operculum. All are hermaphroditic and feed mainly on plants.

GASTROVERMS

Class **Monoplacophora**

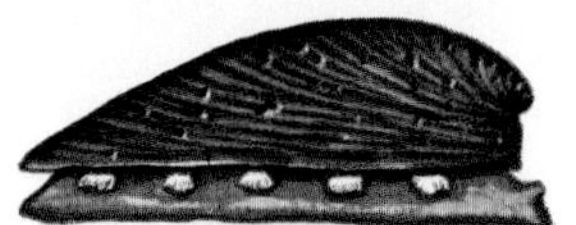

Gastroverm
Neopilina

CHITONS

Class **Amphineura**

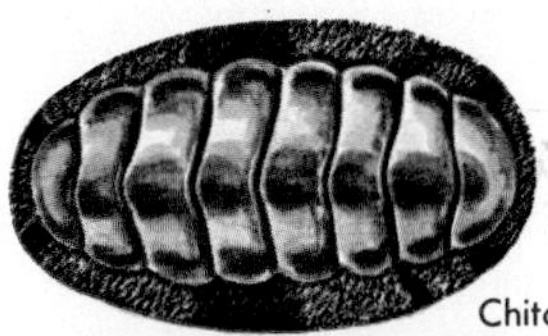

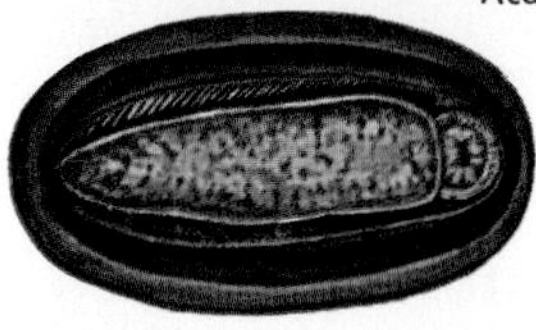

Chiton
Acanthopleura

UNIVALVES Class **Gastropoda**

Subclass **Prosobranchia**

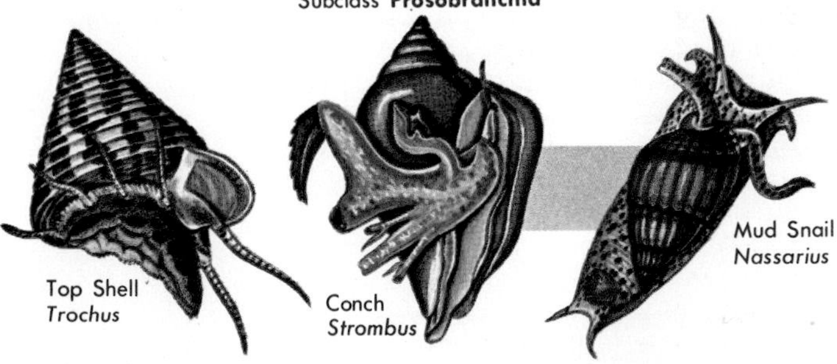

Top Shell
Trochus

Conch
Strombus

Mud Snail
Nassarius

Order Archeogastropoda

Order Caenogastropoda

Subclass **Opisthobranchia**

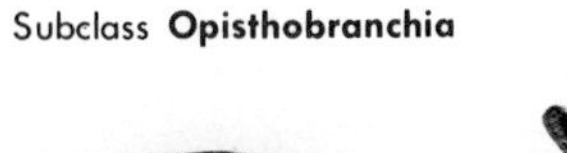

Sea Hare
Aplysia

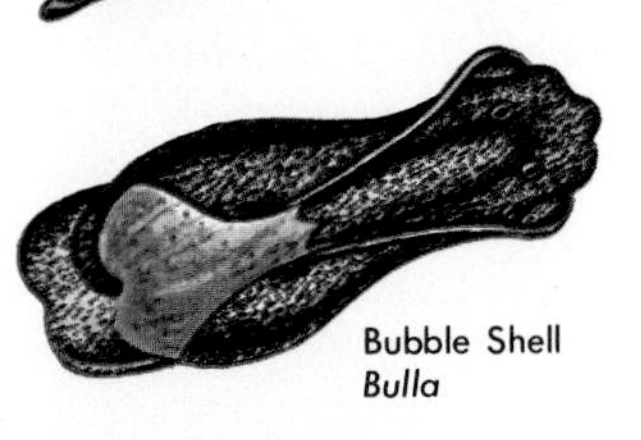

Bubble Shell
Bulla

Subclass **Pulmonata**

Pond Snail
Lymnaea

Garden Snail
Helix

The class PELECYPODA, containing the bivalves, is the second largest group of mollusks and has about 20,000 living species. Most are marine, but many live in fresh water. The soft parts are enclosed within two hinged shells, or valves, hence the term "bivalve" for this group. The valves are pulled together by one or two large adductor muscles and are kept ajar by a chitinous, elastic ligament. Interlocking teeth in the hinge of each valve add strength to the connection and prevent the valves from slipping. The foot is well-developed in clams that burrow, but it is small in attached forms, such as the oysters. Some groups, like the mussels and pen shells, spin a byssus, a clump of chitinous threads that serves as an anchor. Bivalves do not have a head or radula. Eyes sometimes occur along the edge of the mantle, as in the scallops and thorny oysters.

The posterior section of the mantle is often modified into two tubular extensions, the siphons, one of which brings water and algal food into the mantle cavity, and the other of which expels wastes. The siphons are long in species that burrow deeply into sand, rock, or wood. Food particles stick to mucus that is transported by many hairlike cilia on the gills to a groove near the mouth. A pair of fleshy pads, the palpi, then push the food into the mouth.

In most bivalves the sexes are in separate individuals, although scallops and others may be hermaphroditic. In some oysters, a female phase is later replaced by a male phase. Eggs and sperm of bivalves are shed into the open sea water. The developing young are free-swimming for many days or weeks before settling to the bottom. In fresh-water mussels, the larvae, called glochidia, attach themselves to the gills of fish and suck the blood of the host until they are ready to drop to the bottom as young mussels.

Most bivalves move freely through the sand or mud substrate, but many, such as the oysters and jingles, are attached permanently to rocks or wood. Many bore into wood and rocks.

Both the gills and shell hinges have been used in the higher classification of the bivalves, but no system has been universally accepted. There are four main types of gills, as illustrated on the opposite page. The *protobranch* gills, found in the order Protobranchia, are flat and platelike. They are regarded as the most primitive. The *filibranch* gills, found in arks, mussels, oysters, and scallops (order Filibranchia), are long curtains folded back against themselves. The *eulamellibranch* gills, in cockles and clams (order Eulamellibranchia), are similar to filibranch gills except that the curtains are united by cross-channels. The *septibranch* gills, in dipper clams (order Septibranchia), are mere slits in the thin wall that closes off the respiratory chambers. The Palaeoconcha, containing the awning clams, have filibranch-like gills, but this order is considered unique because of its toothless hinge, lack of siphons, extensive periostracal outer covering, and flat-ended foot.

BIVALVES Class **Pelecypoda**

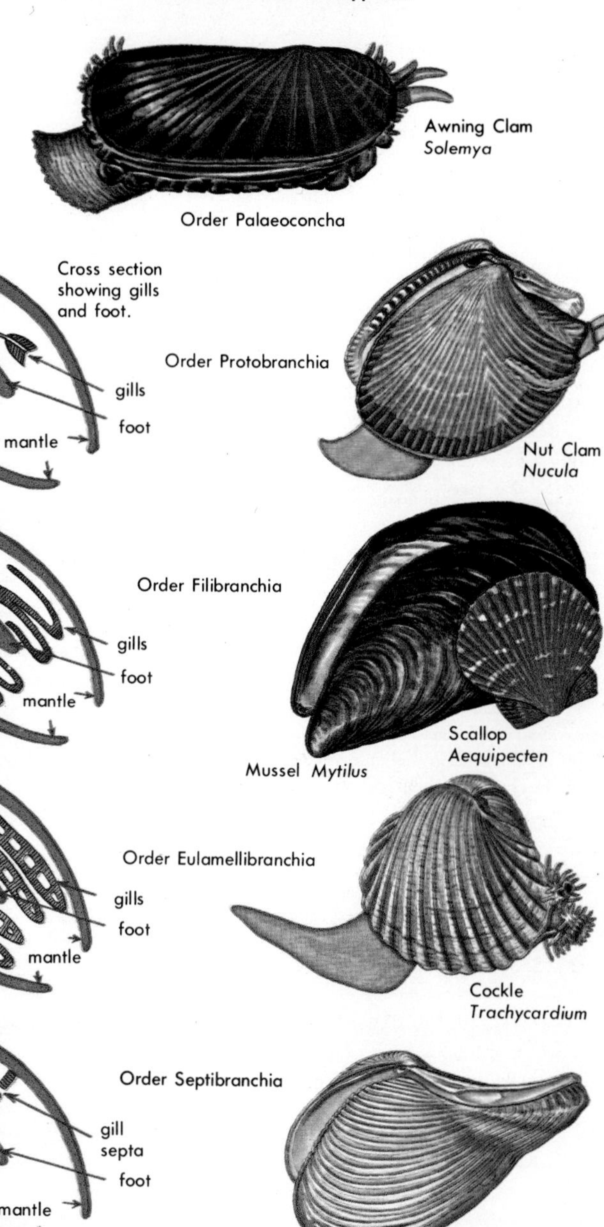

Cross section showing gills and foot.

The class SCAPHOPODA contains the tusk, or tooth, shells. These mollusks live in marine waters buried in mud, with the narrow posterior end of their shell protruding into the water. The shell, from 0.5 to 5 inches long, is tubular and open at both ends. A conical foot projects from the larger posterior end. Numerous, small, tentacle-like filaments, the captacula, capture foraminifera and other small organisms and pass them to the mouth, within which is a radula. No true head, eyes, or gills are present. Water passes in and out of the posterior end, with respiration taking place through the walls of the mantle. There is no heart or pericardium. The blood circulates through contracting sinuses. The sexes are separate, and the sperm or eggs are passed into the sea from two small kidney openings. Most of the 300 living species live at depths of from 10 to 100 feet, but some exist at a depth of 3 miles.

The class CEPHALOPODA, which is entirely marine, includes the most active of the mollusks and, indeed, of all the invertebrates. One species of squid is the largest known living invertebrate. The squids, octopuses, and nautiluses have a cartilaginous brain case and a well-developed nervous system that permits great speed, strength, and alertness. The large eyes reach a perfection found nowhere else in the invertebrates. A series of prehensile tentacles surround the mouth, which is armed with a parrotlike, chitinous beak and, within the pharynx, with a strong set of radulae. Most of the 400 species are predatory carnivores. The sexes are separate, and the heavily yolked eggs are laid in grapelike festoons of jelly material. There is no trochophore nor is there a veliger larval stage, as there is in other mollusks. Many forms have the ability to squirt out clouds of brown or purple ink, evidently as a defense mechanism. Some species display vivid and rapid changes of color by the muscular control of chromatophore pigment cells. Propulsion is accomplished by squirting water from a funnel-shaped fold of the mantle edge. Deep sea species bear luminescent organs of various colors, and the ink of some may give off a slight glow. In most species, when mating takes place, the male transfers bundles of sperm to the female by means of a modified arm called the hectocotylus.

The shells of the cephalopods are very diverse. In the subclass **Nautiloidea,** represented today by only one family in the Pacific, the shell is large, external, and divided into numerous gas-filled chambers. In the subclass **Coleoidea,** which includes the squids and octopuses, the shell may consist of an internal, chalky cuttlefish "bone," or a thin, flat, rodlike support, as is found in squids. The octopus does not have a shell. The white, parchment-like "shell" of the paper nautilus, or *Argonauta,* is actually an egg case secreted by two flat arms of the female. The *Spirula* deep-sea squid has a chambered, coiled shell resembling a miniature ram's horn.

TUSK SHELLS Class Scaphopoda

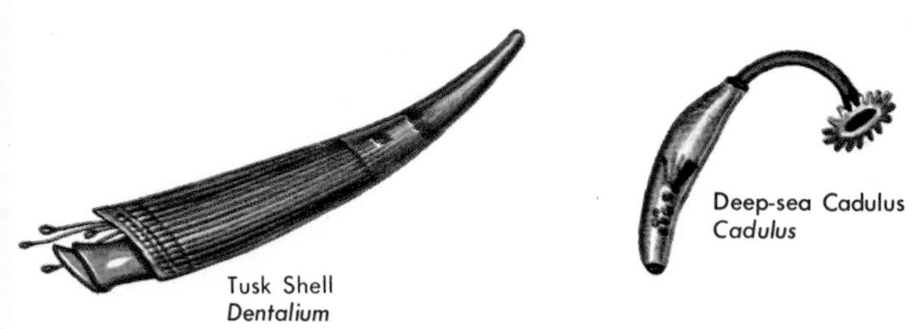

Tusk Shell
Dentalium

Deep-sea Cadulus
Cadulus

SQUIDS AND OCTOPUSES
Class Cephalopoda

Subclass Coleoidea
Order Decapoda

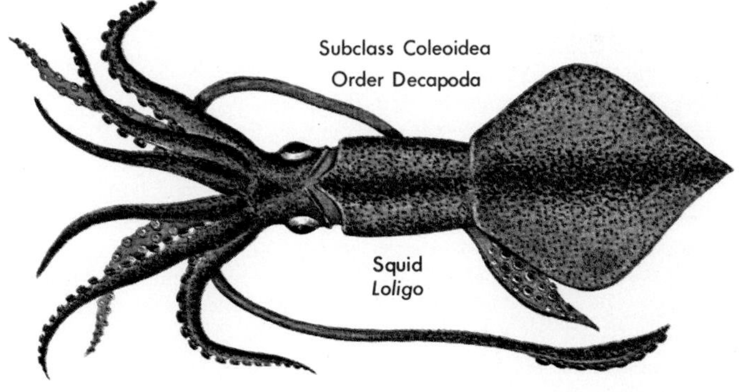

Squid
Loligo

Subclass Coleoidea
Order Octopoda

Subclass Nautiloidea

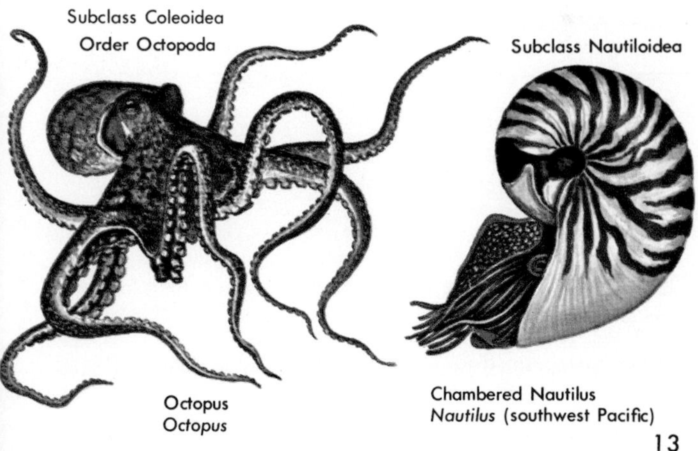

Octopus
Octopus

Chambered Nautilus
Nautilus (southwest Pacific)

EVOLUTIONARY HISTORY

The origin and the early history of the mollusks are shrouded in mystery because not many fossil remains are known from the Pre-Cambrian period, when this phylum probably came into being, more than 600 million years ago. By examining the numerous fossils of the Cambrian and later periods, and by making comparative studies of the anatomy and embryology of living forms, reasonable deductions may be drawn. The immediate ancestors of the first mollusks were probably simple, shell-less animals having segments like those still found in today's annelid worms and arthropods. The development of the mantle and the production of a protective calcareous shell set the stage for the evolution of the greatly diverse classes of mollusks.

The univalves and bivalves appeared in the Early Cambrian, about 600 million years ago. Evidently, it took some of the bivalves another 200 million years to enter fresh water, and it took certain univalves at least 300 million years to evolve into land and fresh-water forms. The chitons and tusk shells appeared fairly soon after the univalves, but the production of new forms and attempts to try new habitats were largely unsuccessful. The gastroverms, represented today by only five known species of *Neopilina,* are the closest known forms to the ancestral mollusks, but from the Cambrian to Recent times they have varied little and given forth few species.

The cephalopods have a long, well-documented, and interesting history. The now extinct ammonoid subclass of huge, coiled shells first appeared in the Lower Devonian, some 400 million years ago. They flourished in great numbers until the Cretaceous, when suddenly and mysteriously they all vanished. The nautiloid subclass first came on the scene as early as the Upper Cambrian and abounded in the seas for about 200 million years. Gradually they waned in number and today they are represented only by four or five living species of Chambered Nautilus in the southwest Pacific Ocean. Most of today's cephalopods, such as the squids and octopuses, belong to the subclass of two-gilled Coleoidea. Perhaps because many were shell-less, their remains in the fossil records do not appear until the late Carboniferous, some 265 million years ago.

Most of today's genera of marine univalves and bivalves blossomed into dominance about 65 million years ago when great geologic and biologic changes heralded the end of the Cretaceous period. With the advent of the Tertiary period, the ammonoids died out, many primitive families of univalves disappeared, and such genera as *Conus, Cassis, Strombus, Venus,* and *Ensis* came into being. Some groups, like the cones, exploded into hundreds of species, while others, like the *Strombus* conchs, produced but few new kinds. In the last million years the land and fresh-water mollusks have evolved very rapidly, but the marine forms seem to be decreasing in the number of species.

GEOLOGIC HISTORY OF THE PHYLUM *MOLLUSCA*

Major Groups of Mollusks

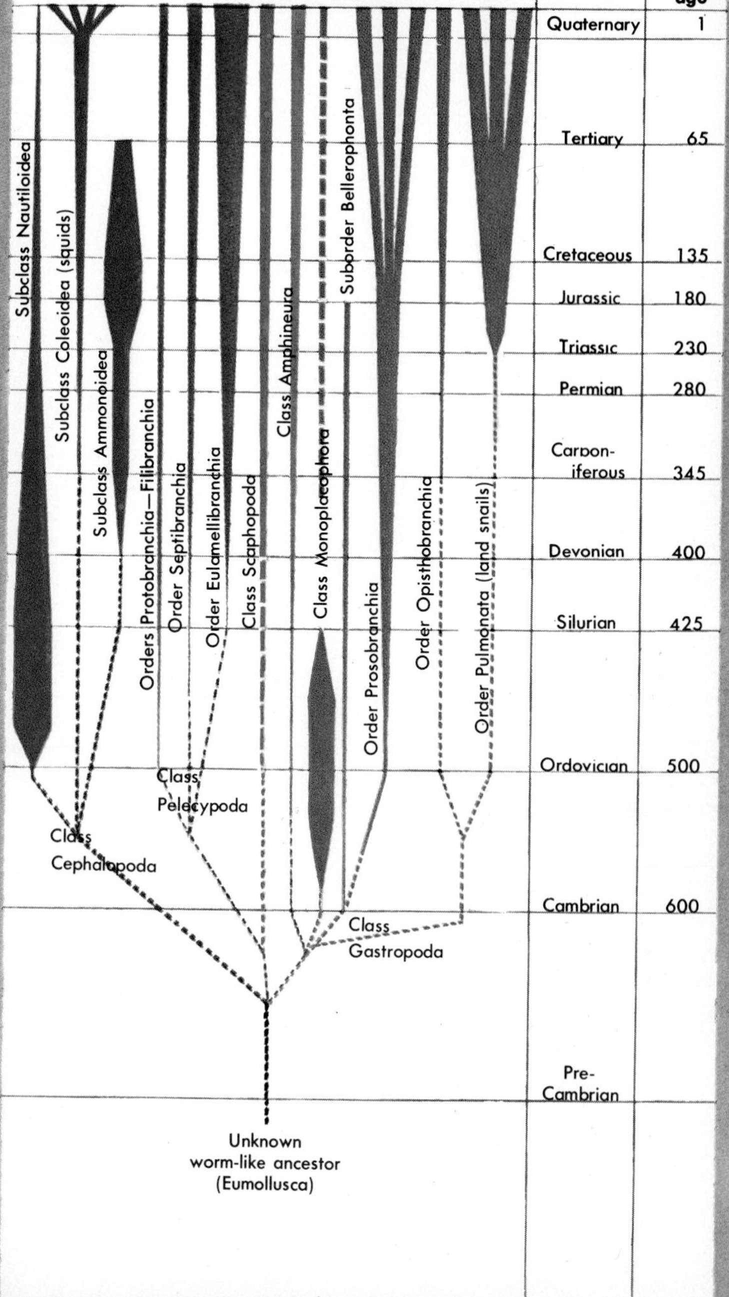

Subclass Nautiloidea
Subclass Coleoidea (squids)
Subclass Ammonoidea
Class Cephalopoda
Orders Protobranchia—Filibranchia
Order Septibranchia
Order Eulamellibranchia
Class Pelecypoda
Class Scaphopoda
Class Amphineura
Class Monoplacophora
Suborder Bellerophonta
Order Prosobranchia
Order Opisthobranchia
Order Pulmonata (land snails)
Class Gastropoda

Unknown
worm-like ancestor
(Eumollusca)

Geologic Period	Millions of years ago
Quaternary	1
Tertiary	65
Cretaceous	135
Jurassic	180
Triassic	230
Permian	280
Carbon-iferous	345
Devonian	400
Silurian	425
Ordovician	500
Cambrian	600
Pre-Cambrian	

REPRODUCTION

The manner of sexual reproduction and larval development varies considerably among the various classes and families of mollusks. In all chitons and tusk shells and in most cephalopods and bivalves the animals have separate sexes. In some bivalves, such as the scallops and cockles, and in many gastropods, including all of the pulmonate snails, the individuals are hermaphroditic, each with functional male and female organs. In some fresh-water snails, the eggs may develop parthenogenetically without being fertilized by sperm.

Sperm or eggs produced in the gonads pass down a separate duct, which exits either into the kidney tube or directly into the mantle cavity. The eggs may be fertilized in the open sea water or within the oviduct of the female. In some families of gastropods there are two types of sperm—the very small, numerous normal sperms which fertilize the eggs and the large spermatozeugma, which is a swimming, undulating carrier of the small sperm. The latter type is found in such genera as *Epitonium, Janthina,* and *Cerithiopsis,* which lack a penis and require the carrier-sperm to gain entrance into the oviduct.

In the cephalopods, a torpedo-shaped tube of chitin containing a dense mass of sperm is plucked by the male from his own genital opening by a specially modified arm, known as the hectocotylus. These spermatophores are then placed inside the mantle cavity of the female, where they eventually work their way into the oviduct to fertilize the eggs.

Some snails, particularly the land forms, practice a simple form of courtship in which they entwine themselves before pairing. A calcareous shaft, or "love dart," is exchanged by the partners in order to stimulate sexual behavior. In some snails the penis bears numerous hooks. In the hermaphroditic opisthobranchs, such as *Bulla* and *Hydatina,* chains or rings of mating individuals may be found at certain seasons.

Fertilization

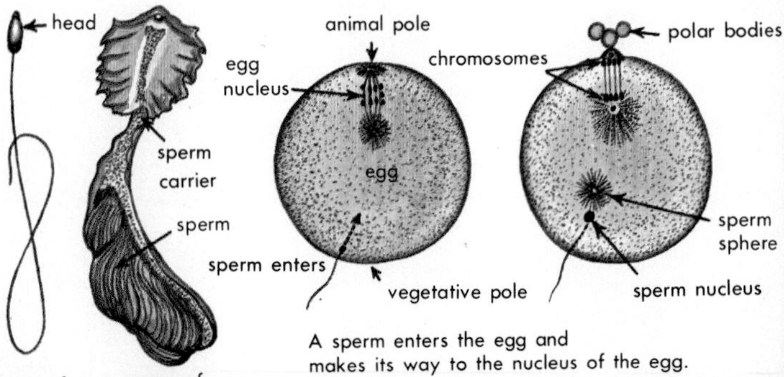

A sperm enters the egg and makes its way to the nucleus of the egg.

sperm of *Trochus*

sperm of *Cerithiopsis*

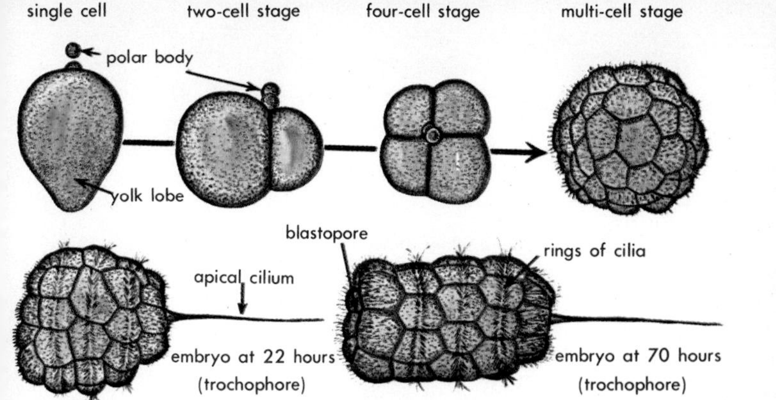

Cleavage of a Fertilized Egg

DEVELOPMENT

In all mollusks except the cephalopods the fertilized egg and the subsequent cells formed from it divide completely in half. However, the cleavage is oblique, not parallel to the polar axis of the egg. This results in a spiral arrangement of the dividing cells, a form common to flatworms and mollusks. In the cephalopods, the cleavage does not divide the egg completely; rather, it forms a cap of divided cells at the top third. The remainder of the egg becomes the yolk, which later develops into a yolk sac. Under ideal conditions, the first cleavage of the egg takes place within three or four hours after the sperm enters.

The size of the egg differs from one species to another, even within the same genus. Larger eggs contain a greater amount of nutritive yolk and usually complete their development within the egg capsule. Those species with smaller eggs generally hatch sooner and give rise to a free-swimming larval stage.

The egg is bathed in a sphere of albumen and is protected by a thin, clear membrane. In most mollusks, all development takes place inside this sphere; later stages in the gastropods, which produce free-swimming larvae, take place outside. As in the annelid worms, the egg first develops into a trochophore, a cylindrical larva ringed with girdles of ciliated cells.

In most mollusks the trochophore breaks out of the egg sphere and swims at large. This stage of larval development is usually very critical for marine species, for the delicate organisms are easily killed by changes of water temperature and salinity or by unduly rough waters. In sea snails and chitons, which brood their young within the mantle cavity, the trochophore remains in the egg sphere, where it transforms into a more advanced larval stage. The trochosphere stage usually lasts for only a few hours and may be absent or very fleeting in many gastropods.

Gastropod Free-swimming Veliger

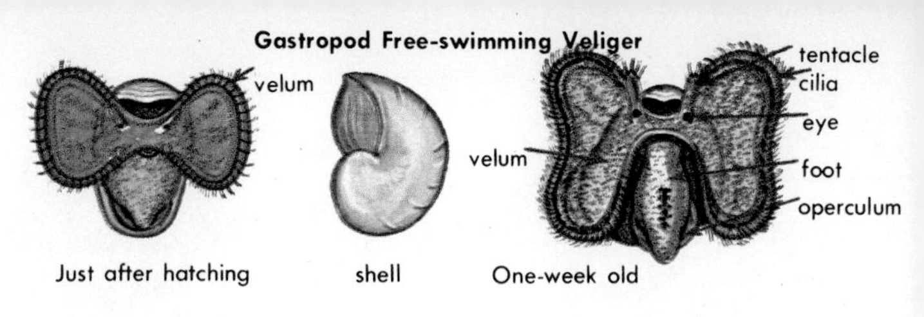

Just after hatching   shell   One-week old

The most advanced larval stage in mollusks is the veliger, a developmental phase much more advanced than the trochophore. The foot, tentacles, eyes, and the tiny early shell (prodissoconch) appear. The most characteristic feature is the large paddle-like lobes of the velum, a skin flap endowed with long cilia. The undulating velum and beating cilia are the means by which the young mollusk swims. In some gastropods the veliger stage is spent within the egg capsule, while in others it hatches out and may spend from a few days to several weeks swimming about freely. The velar lobes grow in size with the rest of the embryo, but finally are shed. Most species of mollusks have veligers of characteristic shape and coloration.

Veligers feed on microscopic algae, and when they have reached an advanced stage of development the velum is cast off and the mollusk begins its life on the bottom of the sea. Snails begin to crawl, and clams either dig into the sand or attach themselves to some hard object.

The advantages of a free-swimming veliger stage are obvious, for this stage permits the young to be carried by currents and tides to new areas. The mortality rate of veligers is very high because of predatory shrimp and fish and because of abruptly changing water conditions.

In a few genera, such as *Trivia* and *Capulus,* a second specialized shell, the echinospire shell, surrounds the veliger. It is made up of conchiolin reinforced with calcareous matter. Some echinospire shells are beautifully ornamented or strongly keeled. The extra shell is for protection and for helping to keep the animal afloat. It is shed later.

During the early veliger stage of gastropods, the animal twists itself so that the shell and visceral mass are turned 180° in relation to the head and foot. This *torsion* is characteristic of snails and accounts for the anteriorly facing anus and goniducts.

Development of a Bivalve Veliger

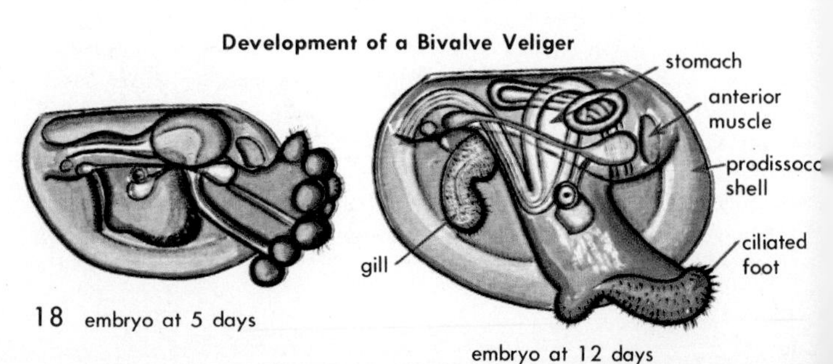

 embryo at 5 days

embryo at 12 days

GROWTH

Growth is accelerated as soon as the veliger stage is past. Often there is an abrupt change in the nature or sculpturing of the shell. In snails the early embryonic whorls are called the protoconch or nuclear whorls. In bivalves, they are called the prodissoconch. The latter may be seen with the aid of a hand lens at the tip of the umbo. In rough oysters it can be seen as a tiny, glistening, perfect clam valve.

The mollusk grows by producing additional shell material exuded from the glands in the edge of the mantle. The shape of the shell is largely predestined by inheritance, although environmental influence may modify the final results (p. 49). Most mollusks attain their mature size in one to six years. The Bay Scallop, for instance, reaches its maximum size at the end of the second year and rarely survives a third season. Many nudibranchs live only one year. *Littorina* periwinkles have been kept in captivity for over 10 years. The *Tridacna* clam of the southwest Pacific, which may reach a length of over four feet and a weight of 500 pounds, is believed to live for over 50 years. Growth rings in scallops and clams are the result of changes in salinity and temperature, and, although they may coincide with seasons, they do not necessarily indicate the age of the individual. In general, the warmer the water, the more rapid will be the deposition of new shell material.

Some gastropods produce a radically different shape upon reaching maturity. In the cowries, the outer lip is gradually curled inward, thickened, and supplied with numerous shelly ridges, or teeth. The *Strombus* conchs have a thin outer lip in their immature or ''roller'' stage, but at maturity they produce a thick, flaring, winglike outer lip. In addition, a special stromboid notch is formed through which the tentacle and eye may protrude. Many other snails keep growing until death and never modify the outer lip. In senile specimens, beyond the age of active reproduction, the shell may become greatly thickened and covered with marine growths. The delicate coloration in the lip of the snail shell may be covered over with a thick gray glaze.

Growth Stages of the Cowrie and Conch Shells

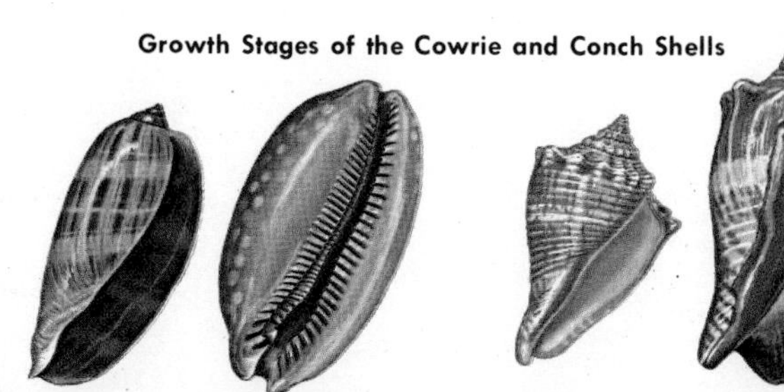

mmature vith fragile lip

adult cowrie with toothed aperture

immature with thin lip

adult conch with flaring, thick lip

SEXUALITY

Although the sexes are separate in a great number of mollusks, hermaphrodism and sex reversal occur in many snails and clams. There are various degrees of hermaphrodism. In some scallops the gonads produce both sperm and eggs at the same time, and self-fertilization may take place. In oysters, some individuals may first be males, then become females for a few months, and finally return to a sperm-producing phase. In the hard-shelled clam *Mercenaria*, all individuals are male the first year, but the following season about half of a local population will turn female and remain so for the rest of their lives. Many factors influence the sex cycle of bivalves, including nutrition, climate, salinity of the water, and the presence or absence of hormones given off in the water by the opposite sex.

Among gastropods, sex change takes place in a number of species. Limpets, such as *Acmaea* and *Patella*, are male when they first mature and change to female at the age of 12 to 18 months. Most inch-long limpets are males, and those over two inches are all females. Occasionally, medium-sized limpets are hermaphrodites.

The *Crepidula* slipper shells are normally protandrous hermaphrodites—that is, the younger and smaller specimens are males but later change to females. These snails, as illustrated below, live one upon the other in clumps or chains of up to 12 animals, each clinging to the shell of the one beneath. As long as the larger female snails at the bottom of the clump give off hormones, the top snails remain male. Removal of the female starts the male changing its sex. Sperm is no longer produced, and the penis is gradually absorbed. Finally, an oviduct is formed and the gonads produce eggs.

Sexual dimorphism is sometimes expressed in the shape or sculpturing of the shell. Female clam shells are apt to be more obese in order to accommodate the swollen gonads or the numerous young brooded in the mantle cavity. Female snails are usually much larger than males. The shells of male conchs and helmets are more likely to have longer but fewer knobs on the last whorl. These differences are termed sexual dimorphism. They may also be expressed in the size, shape, and number of radular teeth in certain marine snails.

Sex Reversal in the Slipper Shell, *Crepidula* (shell removed)

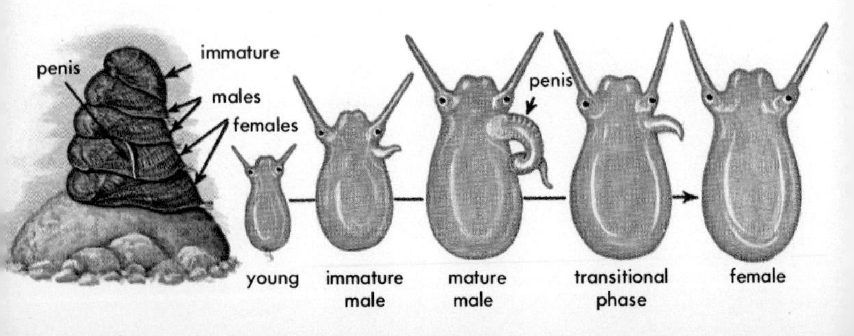

floating capsule of *Littorina* periwinkle (x 50)

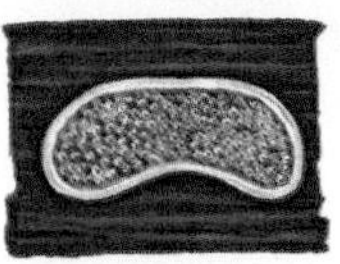

jelly mass of *Lacuna* on weed (x 10)

chain of leathery capsules of *Busycon* whelk (1″ diam.)

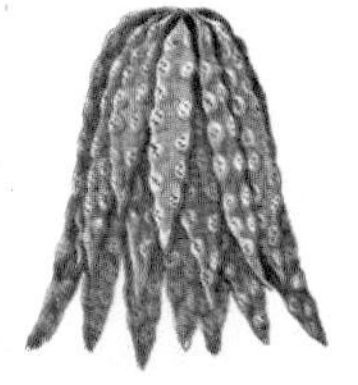

festoon of jelly masses of squid (5″)

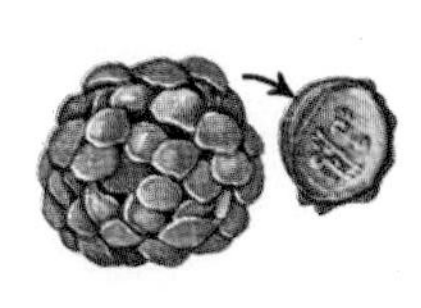

ball (3″) of leathery capsules of *Buccinum* whelk

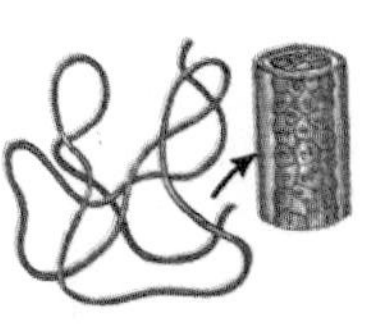

entangled egg strands of *Aplysia* sea hare (0.1″ diam.)

solitary capsules of oyster drill snail (x 3)

EGG PROTECTION

In most bivalves, chitons, and tusks and in a few primitive snails, the eggs are shed freely into the water with no extra protective coating. Some chitons, a few bivalves, and such snails as *Planaxis* and *Littorina* may brood their young within the mantle cavity or within a greatly enlarged oviduct. The female cowrie sits on her clump of eggs for many days until they hatch. Soft egg capsules are retained within the mantle cavity of *Crepidula* slipper shells, *Coralliophila* coral snails, and other univalves that remain attached to the substrate. The octopus cleans and circulates fresh sea water over her long strands of eggs.

Some gastropods bury their eggs in sand; others produce protective jelly masses, roll their eggs into sand collars, or place them in horny capsules. The horny capsules are produced by a gland within a special pore on the sole of the foot. In such genera as *Fasciolaria* and *Buccinum* as many as 100 eggs may be placed in one leathery capsule, but the unhealthy or smaller embryos, known as nurse cells, are eaten by the more aggressive young snails. Many kinds of capsules have an escape hatch or weak section through which the crawling young emerge. Egg strands are usually firmly anchored to stones, to other shells, or under protective rock ledges. The jelly masses of herbivorous snails are usually attached to the stems of seaweeds.

Eggs require varying lengths of time in which to develop and hatch, and they must remain submerged in water at a particular temperature. Once dried or washed upon a beach, the eggs will not survive. Some mollusks produce many millions of eggs during one breeding season. This serves to offset the high mortality rate due to adverse environmental conditions. Some snails congregate and contribute their egg capsules to a large community mass of spawn.

SHELL STRUCTURE AND PIGMENTATION

The molluscan shell is made up basically of crystals of calcium carbonate. The blood of the mollusk contains a high concentration of this salt in liquid form, and the fleshy mantle is capable of concentrating the calcium through an osmotic process. Crystallization takes place in a protected space between the mantle and the old shell layers. Solid crystals form either on other crystals or on a matrix or lattice work of conchiolin. The conchiolin is a brown, soft material consisting of proteins and polysaccharides. This is the same material of which the horny opercula of snails and the outer periostracal layers are formed.

Depending upon the degree of acidity and the chemical content of the blood, the crystals of lime form into calcite, as they do at the very edge of the mantle, or into aragonite, a heavier form of lime that may give a nacreous, or mother-of-pearl, effect. Several types of crystal layers may be laid down, each by a specialized portion of the mantle. Among the commonest types are a prismatic structure, a sheetlike laminar structure, and a mixed structure of lamellae crossing each other. The crossed lamellar layers give a high degree of strength to seemingly thin, fragile shells.

A vertical cross section of a piece of shell, magnified about 7,000 times, looks like a brick wall, with the calcite crystals representing the bricks and the protein conchiolin the mortar. In transverse section the crystals resemble shingles on a roof. The size of the crystals is determined by the rate at which their formation takes place, and this, in turn, is determined by the temperature of the water.

Structure and Layers of a Clam Shell

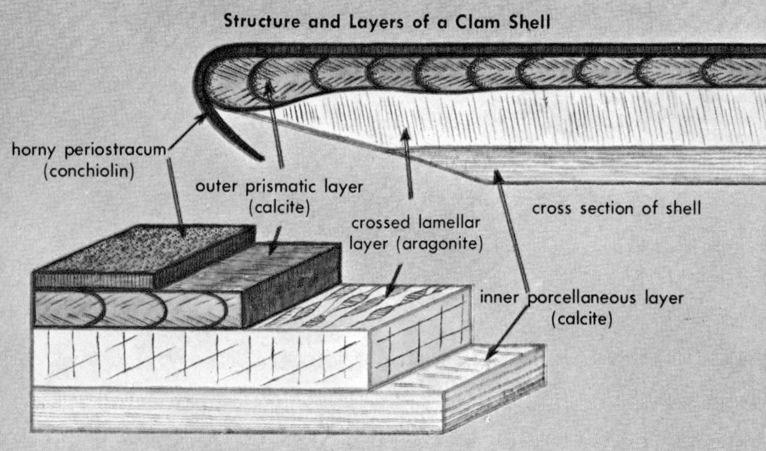

Pigments of many types are common in both the soft and hard parts of mollusks. They are obtained from the food of the mollusk and are distributed by the blood system. Pigment cells throughout the body and especially in the mantle concentrate the pigments and contribute to the colors of the shell material. Among the leading pigments are the orange to yellow carotenoids, the black to tan melanins, the prophyins (which include the hemoglobins), and the indigoids. The last include indigo blues and indigo reds that contribute to the coloration of the Royal Tyrian Purple of the ancients. This fabric dye was obtained from the accessory glands of several species of Mediterranean *Murex* rock shells.

The brilliant iridescent colors of the inside of abalones and trochid snails are the result of the physical structure of the shell and not of pigments. The wavelength of light is altered when the entering and reflected rays pass through the different layers of calcite.

Various patterns of pigmentation, such as dots, circles, and stripes, are produced in the shell by the migration of color-depositing cells along the edge of the mantle and by the intermittent production of pigment. If a pigment center on the edge of a snail's mantle remains in the same position and continuously colors the newly deposited shell, a simple stripe will be formed. If the pigment cells wander across the edge of the mantle, an oblique stripe will be produced. The shade or intensity of a color may change as the mollusk grows, thus giving a further modification to the general color pattern. Such changes may be due to varying diet or changing water salinities.

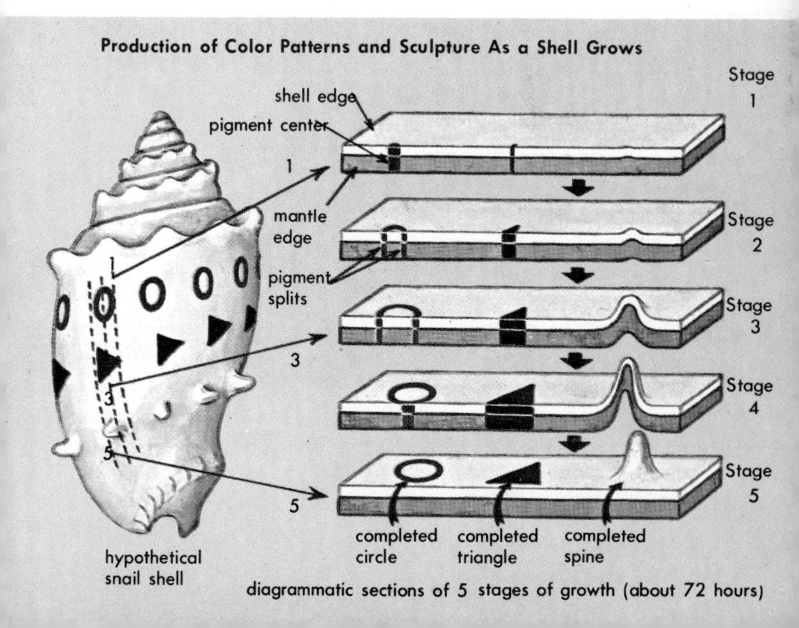

FEEDING

The food of mollusks includes many forms of animals and plants. Most bivalves and most chitons are herbivorous. All cephalopods and over half of the marine gastropods are carnivorous. A number of mollusks are external or internal parasites of other marine creatures.

In the bivalves the gills have assumed a major function in feeding. The hairlike cilia create currents of water that come in through the tubelike inhalant siphon and pass over the gills. Tiny food particles, such as planktonic algae and protozoa, adhere to a mucous film that passes slowly, in tiny strands, along the ciliated food gutters to the clam's mouth. Some clams, the *suspension feeders,* like the scallops, mussels, and venus clams, draw in food-laden sea water. The *deposit feeders,* like the tellin clams, use their long, movable siphons to suck up food lying on the bottom. The primitive protobranch clams belonging to such genera as *Nucula* and *Yoldia* have a pair of long, muscular proboscides that sweep food off the bottom and carry it along a ciliated groove to the mouth. The highly specialized septibranch clams, like *Cuspidaria,* pump in small crustaceans and worms, which are seized by small muscular lamellae and carried to the mouth. The stomach of the septibranchs is lined with tough chitin and acts as a crushing gizzard.

Squids are rapacious feeders, darting quickly after fish, ensnaring them in their tentacles, and killing them by biting a triangular section out of the fish's neck with a strong, parrot-like beak. The octopus feeds on mollusks and crustaceans. It can enwrap a crab and crush its shell within a few seconds. An octopus lair, usually a hole in the rocks, can be recognized by the mound of empty shells outside the entrance. Most chitons are herbivorous, but a few species capture and eat shrimp.

The sluggish scaphopod tusk shells that burrow slowly through the sandy mud substrate feed upon single-celled animals, such as foraminifera, by ensnaring them with threadlike tentacles.

Gastropods are very diverse in their methods of feeding and choice of food. Some are carnivorous or scavengers, others are herbivorous. The majority have a *radular ribbon* of many small, hard teeth set in the mouth, which acts somewhat like a rasp.

Boring Mechanism in Snails

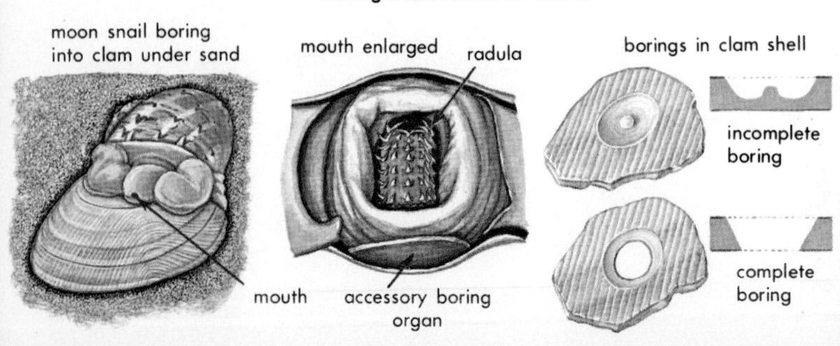

Parasitic snails, which suck the blood of bivalves or feed upon the soft tissues of other invertebrates, lack a radula. Some carnivores have a remarkably long, tubelike extension with the radular apparatus at the end. When not in use, this proboscis is retracted within the head. Some univalves, like the *Natica* moon snails, drill small, round holes through the shells of their molluscan victims. In some cases, the action of the radulae is aided by an acid secretion that softens the shell. Large *Busycon* and *Murex* snails open large clams by exerting a steady pull with their foot and prying open the valves of the clam with the edge of the outer lip of their shell.

Most bivalves and many herbivorous snails possess a unique *crystalline style* in the stomach. This is a firm, gelatinous protein rod that rotates and from its end liberates enzymes, making possible the digestion of vegetable matter. Algae-eating *Strombus,* filter-feeding *Crepidula,* and mud-eating *Ilyanassa* are among the few snails that possess a crystalline style. In *Strombus,* both the pouches in the esophagus and the salivary glands produce amylase and cellulase necessary for the digestion of algae.

RADULAE, the teeth found in all classes of mollusks except bivalves, are attached to a ribbon, or odontophore, in the mouth. They are used as a rasp, or rarely as a stinger. Individual teeth are almost microscopic, and the radular ribbon itself is small. The teeth are made of hard chitinous material. There may be from 20 to over 300 rows of teeth in a ribbon, depending upon the genus or family. A few of the major types of radulae found in snails are shown here:

(1) *rhipidoglossate,* found in the archaeogastropods (abalones, top shells, nerites, and keyhole limpets)—each row has upward of 100 teeth;

(2) *taenioglossate,* having only seven teeth in each row and being, on the whole, rather delicate—found in less advanced gastropods like the periwinkles, conchs, and cowries;

(3) *rachiglossate,* usually having only three teeth, all large and heavy, per row—found in carnivorous families, such as *Murex* rock shells, volutes, miters, and whelks;

(4) *toxoglossate,* having hollow, harpoon-like teeth—used individually to inject poison into prey, radulae of this type are found in cones, terebras, and some turrids.

Types of Radular Teeth in Prosobranch Snails

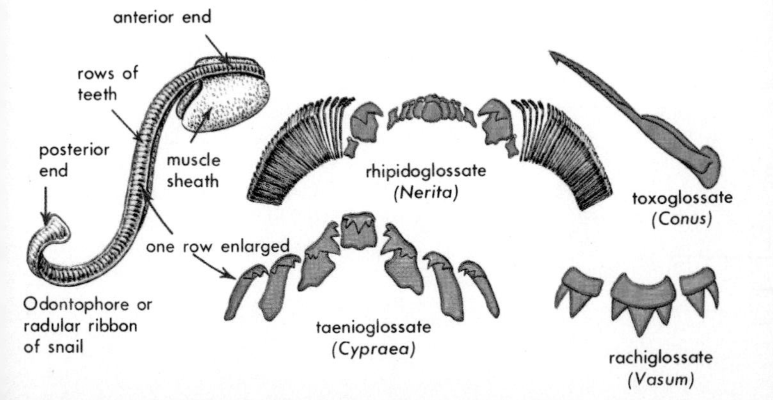

LOCOMOTION

There are many modes of locomotion among the mollusks. Many crawl, swim, burrow, or dig; a few walk (the octopus), and one even "flies" (an air-gliding squid). During their free-swimming larval stage many snails and clams remain suspended in the ocean currents by beating the tiny oarlike cilia that girdle their body. Some snails, like *Janthina,* float all their lives at the surface of the ocean and are wafted about by wind.

The foot of the snail is well suited for locomotion. In most forms with a fairly broad sole, crawling is accomplished by a constant series of muscular waves that start at one end of the foot and move toward the other. From a narrow slit along the front edge of the foot, a gland secretes a slimy mucus that assists the snail over dry or coarse bottoms. In some snails the sole of the foot has numerous microscopic cilia that beat rapidly and aid in the creeping motion. In the *Strombus* conchs, the foot is long, narrow, flexible, and used like a vaulting pole to flip the animal forward. Large flaplike extensions of the foot, when undulating, permit the *Aplysia* sea hares to swim slowly but effectively.

Clams are able to dig through mud and sand by manipulating the pliable muscular foot, which can be extended out in a long, narrow projection. The tip of the foot is slowly filled with blood and spread muscularly to form a swollen, anchoring bulb. Then the clam pulls itself forward.

Some bivalves, such as *Pecten* and *Lima,* can swim by snapping their valves rapidly and thus creating a jet stream of water. Other clams, like the date mussels and piddocks, bore into clay, coral rock, or wood, usually by rotating their shells and, less commonly, by producing acid. When a cockle has been uprooted from the sand, it can hop and take evasive action by lashing its long muscular foot against the bottom.

Many species of bivalves that attach themselves to rocks and pilings have a much-reduced foot. They anchor themselves by a byssus, or clump of chitinous threads, from a pore in the foot. The mussel and lima clams are able to move from one place to another by abandoning the old byssus and spinning a new one a few inches away.

The tusk shells dig through the sand in the fashion of most clams, since their foot is also hatchet-shaped. The chitons with their broad, flat soles creep along in the fashion of most snails.

The cephalopods have the most advanced type of locomotion. Swimming speeds up to 30 miles per hour are attained by some squids. The motive power comes from a strong jet of water forced from the mantle cavity by sudden muscular contractions of the mantle. The exhalant funnel, when turned to one side or another, assists in determining the directions of swimming. The undulating fins of the squid are mainly used as stabilizers.

NERVOUS SYSTEM AND SENSE ORGANS

The nervous system in mollusks varies from a primitive condition of nerve-connected ganglia, found in scaphopods and chitons, to the highly developed "brain," or fused ganglia, of the cephalopods. The esophagus in all mollusks runs through the center of the mass of the pairs of cerebral, buccal, and pedal ganglia. In the chitons and in some univalves, the ganglia are mere swellings in the nerves. In gastropods, the two pedal ganglia, serving the foot and columellar muscle, are usually larger than the cerebral ganglia that serve the eyes and tentacles. The largest nerve fibers among animals occur in the mantle of the fast-swimming squid. Their diameter is 1/20 inch, about 50 times thicker than those in most other animals.

Cephalic eyes are present in most gastropods and in nearly all cephalopods. A few pelagic snails, *Janthina,* and fresh-water cave snails are blind. The mantle edge and shelly plates of certain chitons may possess aesthetes, bundles of light-sensitive cells that are clumped together to form a type of primitive "eye." Ocelli capable of detecting light are found along the mantle edge or even on the siphons in some bivalves. The approximately 100 mantle eyes of scallops are about 1 mm. in diameter and are bright blue. The eyes of gastropods are usually located at the base or near the end of the cephalic tentacles. The eyes of squids and octopods are able to form an image. Cornea, iris, lens, and retina are present, and focusing is accomplished by moving the lens forward or backward, rather than by changing the shape of the lens.

The sense of balance and orientation is believed to be located in a pair of pouchlike statocysts usually attached to the base of the pedal ganglia. A small calcareous statolith, or "ear stone," rolls about and activates the nerve endings in the walls of the statocyst.

Touch and smell are made possible by specialized tactile and chemoreceptor cells. These cells may be concentrated in various parts of the mollusk's body, the tentacles of snails and the mantle margins of bivalves being the commonest sites. In snails, an osphradium, or elongate, filamentous organ, lies adjacent to the gills. It is highly sensitive to sediments and changes in the salinity of water. The plate-like tentacles, or rhinophores, found in nudibranchs have been shown to be very sensitive to chemical changes in sea water, including odors.

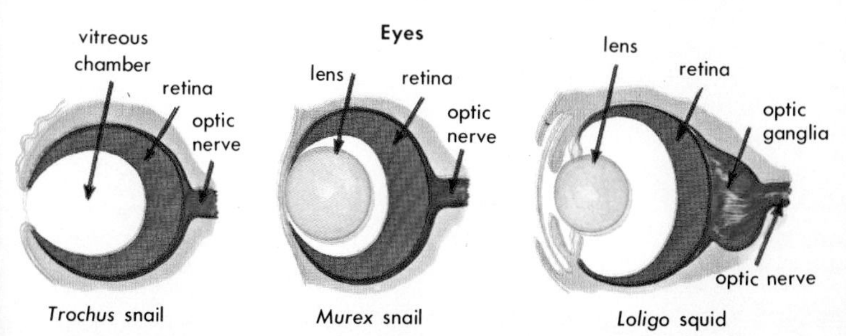

COMMENSALISM AND PARASITISM

The molluscan phylum has many examples of close associations with other animals. In some cases, such as in commensalism, the two organisms live together and benefit from the association, without one harming the other. In other instances, such as in parasitism, one of the associates suffers as a result of the feeding activities of the other. There are many gradations between commensalism and parasitism.

Among the cases of commensalism, or "feeding at the same table," is the existence of certain copepods, pea crabs, and hydroids within the mantle cavity of clams, where they feed upon surplus food supplies without injuring the bivalve host. Some hydroids, such as *Stylactis,* can survive only if they are growing on the outer shell of the New England Mud Snail *Ilyanassa.* The mantle cavity of the pearl oyster, *Pinctada,* and the pink conch, *Strombus,* serves as a welcome protective chamber for certain kinds of *Apogon* fish and shrimps. Certain kinds of limpets and parchment worms may live exclusively on the underside of or within the shell umbilicus of herbivorous snails, where they receive protection and feed upon the feces of the host mollusk.

Mollusks themselves act as parasites and also, in other cases, serve as the host of other parasites. The trematode worm parasites of sea birds must spend the larval period of their lives in specific species of intertidal periwinkles, *Littorina.* The digestive gland in the apex of the snail is the site where the larval sporocyst worm multiplies and transforms into thousands of cercariae that, in turn, burst forth and swim freely in the sea to infect swimming gulls and ducks. If a man's skin is attacked, a discomforting but harmless rash, or "swimmer's itch," is caused. Unusually heavy parasitic infections can cause drastic decreases in the mollusk population.

Many mollusks, including about 20 genera of snails, act as parasites. *Stylifer,* a snail, burrows into the spines of the club-spined sea urchin and causes a large gall to form. *Entocolax* and *Entoconcha* snails live embedded in the interior walls of the intestines of holothurian sea cucumbers.

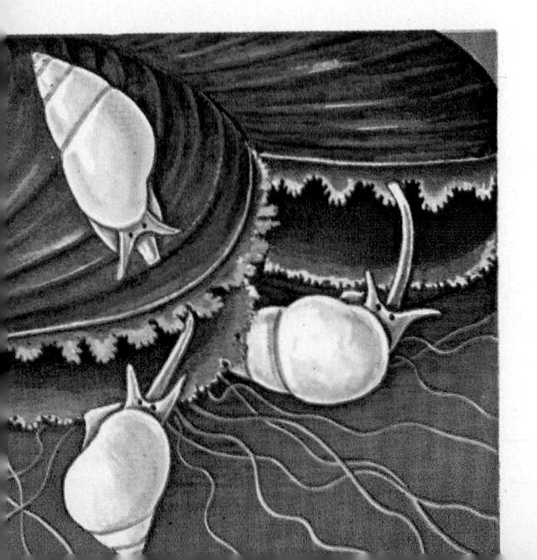

The snail family Pyramidellidae includes many species that are ectoparasites, living on the blood of marine worms, bivalves, and other snails. The proboscis of the parasitic snail is modified into a long, hoselike tube bearing a sucker for attachment, a stylet for piercing the flesh of the host, and a special pump to suck blood.

MARINE COMMUNITIES

Mussel community on wharf piling

Throughout the world's oceans there are countless communities of animals and plants living in close association and maintaining a characteristic makeup. In many instances, mollusks may dominate the scene, although other forms of life are present that contribute to the balance of the community. The stability of these communities depends upon food availability, the nature of the bottom, growth and mortality rates, and other basic biological conditions. Similar marine communities are found in corresponding environments in various parts of the world. A wharf piling has much the same general kinds of life in New England or New Zealand.

Many marine communities are best identified by or named after the organism that dominates the scene. The *Mytilus* mussel communities of New England and the northwestern United States are characterized by the myriad of other animals found among the clusters of mussels. Hydroids, bryozoans, barnacles, brittle stars, ascidians, small crabs, and worms live together there and feed upon each other. *Littorina, Lacuna, Odostomia,* and *Anachis* are a few of the associated gastropods.

The *Terebra-Donax* community is a common sea-beach association in which single-celled algae are washed ashore on the slopes of the beach and are fed upon by the *Donax* clam. Almost invariably a predacious gastropod is present in such communities. In Florida, it is *Terebra dislocata;* in the East Indies, *Terebra hectica;* and in southern Asia, *Terebra duplicata.*

Macoma clam communities are common in many parts of the world. They occur at depths from about 6 to 100 feet, under estuarine conditions, where the bottom is a mixture of sand, silt, and mud. Several species of *Macoma* may be present in one area, such as in the *Macoma nasuta-Macoma secta* community of the Pacific Northwest. *Arenicola* worms and *Dentalium* tusk shells are usually present.

The *Batillaria-Cerithidea* snail community of southeast Florida and the West Indies is characterized by a hot, brine environment in the area of mangrove trees. These horn snails are mud-detritus feeders, and they serve as food for aquatic birds, shallow-water fish, and crabs. The snails grow in great profusion, and only a constant and rapid growth of diatoms and green algae will sustain their numbers.

The *Cyphoma-Coralliophila* snail community of the Florida Keys and the West Indies is restricted to the very limited environment of the sea fan. These snails live on, feed upon, and attach their eggs to these fans. Other genera sometimes associated with them are *Neosimnia* and *Triphora.*

ENVIRONMENTAL NICHES

Most forms of life have become adapted to special living conditions, and the sea offers many environmental niches or specialized habitats. Marine mollusks may be divided into three major ecological or behavioral groups: (1) the *benthic* forms, which burrow into, creep on, or are attached to the bottom; (2) the *planktonic* forms, which as adults or as larval forms float about passively in the ocean currents; and (3) the *nektonic* forms, or strong swimmers, which operate freely in open waters.

The seashore and deeper parts of the ocean are divided by ecologists into arbitrary zones, each of which is characterized by conditions of light, depth of water, tidal fluctuation, or the kinds of animals or plants dominating the scene. Some, like the splash zone where waves keep the upper shore rocks wet, support colonies of limpets and periwinkles in most areas of the world.

Open Ocean

The planktonic forms found in the pelagic, or open-sea environment are quite abundant in species and numbers, especially because most bivalves and many snails go through a temporary floating veliger stage. The pteropod snails, however, maintain throughout life a pelagic existence, usually floating and swimming in schools at depths of a few to 100 feet. At night they rise near the surface where they feed on microscopic animals, but with daylight they sink because of the strong sunlight. The *Argonauta,* the *Janthina,* and the *Recluzia* snails with their bubble floats, several nudibranchs, and the *Carinaria* Glass Snail all live permanently just below or at the surface of open seas. A veritable jungle of organisms dwells among the floating mats of *Sargassum* weed in the tropical Atlantic, including the weed-colored *Scyllaea* nudibranch and the brown *Litiopa* snail, which dangles from the weed stems by means of mucous strands.

Rocky Shorelines

Just above the normal reach of the high tide is a zone in which a few hardy marine snails and several salt-loving land pulmonates are able to survive. Intermittent rains and occasional salt-water drenchings during storms give rock cliffs enough moisture for the growth of lichens, fungi, and microscopic algae. *Tectarius, Nodilittorina,* and *Acmaea* limpets dominate this inhospitable area of extreme temperatures and variable humidity. Below this zone are found the unique and magnificent tidepools that are isolated at low tide and are flushed with new sea water at high tide. Protective crevices and submerged rocks encourage a rich growth of seaweeds, sponges, sea anemones, and other invertebrates. In temperate climates, tidepools support chitons, *Acmaea* limpets, *Nucella* dogwinkles, and *Mytilus* mussels. In tropical areas, *Nerita, Engina* and *Cantharus* snails, *Brachidontes* mussels, and *Thais* rock shells are the common inhabitants.

Sand Beaches and Mud Flats

Sand beaches, depending upon their degree of slope and the size of the sand grains, offer a suitable habitat to only a few mollusks, among them being *Donax* wedge clams and a few kinds of carnivorous *Terebra* augers. In contrast, nearby sand or mud flats are usually rich in mollusks. Again, the physical conditions, such as the stability of the sand bars, the salinity and temperature of the water, and the length of intertidal exposure, will influence the number and kinds of mollusks. Brackish water and a muddy sand are favorable to burrowing *Macoma* and *Mya* clams and to *Nassarius* and certain *Cerithidea* snails. Add a rock bottom, and oysters and *Urosalpinx* snails become dominant. The addition of eelgrass favors scallops and *Crepidula* slipper shells. In waters nearer the open ocean, the *Macoma* clams are replaced by *Trachycardium* cockles, *Tellina, Ensis,* and the surf clams *Mactra* and *Spisula. Polinices* and *Natica* moon snails replace the mud snails in deeper water.

Sublittoral or Shelf Bottoms

From the zone below the low-tide mark out to a depth of about 300 feet is a rich continental shelf area of varying types of bottom—sand, ooze, rock, coral rubble, and combinations of these. Mollusks may be abundant or very scarce, depending upon many environmental factors. Many normally intertidal species are quite capable of surviving in depths down to 100 feet. Other species, however, especially the *Neptunea* whelks, *Mitra, Conus, Yoldia* and *Nucula* nut clams, turrid snails, and *Calliostoma* top shells, usually thrive best in the sublittoral zone where tidal currents are less felt. Many diversified ecological niches exist offshore, from sponge beds to rock ledges in submarine canyons and from submerged wooden logs to swaying sea fans on a coral reef. Many mollusk species can survive only in very limited areas where ideal conditions may exist. The *Cyphoma* flamingo tongues and *Neosimnia* are found only on sea whips and sea fans; *Epitonium* Wentletraps are found near sea anemones or certain corals; *Architectonica* sundials are associated with sea pansies, and their *Heliacus* cousins feed and live on soft polyp-like zoanthids.

Deep Sea, or Abyssal Zone

The Abyssal Zone consists of animals living at depths greater than about 6,000 feet, or about 1 mile. Although the 1,200 species of mollusks living at these great depths are somewhat the same throughout the world, there is a strong relationship between the mollusks of an ocean basin and its bordering continent. Most shells are 1 inch or less in size, whitish or gray, thin, and without pronounced ornamentation. Most feed on animal matter, but about one fourth of the species feed on plant detritus or are parasitic. The water is very cold and sunlight is absent at these depths. Only rarely do abyssal mollusks venture into shallow waters.

DISPERSAL AND ABUNDANCE

Every species of marine mollusk has its own natural geographical distribution that is controlled by an interaction between the biological nature of the individual and the oceanic environment. On occasions, a species may expand its normal domain. Free-swimming larval stages in ocean currents or adults aboard floating objects, such as logs, help to disperse each species.

Sea temperature, salinity, and bottom conditions are always changing, particularly from one season to another, and sometimes from one decade to another. If conditions change sufficiently, a normally cold-water species may spread southward along the coast for several hundred miles in a matter of a few years. Likewise, warm-water species may spread northward if there is a warming trend at the north end of the range of the species. Areas of severe pollution may locally exterminate populations.

Sometimes other environmental changes may affect the abundance of certain mollusks. The death of eelgrass beds, *Zostera,* along the east coast of the United States in the early 1930's caused a temporary disappearance of the Bay Scallop *Aequipecten,* which lives in that type of protective habitat. Within the next decade or so the eelgrass returned, and with it the scallop.

There are natural, short-range migrations during the lifetime of individual mollusks. Many species of *Strombus, Oliva,* and nudibranchs come into shallow, warmer water to breed at certain seasons of the year, especially the spring. Schools of scallops move around, evidently to better feeding grounds or possibly to areas with fewer predatory schools of fish. Certain intertidal species, particularly limpets and periwinkles, make short nocturnal migrations in search of unbrowsed growths of algae, and they usually return to the same resting location during the day.

Probably the only true examples of long-range migrations among mollusks would be found in the fast-swimming squids that roam the seas along our shores, but little evidence about their movements has so far been obtained.

There can be sudden appearances of mollusks at unpredictable times. Favorable currents may seed a stretch of beach with young *Donax* clams and result in a very large population for the next few months. In certain areas there have been "plagues" of *Octopus,* presumably due to especially ideal conditions for reproducing. They have been sufficiently abundant to eliminate most of the large lobsters in an area for several years. Onshore winds may bring ashore schools of floating *Janthina* in the spring months and, at times, litter the beaches with these purple snails. Cold currents or very cold winds at low tide can kill off large populations of clams and snails, and sometimes result in thousands of tons of dead mollusks being thrown up on the beaches.

INTRODUCED SPECIES

Man has introduced certain marine mollusks, either accidentally or purposely, to various parts of the world. Many of the introductions have occurred when commercially important bivalves, especially oysters, have been taken from one ocean to another. Eggs and tiny young of other mollusks have been attached to the outside of the live oysters. Several Japanese and eastern American species, including the mud snail, *Ilyanassa,* the drill, *Urosalpinx cinereus,* and the *Busycon* whelk, are now well established on the Pacific coast of the United States. Conversely, American species have been introduced to foreign countries. The Common Atlantic Slipper Shell, *Crepidula,* for example, was added to the European fauna about 80 years ago. It has multiplied so rapidly that its presence in the estuaries of eastern England hampers the raising of the native oysters. In France, it is becoming an edible shellfish product.

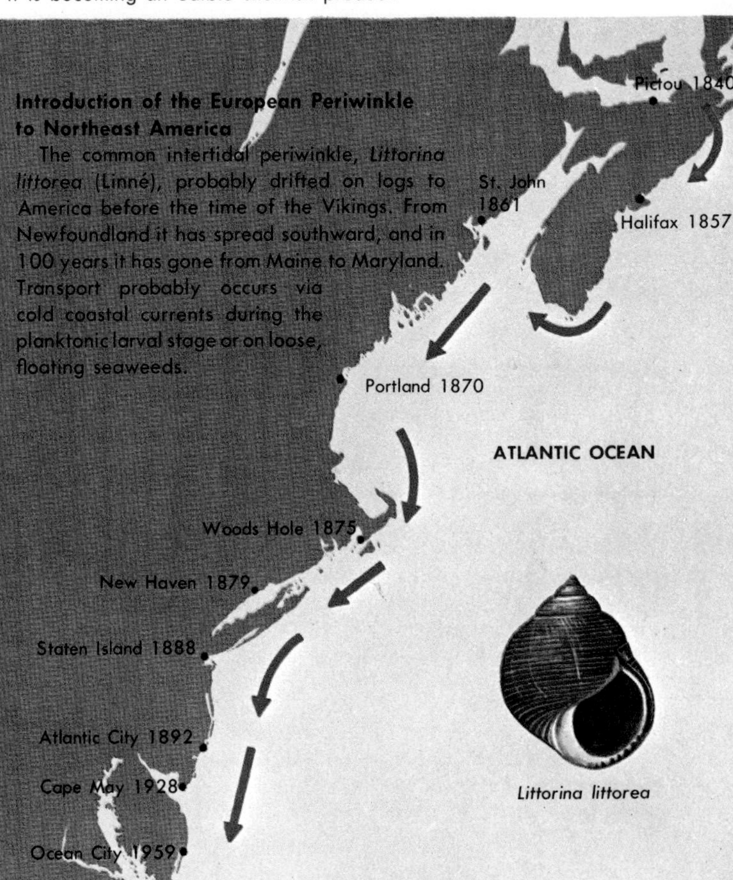

Introduction of the European Periwinkle to Northeast America

The common intertidal periwinkle, *Littorina littorea* (Linné), probably drifted on logs to America before the time of the Vikings. From Newfoundland it has spread southward, and in 100 years it has gone from Maine to Maryland. Transport probably occurs via cold coastal currents during the planktonic larval stage or on loose, floating seaweeds.

Littorina littorea

MARINE FAUNAL PROVINCES

From the icy waters of the Arctic, which surround the northern shores of North America, to the warm tropical waters of the Caribbean and the Gulf of Panama a number of more or less natural marine faunal provinces support their own particular kinds of mollusks. The inshore, shallow-water areas are characterized by unique sea temperatures, not only from the standpoint of the mean annual temperature, but also, and perhaps more importantly, from the minimum winter and maximum summer temperatures.

Many factors control these inshore temperatures—the general mass of surrounding seas, surface currents, deep-water upwellings, seasonal oceanic and continental winds, the sun's rays on shallow bay waters, and the amount of rainfall. Although several faunal provinces are recognized, their limits and even their centers are difficult to define, not only from physical oceanic data, but also from the kinds of animals and plants that are present. Overlapping and mixing are commonplace.

Some mollusks are constantly extending their distributions, especially while in their free-swimming larval stages. The present-day ranges of species are kept about the same by the adverse temperature conditions bordering the mollusk's normal distribution. A warm-water Caribbean species, such as the cowrie *Cypraea zebra,* survives with ease in the Lower Florida Keys and the Caribbean, but farther north, along the Carolinas and southern Texas, colder waters make breeding impossible or may cause their special food to be scarce or absent. Northern species migrating southward are unable to survive because of the overly warm sea temperatures.

Since ideal temperature conditions may exist temporarily in boundary areas for a few years, some species may invade these new areas and continue to exist successfully until adverse conditions exterminate them. A series of mild winters or periods of low rainfall along the southwest shores of Texas are believed to be responsible for the occasional and temporary presence of Caribbean species along the beaches. The stunted Fossor form of the Coquina Clam from Florida turns up on New York beaches in the summer.

Many of today's faunal provinces have existed for millions of years in approximately the same positions on our globe that they now hold. In the Pliocene and Miocene beds of Florida and California we find shells not unlike those of today. While many fossil species are extinct, others have remained unchanged and are still living today in the same areas. Some different, present-day faunas, like the Caribbean and the Panamanian, show a remarkable similarity because in former geologic times they existed in a common sea. The presence of fossil remains of tropical species in northern regions of North America suggests that warmer provinces formerly existed in today's arctic and boreal regions.

MOLLUSCAN FAUNAL PROVINCES
OF NORTH AMERICA

The ARCTIC province is circumpolar in extent and includes not only the cold waters of northern Canada, but also those of northern Europe and Asia. The shelled mollusks of this polar region are characterized by dull brown or chalky-white shells covered with a thick brown or blackish periostracum. There are relatively few species, about 350, and all live well below the low-tide line beyond the destructive reach of grinding ice. The dominant bivalves are *Astarte, Yoldia, Nuculana, Hiatella,* and *Mya,* and the most prevalent univalves are in the genera *Buccinum, Colus, Margarites* and *Trophon.*

The BOREAL province lies to the south of the Arctic Sea and extends southward. In it there is an increase in number of species, but at the same time a slight differentiation in species, depending upon whether the province borders the shores of Alaska or Labrador. Since these areas have predominantly rocky shores, a number of cold-water forms, such as *Mytilus, Modiolus, Littorina,* and *Acmaea,* supplement the otherwise Arctic-like species of this province. A number of species, generally belonging to warm-water genera, such as *Columbella, Trivia,* and *Epitonium,* are prevalent in the southern limits of the Boreal province. In western America, this warmer Boreal region extends along the southern shores of Alaska to northern British Columbia, and is known as the ALEUTIAN subprovince. *Trophon, Littorina,* and *Patinopecten* are abundant. In eastern America, a comparable area, the ACADIAN subprovince, extends from the southern half of Newfoundland to the Gulf of Maine. The latter subprovince has a unique warm-water pocket trapped in the Bay of Chaleur around Prince Edward Island. In this MAGDALEN pocket are a number of species, like the *Mercenaria* Quahog and the Eastern Oyster, that thrive normally in the Virginian subprovince to the south of Cape Cod.

The VIRGINIAN subprovince, extending from Cape Cod to Cape Hatteras, has basically a Boreal fauna, but also has many elements, such as the whelk *Busycon carica,* the drill *Eupleura caudata,* and the scallop *Aequipecten irradians concentricus,* that make it a rather unique area. To some extent, this subprovince resembles the European CELTIC province of England and France, as well as the OREGONIAN province of western United States.

The CAROLINIAN province extends from Cape Hatteras to Texas. It is richer in species than the Boreal and is greatly influenced in the south by a slight mixture of tropical species. Among its dominant or unique species are *Littorina irrorata, Thais haemastoma floridana, Busycon spiratum,* the Angel Wing *(Cyrtopleura costata),* the Atlantic geoduck *(Panopea bitruncata),* and the Southern Quahog *(Mercenaria campechiensis).* This province is separated in its middle by Florida, so that some of the species have a discontinuous range. The semitropical waters of southern Florida act as a temperature barrier. The separation is geologically too recent to have caused any significant subspeciation.

The CARIBBEAN province is a typical tropical environment supporting reef corals and about 3,000 species of mollusks. About one fourth of these are colorful shallow-water species, the others living below 300 feet. The province extends from Brazil northward throughout the West Indies to Bermuda and to the shores of the Lower Florida Keys. Warm-water currents carry Caribbean species on to the offshore sea mounts of Texas and the outer banks of the Carolinas. Cones, cowries, miters, olives, tellins, and other tropical genera dominate the Caribbean fauna. While many of the species are associated with clear, oceanic-like waters, there are quite a few which are found only in the muddy waters of mangrove-bordered bays. The province extends over such a wide area that the individuals of, say, Bermuda species are somewhat different from those in Brazil. Most experts do not consider them as separate species.

The PANAMANIAN province, not within the bounds of this book, is the tropical Pacific counterpart of the Caribbean. This rich fauna, extending from the Gulf of California to the shores of Ecuador in South America, bears a close relationship to that of the West Indies because of their common connection during the pre-Pliocene times. *Conus, Terebra,* and *Mitra* species are abundant. Many unique species are limited to this province, such as the giant Tent Olive, the Grinning Tun *(Malea),* and the Radix Murex. Very few of the approximately 2,400 Panamanian species resemble those from the central Pacific, although most of the genera are the same.

The CALIFORNIAN province, to the north, extends from Baja California to Point Conception in central California. Some species native to this cool-water area are also isolated in the upper regions of the Gulf of California. Several large, handsome kinds of abalones *(Haliotis), Calliostoma* top shells of the kelp beds, the Great Keyhole Limpet *(Megathura),* and the only West Coast species of *Conus* are common in this temperate to subtropical area. The fauna of this province is comparable to that of the Atlantic Carolinian. Although primarily made up of about 1,200 cool-water species, there are a few tropical and Boreal species that have invaded the province.

The OREGONIAN province, a cool- to cold-water region extending from central California to southern British Columbia and having its center around Puget Sound, is dominated by rock-loving *Acmaea* limpets, *Nucella* dogwinkles, and numerous large chitons. Here is the northern limit for oysters, for the *Solen* Jackknife Clam, and other temperate-water mollusks. The largest clam in America, the Pacific Geoduck, is fairly common in this province. Offshore, the mollusks are mainly cold-water whelks of the genera *Colus, Trophon,* and *Buccinum.* The clams are arctic-like and belong to such genera as *Astarte* and *Nuculana.* Some of the Oregonian species extend in range to northern Japan. This province is not as rich in unique species as is the Aleutian subprovince to the north.

GUIDE TO COLLECTING AREAS

Although the saying "shells are where you find them" is probably a sage maxim, nonetheless any guide to good collecting areas must attempt to help collectors locate rich, productive, and exciting shelling grounds. A very excellent collecting area can produce a disappointingly "next to nothing" catch if the weather is windy, the water muddy, the season wrong, or the collector inexperienced. By contacting shell club members in various parts of the country you can obtain important information on the most productive months and, indeed, the best bays and reefs to visit. Honor this cooperation by collecting in very modest quantities and practicing conservation. Clubs are listed in *Van Nostrand's Standard Catalog of Shells* (1967), edited by R. J. L. Wagner and R. T. Abbott. Most seaboard states have shellfishery laws, but these largely apply to commercially important oysters and to large species of clams and scallops. In California, a sport-fishing license is required of those over 15 years of age who wish to collect live mollusks.

East Coast

Canada—The shores from Quebec to Nova Scotia support a moderately rich, cold-water molluscan fauna consisting of about 150 readily obtainable, intertidal species. A third of these are arctic in origin, the remainder being more southerly species. The Northumberland Strait, south of Prince Edward Island, is richest in temperate-water species. Consult E. L. Bousefield's *Canadian Atlantic Sea Shells,* Ottawa: Queen's Printer, 1960.

New England—Collecting north of Cape Cod is excellent in rocky tidepools, especially at low tide. In Maine, several large *Colus, Neptunea,* and *Buccinum* occur in shallow water. Contact scallop fishermen and those who use baited traps. South of Cape Cod the warmer waters of protected bays have more easily collected species. Dredging is productive in eelgrass patches. Areas near large cities are usually badly polluted.

New York to Georgia—The Middle Atlantic States offer a variety of habitats, from the rocky points and beaches of Long Island, New York, to the tidal marshlands of the Chesapeake Bay. Over 40 temperate-water species are readily collected near New York City. (Consult M. K. Jacobson and W. K. Emerson's *Shells of the New York City Area,* Argonaut Books, N.Y., 1961.) The inside bays of the New Jersey Coast are better than the open beaches facing the Atlantic Ocean. Both sides of the southern end of the Delaware-Maryland peninsula are good. Farther to the south in the Carolinas and Georgia semitropical, sand-dwelling species appear. Marsh collecting is poor. Dredging off this area results in a few tropical species. Sometimes after severe storms, Cape Hatteras is strewn with such Floridian species as the *Phalium* Scotch Bonnet. Visit scallop-shucking dumps in the Carolinas, where discarded, dredged species can be found.

Florida—Probably the best collecting within the continental United States is in southern Florida. The Lower Florida Keys have a high percentage of colorful West Indian species. Night collecting at low tide on the Gulf Stream side of the Keys can be very productive in the late spring and summer. It is against the law to collect live mollusks in the John Pennykamp Underwater State Park, which borders the east side of Key Largo, or around the shores of Dry Tortugas. The west coast of Florida, from about Tarpon Springs south to Marco, is renowned for its abundance of beautiful shells. Beachcombing is best, particularly at Sanibel Island and Naples, three days following a strong northwest blow in the winter. (Consult L. M. Perry and J. S. Schwengel's *Marine Shells of the Western Coast of Florida,* Paleontological Research Institute, Ithaca, N.Y., 1955.)

Gulf Coast States—From northwest Florida to Texas the collecting is moderately good, but requires careful search for suitable habitats. Galveston Bay, San Antonio Bay, and Aransas Pass are favorite collecting spots. Tropical species make an appearance at the south end of South Padre Island, not far from Port Isabel. Dredging off here and to the south results in choice deep-water material, and after storms the beaches are sometimes littered with shells.

West Coast

Alaska and British Columbia—As a whole, the entire Pacific coast is richer in species and number of individuals than is the Atlantic coast of the United States. Although the water is cold, the protected bays and estuaries of southern Alaska and British Columbia offer a wide variety of habitats. Shore and shallow-water collecting is excellent, particularly in the summer.

Washington to Central California—Collecting improves farther to the south with an increase in the number of species. Rocky shore and tidepool collecting are excellent, with a varied selection of chitons, limpets, and nudibranchs. Protected bays, sand bars, and grassy areas at low tide are fairly rich. (Consult E. F. Ricketts and J. Calvin's *Between Pacific Tides,* Stanford University Press, Calif., 1952.)

Southern California and Mexico—From Point Conception southward, semitropical species make an appearance. Diving and dredging result in excellent catches. The several offshore islands, such as Santa Catalina and San Clemente, offer good collecting in the surrounding shallower areas. Improved roads, tourist facilities, and convenient air travel have made the rich waters of Baja California and mainland Mexico available to shell collectors. The area around Guaymas is rich in tropical, shallow-water species. Mazatlan, Acapulco, and Salina Cruz are favorite collecting spots, where the less windy months from May to August seem best. The farther south one goes, the more numerous become the tropical genera and species. (Consult A. Myra Keen's *Sea Shells of Tropical West America,* Stanford University Press, Calif., 1958.)

COLLECTING MOLLUSKS

The search for specimens of marine mollusks requires the fascinating combination of skill, knowledge, and luck, found in any hunt or exploration, considering the vagaries of weather, the temperament of the sea, and a number of other factors. It may be true that "shells are where you find them," but the novice can soon become a collecting expert if he is armed with a general knowledge of the life habits of mollusks and provided that patience is liberally mixed with common sense.

Some beaches, on rare occasions, are strewn with freshly cast-up shells, but the upper shore line is usually a poor collecting area. Most unbroken and colorful shells are obtained by collecting living mollusks in their natural habitats. Since mollusks are mainly nocturnal, night collecting at low tide with the aid of a strong light is usually productive. Daytime collecting can also be rewarding, but then protective rocks should be overturned, crevices investigated, seaweeds pulled and shaken, and sand trails followed. If you are a good swimmer, a face mask and snorkel are useful equipment in water from 2 to 12 feet deep. Remember that living mollusks are seldom as clean and polished as museum specimens. Growths of algae and smaller marine invertebrates can make some gastropods hard to see in their natural habitats.

In addition to a bathing suit and canvas sneakers, little collecting gear is necessary. A cotton or nylon-mesh collecting bag, about 5 by 12 inches, will hold most of your catch. A plastic pail will keep specimens alive. Vials or jars can be used to protect fragile or very small shells. There is a great temptation to overload oneself with equipment. On occasion, gloves may be needed for handling jagged coral rocks, tweezers for getting into tiny crevices, sieves for straining sand, a crowbar for breaking up coral or wood for boring mollusks, a knife for dislodging limpets, and a shovel for digging up long-siphoned clams. Experience and the occasion will dictate what you will most need.

In some areas, baiting carnivorous snails with dead fish or meat is fairly successful. Commercial lobster and crab fishermen often find rare snails in their nets and traps. The stomachs of recently caught fish are a good place to look for shells that are otherwise rarely obtained.

With today's increase in shell collectors, it is essential to practice conservation. Do not disturb a habitat more than absolutely necessary. Roll back the rocks to where you found them, and always leave some specimens behind. If you find collecting successful, it may well be because a considerate collector was there before you! Most maritime states have shellfishery laws, and a few require a sport-fishing license for collecting certain live mollusks. Always practice safety in the field. Respect the sea, beware of tidal currents, protect yourself from severe sunburn, and always wear shoes. These are rules that experienced collectors always keep in mind.

Dredging, or trawling, is perhaps the most effective way of obtaining choice specimens. A small rectangular dredge, with a mouth no more than 20 by 8 inches, can be handled by one person and can be hauled with a quarter-inch line from the stern of a rowboat or an outboard motor boat. Any larger dredge will require steel cable, a winch, and perhaps a davit to get the heavy load aboard. The correct ratio between the depth of the water and the length of line let out should be about 1 to 3—that is, in 50 feet of water use 150 feet of line. Rope tends to float, and this will lift the "biting" edge of the dredge off the bottom. Weights of 10 to 20 pounds should be attached to the pulling line about 15 feet in front of the dredge. By feeling the line, you can tell whether the dredge is skipping over a rocky bottom, fluttering uselessly in midwater, or digging into oozy mud. Excess mud and sand can be sloshed away just before the dredge is brought on board. Dump the contents on a wooden sorting board or in a large tub. Place the fragile specimens in jars or tins. Save a large sample of sand and rubble in a cloth bag. At home the dried bottom sample may reveal many choice miniature shells. Be sure to record the exact locality and depth of water for each dredge haul.

Small Hand Dredges

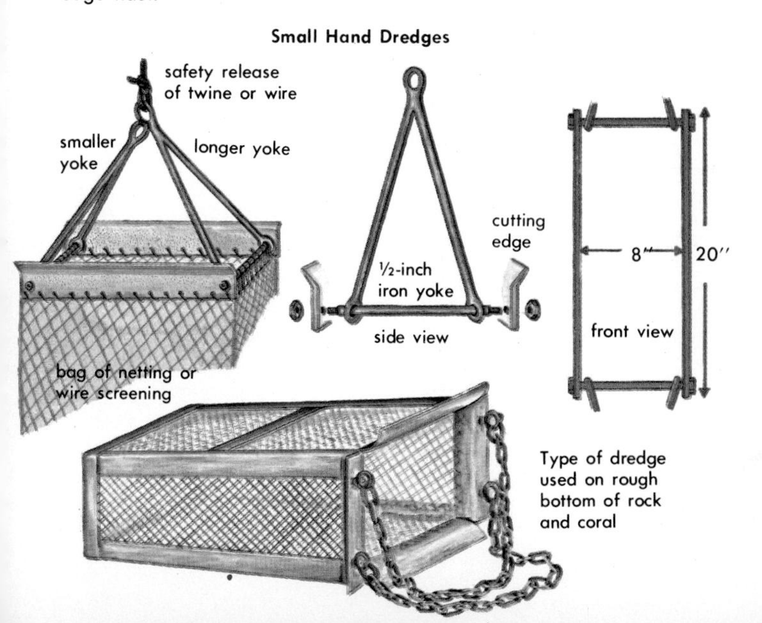

CLEANING AND PRESERVING

Unless live-collected mollusks are properly cleaned or preserved, the soft parts will soon deteriorate and produce offensive odors. The simplest and usually the most successful method is to boil the shell and its soft tenant in fresh or ocean water. In order to avoid damaging glossy shells, start with cool or warm water, and boil for about five minutes. Then let it stand awhile, or add small amounts of cold water to permit gradual cooling. The bivalves will gape, and most of the snails can be hooked and wound out, corkscrew-fashion, with a bent safety pin, an old dental probe, or an ice pick. In emergencies, shells under two inches can be preserved and brought home in layers of table salt.

Some genera, like *Terebra,* are almost impossible to clean by boiling and pulling. Sometimes *Terebra* and other snails can be cleaned alive, if the animal has slightly extended itself. Quickly thrust the prongs of a fork into the columellar muscle and with a twisting motion give a sharp yank. Jet-stream flushing of dead flesh is worth trying, if you plan to do a lot of cleaning. Screw the nozzle of a garden hose to the spigot in the laundry tub or at the side of the house. Set the nozzle for the strong jet. Hold the snail just under the surface of the water in a full bucket in order to prevent annoying spray. Some collectors solder a tubelike reducer to the screw part which fits on the spigot. The "liver" of boiled or two-day-old small snails can usually be flushed out. Large *Strombus, Cassis,* and *Buccinum* may be placed in a deep freezer for a few days, thawed for a day or more, then pulled and flushed. Some heavy snails may be "hung" alive by tying the foot with string and hanging to an overhead branch. The weight of the shell will gradually pull the snail free in a day or so. In certain tropical areas some species of ants will clean the meat out in a few days.

Museum expeditioners place most of their live mollusks in 70 percent grain or ethyl alcohol or in 50 percent methyl or isopropyl alcohol. Formaldehyde and water, *only* if buffered with one teaspoon of bicarbonate of soda per quart of fluid, will do. The druggist sells 40 percent formaldehyde. Add 8 parts of water to get 5 percent formalin. If soaked in one of these preservatives or in strong Jamaican rum for more than a week, snails can be pulled with ease. Very small shells, soaked in alcohol, need only be dried in the sun or oven. If time permits, shells may be buried in soft, dry sand where they will rot out in several weeks. With cowries, make sure that the aperture faces down or the trapped rotting liquid will discolor the shell. In all cases, save the operculum with each snail. A small plug of cotton will keep it in place inside the dry aperture. Small, delicate snails, like *Trivia* or *Marginella,* may be soaked in fresh water for at least 48 hours, with a couple of water changes, until the soft parts have rotted enough to be flushed out.

Some groups of mollusks require special preserving techniques. Squid and octopods may be initially preserved in a 5 percent solution of formalin for a day, and then transferred to 70 percent alcohol. If you do not wish to preserve the whole beast, you may want to cut out the squid's pen or cuttlefish bone or the two horny, parrot-like beaks. Live wood borers, such as *Teredo* and *Bankia,* should be carefully dissected from the boards or logs in which they live and placed in jars containing 4 parts of 70 percent alcohol and 1 part of glycerine. The latter will keep the periostracal margins of the pallets, so important in identification, soft and pliable. Permanent microscope slides, for later consultation, may be made by mounting the pallets in "Diaphane" or "Euparal."

No satisfactory method is known for preserving the natural colors of the soft parts of mollusks, nor for holding the color and shape of the beautiful nudibranchs. If a freezer is available, nudibranchs may be placed in small plastic compartments of ocean water where the snails will freeze in an expanded position. Place the solid ice cube in 50 percent grain alcohol and when it has melted, transfer the snail to 70 percent alcohol.

Various mollusks react differently to such narcotizing agents as menthol crystals floated on top, pinches of epsom salts, a few drops of alcohol, or a 1 percent solution of propylene phenoxetol. The latter works well on most bivalves, cones, and olives.

Unless chitons are specially handled, they will roll up into armadillo-like balls and make unattractive, useless specimens. Allow the live chiton to crawl about in a dish of sea water, preferably on the flat surface of a piece of wood, plastic, or glass. While the chiton is in a flattened position, press your thumb on its back and at the same time pour off the sea water and add 70 percent alcohol. Hold the chiton flat in this solution for about a minute. It will now remain in this flattened position without being held down and may be soaked in the preservative for a few days. The foot and viscera may be removed. Do not damage or remove the girdle. At a later date, if you wish, one specimen of each species may be soaked in warm water and detergent for several hours and all of its valves removed and mounted on cardboard.

Some specimens of the more attractive shells require a certain amount of "face-lifting," although most of this operation is simple scrubbing in warm, soapy water. Shells may be soaked in strong lye or laundry bleach for a few hours to remove algae, stains, and some growths. Following the bleach treatment, some collectors dip the shell for a few seconds in 10 percent muriatic acid, then douse the shell in cold water. Acid, however, is apt to do more damage than good. It is advisable to save some of the shells with their natural periostracum, otherwise important identification features may be lost. The judicious application of a little neatsfoot or baby oil sometimes hides blemishes.

THE SHELL COLLECTION

A well-kept shell collection is an important focus of conchological studies, for it assists in the identification and classification of species, and enables one to understand the biological relationships of the various families. It also helps in planning future trips into the field in search of species new to the collection. Adequate samples, with individuals showing growth stages, accompanied by geographical and ecological data, will reveal interesting variations within a species.

If the collection is housed in a simple, inexpensive manner with sufficient expansion room in each drawer, adding new samples requires little time and effort. Properly catalogued and numbered specimens are protected against accidental mixing and the loss of valuable field observations. Simplicity, mobility, expansibility, and inexpensiveness are the keys to a worthwhile and rewarding collection.

The type of cabinet shown below is ideal, but there are other, although less desirable, methods of housing shells. Large cardboard boxes, cigar boxes, wooden boxes with glass covers, and papier-mâché egg cartons may be used for storage, but uniformity of size and style is recommended. Secondhand, glass-fronted display cases or old china cabinets may be used to display large shells.

A cabinet of plywood or metal made as dustproof as possible may have wooden or metal runners on which slide wooden or "pressed wood" drawers, uniformly 2 inches deep, 20 inches wide, and 24 inches long. By placing drawers two or three runners apart, room is left for very large shells. The front edge of each drawer should have a removable label on which are listed the genera and species. Leave some room in the drawers for future expansion.

Sample lots of shells (one specimen of one or more species from one locality) are placed with a label in ¾-inch-deep cardboard trays, of which the most convenient sizes are $1\frac{1}{2} \times 3$, 3×3, 4×3, 6×3, and 6×6 inches.

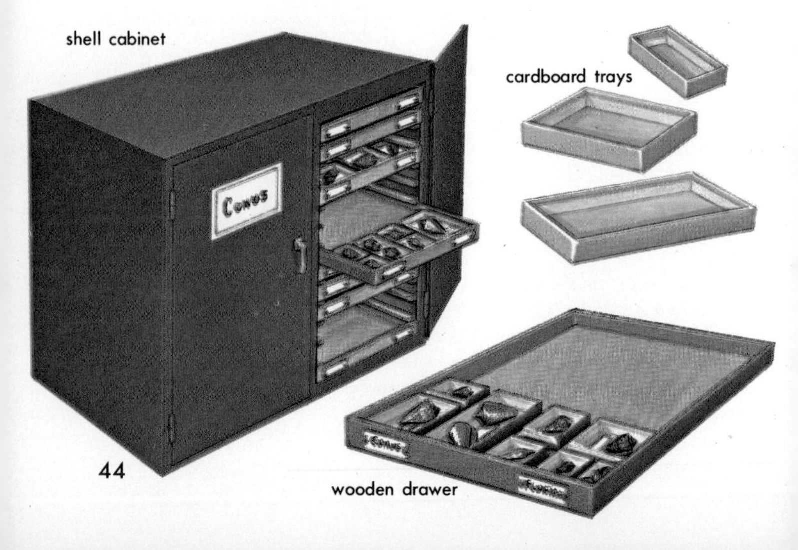

Small shells, on which it is difficult to write the catalog numbers, may be put in glass vials and plugged with cotton, or in small plastic boxes.

Labels of good-quality bond paper should fit easily in the bottom of the smallest-sized cardboard tray and should bear essential data. Catalog numbers should be written in India ink on the label and on the specimens.

The most important data to be placed on a label are the exact collecting locality, date, name of collector, and habitat information. The scientific name of the shell can be added later. Keep old labels; they are a historical record.

shell vials

labels

catalog book

A ledger or loose-leaf book serves well as a catalog, which is basically a numerical recording of the individual lots or samples in the collection. Seven columns across two facing pages suffice to record the following: catalog number; scientific name of shell; locality data (this should be the widest column); number of specimens; collector; donor; and remarks. The last can include special field notes.

A geographical card catalog that cross-indexes the collection serves to show the fauna found at various places. Main divisions (blue cards) may be by states, islands, or collecting stations. One white card per species, with several locality records, will suffice. Buff cards could separate genera or families. A card index of your collection, arranged by families or genera, is not recommended, since your collection, if arranged systematically, gives this information.

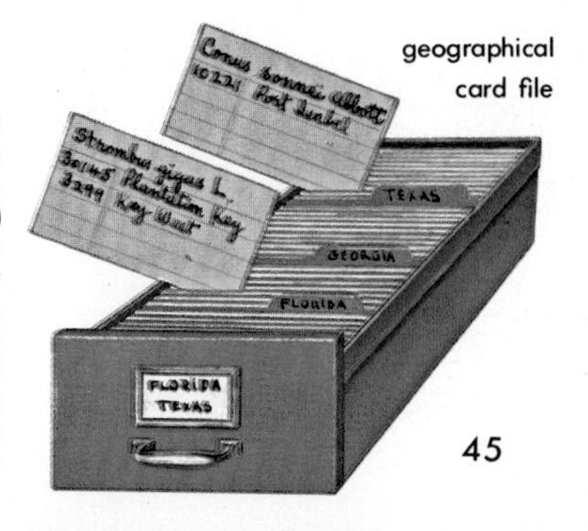

geographical card file

STUDYING AND OBSERVING MOLLUSKS

Although a great deal of scientific research on mollusks is being carried out by professional malacologists in marine laboratories and museums, there remains a vast and untapped field of original observations to be made by amateur investigators. Many avenues of study are open, whether it be making a census of the mollusks of a nearby shoreline or patiently watching to discover the egg-laying methods of a marine snail in a home aquarium.

If you plan to make a study of the molluscan population of a limited area, be sure to keep accurate, methodical notes, and, if possible, do it on a weekly basis for an entire year. Keep samples of the shells of the animals that you observe, so that at a later date you can have your identifications verified by an expert. Make or obtain a good chart or map of the area, indicating the direction of currents, wind, entrances of fresh-water streams, and other physical characteristics that may change and influence the mollusks. In examining a tidepool, for example, here are some of the facts that might be recorded: size and depth of pool (at low and high tide); temperature; clarity; distance from the open ocean; number and kinds of fish, crabs, and starfish; and kinds and distribution of sea plants and algae. In observing the mollusks, for example, find out how many species live in the tidepool. What do they feed upon? What are their enemies? Do the snails move about or stay in about the same place? What happens at night? Does a full moon influence them? Do the *Nucella* or *Thais* carnivorous snails influence the size of the populations of barnacles and mussels? Compare your master tidepool with others nearby. Are they the same, and, if not, can you suggest various types of observations that might explain the differences?

Similar biological detective work can be done in marshlands, mud flats, wharf pilings, beds of eelgrass, or coral reefs. For instance, a cubic foot of mud may be removed, sieved, and analyzed from ten different places from the high-tide mark down to a water depth of about four feet. Does each sample have the same number and kinds of mollusks, algae, and other invertebrates? Does the size of the sand grains differ? Can you deduce why a certain kind of clam or snail prefers deeper water or softer mud? Suggest a reason, then try to prove it by making more observations, always keeping accurate information. Be prepared for unexpected discoveries. You may, for instance, learn why some species of *Oliva* or *Strombus* are found in shallow water only in the spring when mating or egg-laying is taking place. Be on the look-out for interesting feeding or egg-laying habits or methods of defense. If possible, take photographs of living mollusks and keep the shell in your collection for future reference.

Write up your results, if for only your own satisfaction, or prepare them in the form of a formal school research paper. Many shell clubs issue monthly publications and welcome new biological information.

MARINE AQUARIUMS FOR MOLLUSKS

Many fascinating and unexpected happenings can be observed in a home salt-water aquarium. The egg cases, veligers, young stages, and feeding habits of many species of mollusks have been first noted and accurately reported upon by amateur conchologists who were able to maintain simple and inexpensive aquariums at room temperatures.

Beginners invariably make the mistake of overloading their aquariums with too many specimens, too many plants, too much sand, and too much food. The results are polluted water, dead organisms, and a very unpleasant aroma.

Large 50-gallon tanks with filters and pumps, colorful fish, and artistic lighting are for fresh-water fanciers and the very advanced aquarium expert. Pumps for aeration are not necessary if one uses a large, shallow, dishlike aquarium. However, a very inexpensive air pump may be purchased at your aquarium dealer's shop and hooked up by means of small, plastic tubing to several small, salt-water aquariums. Oblong or square plastic, covered, refrigerator containers, sold at most hardware stores, make excellent miniature aquariums. File a round hole where the lid and top edge of the plastic box overlap, so that the aerating tube can fit through into the aquarium water. Use pint- or quart-size aquariums for one or two species at a time, and limit the number of specimens to four or five, depending upon their size. A whelk or conch about the size of an orange can be kept alive for over a year in an aerated gallon glass jar with plastic cover. Do not add sand or algae. Feed the snail a small, live clam every month, being sure to change the water after feeding.

Extra salt water, obtained at the seashore and strained through a few thicknesses of cloth, may be kept in gallon jugs in the dark until ready for use. Artificial sea water, purchased in salt form inexpensively at your aquarium store, has the advantage of being free of organic impurities and microscopic creatures which can multiply and foul the water. Prevent evaporation in your aquarium by covering with a plastic or glass plate, leaving a crack at one end. If the water level goes down, add fresh, not salt, water. Keep small pieces of frozen mussel, clam, or snail in your freezer, and feed it in very small amounts to your carnivorous snails, making sure that uneaten pieces are removed in a few hours. Avoid meat with fat, since this will make a stifling film on the water surface. From the intertidal zone obtain small porous rocks covered with mosslike algae. These and pieces of the green, lettuce-like *Ulva* will give the herbivorous species something to walk and feed upon. A few clams or mussels may be added to deep-water, aerated aquariums, but note that starfish, worms, or carnivorous snails may eat them. Keep accurate notes and make drawings or take photographs. (Consult an article by D. Raeihle in *How to Collect Shells,* published by the American Malocological Union, 1966.)

NOMENCLATURE AND CLASSIFICATION

A stable system of scientific names based on the careful study of species is the backbone of work on molluscan phylogeny, or the study of "family trees." Latinized names are understood and accepted by malacologists in all countries, whereas popular or common names (when they exist at all) may vary from one place to another. A conch in New England is a *Busycon,* while in Florida the name refers to a *Strombus.*

The binomial system of naming a species is somewhat akin to our own method of naming people. The generic name, like the surname of Adams or Smith, is always capitalized—for example, *Pecten* or *Strombus.* The species name, comparable to Ann or Robert, however, is not—for example, *annae, roberti,* or *californiensis.* The name of the scientist then follows—for example, *Strombus gigas* Linné. Parentheses are placed around the author's name if the species is now placed in a different genus. Thus, the Calico Scallop, originally described as an *Ostrea,* is now written as *Aequipecten gibbus* (Linné). If a species name is an adjective, its gender must agree with that of the genus—for example, *Conus albus, Buccinum album,* and *Tellina alba*. The many technical rules of nomenclature are governed by the International Commission on Zoological Nomenclature, located in London.

In order to understand and conveniently discuss groups of related species, taxonomists have set up higher taxonomic categories within the phylum Mollusca. The major division is the class, such as Gastropoda (snails) or Pelecypoda (clams). Next is the family, established to include several closely related genera. An example of a classification below the level of the class Gastropoda is given below:

Subclass Prosobranchia
Order Caenogastropoda
Superfamily Cypraeacea
Family Cypraeidae
Subfamily Cypraeinae
Genus *Cypraea*
Subgenus *Erosaria*
Species *spurca* Linné, 1758
Subspecies *acicularis* Gmelin, 1791.

There is no official pronunciation established for latinized names, and for certain words it may vary from one country to another.

A biological species is a basic unit or kind of plant or animal life, consisting of interbreeding natural populations that are unable to breed successfully with other kinds because of geographical, physiological, and ecological barriers. A subspecies is a geographically isolated series of populations having somewhat different characters but potentially still able to interbreed with the main species.

IDENTIFICATION

The correct identification of a mollusk makes it possible for biologists and shell collectors to seek or share more information about a species. Identification is the foundation of classification and makes it possible to organize and confirm knowledge about mollusks. It is not always easy to identify mollusks, even some of the relatively common forms. Some species exhibit only slight differences, while others may be variable in color, shape, and sculpturing. A constant source of confusion for many beginners are the immature stages of some of the common species, which may differ from the adult stages. Bivalves generally do not show great change as they mature.

The soft parts of mollusks are important in distinguishing species, making it desirable to study the living animal rather than just the shell. The fleshy mantle extending over the shell of the *Cyphoma* snail is characteristic for each species, whereas the shells show few characteristic differences.

Animals living in water and secreting shells are naturally subject to considerable influence by currents, turbidity, temperature, foreign growths, and the nature of their diet. Shells produced in quiet water are usually larger, more fragile, and more spinose than those produced by animals living in exposed, wave-dashed sites. A *Crepidula* slipper shell living on a scallop shell will take on the ribbed pattern of its bivalve substrate. Even after death, the shells of the mollusk may be altered by abrasive sands, by the tiny borings of sponges, and by hermit crabs, which wear a deep, rounded notch on the columella of a snail shell.

Abnormalities caused by injuries are not uncommon among many species. Unusually high spires, extra rows of spines, twisted or double siphonal canals, and "uncoiled" whorls are occasionally made. In some species that are normally coiled clockwise, or dextrally, a sinistrally coiled, or "left-handed," specimen may rarely occur. In normally colorful species, an occasional pure white, or albino, shell may be produced.

Variations in Shell Caused by Environment

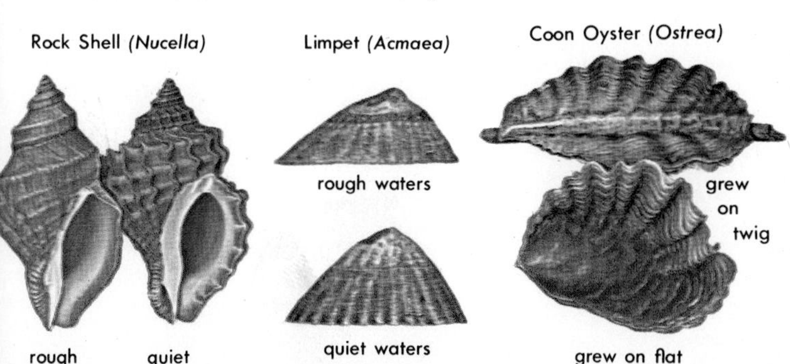

CLASS GASTROPODA (Univalves)

The snail or univalve class includes mollusks with a distinct head, usually with eyes and tentacles, and with a solelike foot adapted for creeping. In such a large class, containing over 80,000 living species and hundreds of genera, there are many peculiar forms that make exceptions to any general definitions. Many gastropods have no shell; many lack a creeping foot; some parasitic forms resemble worms; and one tectibranch snail has two shells. The classification of the gastropods is based primarily on anatomical features, such as the nature of the gills, the heart, and the radular teeth. At the generic level, the radula, shell, and operculum are more important.

An outline of the classification of the living marine gastropods is given below. Only the major groups (most superfamilies and large families) are mentioned.

SUBCLASS PROSOBRANCHIA (auricle of heart is anterior to the ventricle; visceral nerve loop forms a figure eight).

Order Archaeogastropoda (one or two gills; radula with numerous rows of radulae).

Superfamilies: Pleurotomariacea; Fissurellacea; Patellacea; Trochacea; Neritacea.

Order Caenogastropoda (one set of comblike gills; radula with only up to 7 teeth per row. Previously was Meso- and Neo-gastropoda).

Superfamilies: Littorinacea, Cerithiacea, Rissoacea, Epitoniacea, Hipponicacea, Calyptraeacea, Strombacea, Cypraeacea, Naticacea, Tonnacea, Muricacea, Buccinacea, Volutacea, Conacea.

SUBCLASS OPISTHOBRANCHIA (auricle is posterior to the ventricle; visceral loop forms oval).

Order Tectibranchia (gills present on inner right side; usually has shell).

Superfamilies: Bullacea, Aplysiacea, Pyramidellacea.

Order Pteropoda (shell and swimming appendages usually present).

Superfamilies: Cavoliniacea, Peracleacea, Clionacea.

Order Saccoglossa (single row of teeth, preserved in special sac when worn out; shell reduced).

Families: Oxynoidae, Stiligeridae, Elysiidae, etc.

Order Notaspidea (sluglike, with a shell, and true gills on right side).

Families: Umbraculidae, Pleurobranchidae.

Order Nudibranchia (sluglike; gills and shell absent).

Superfamilies: Doridacea, Aeolidiacea, Rhodopacea.

SUBCLASS PULMONATA (mantle cavity modified into "lung"). Land and fresh-water snails.

GLOSSARY OF GASTROPOD SHELL TERMS

Aperture. The opening in the last whorl, providing an outlet for the head and foot.

Apex. The first-formed, narrow end of the shell, usually of several whorls.

Axial. Parallel to the lengthwise axis of the shell; suture to suture features.

Base. The lower, siphonal end of the body whorl; opposite to the apex.

Canal. Semitubular extension of the aperture, either posteriorly or anteriorly, the latter being the siphonal canal.

Columella. The solid pillar at the axis of the shell, around which the whorls grow.

Cord. A coarse, large spiral line or thread.

Dextral. Right-handed or of whorls growing clockwise. Aperture at right.

Growth lines. Irregular, axial lines or blemishes caused by growth stoppages.

Inner lip. The wall on the body whorl opposite the outer lip, or rim, of the aperture.

Operculum. A "trapdoor" attached to the foot that, when withdrawn, helps to seal the aperture. May be corneous (horny or chitinous) or calcareous (hard, shelly). Absent in adults of some families, like abalones, keyhole limpets, and cowries.

Outer lip. Final margin or edge of the body whorl, opposite the inner lip.

Parietal region. The area or wall on the columella side of the aperture; the outer body wall opposite the outer lip.

Periostracum. The layer or coat of horny material covering the outer shell. It may be thin, thick, smooth, rough, or hairy.

Periphery. Part of the whorl projecting or bulging farthest from the axis.

Sculpture. Relief pattern on the shell surface; can be axial or spiral.

Spiral. Sculpturing encircling the whorl parallel to the sutures.

Spire. The whorls at the apical end, exclusive of the last whorl.

Suture. Continuous line on shell surface where whorls adjoin.

Umbilicus. A central cavity at the base of the shell, around which the whorls coil.

Varix. An axial rib or swelling made during a major growth stoppage.

Whorl. A turn or coil of a snail. The *body whorl* is the last and largest. The *penultimate* whorl is the next to the last one.

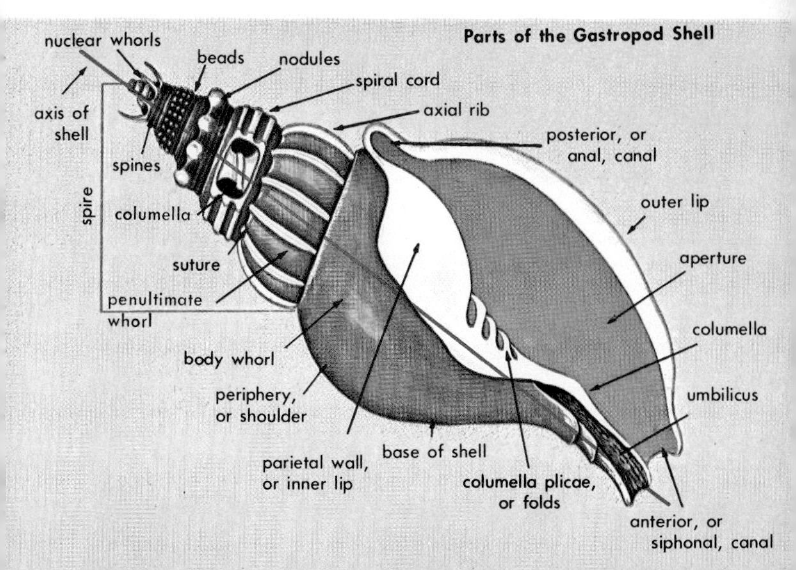

SLIT SHELLS

The primitive slit shells, of the family Pleurotomariidae, live in fairly deep waters in various parts of the world. Although they are abundant as fossils, only about a dozen species are known to be living. There are seven species in western Atlantic waters, and although they are seldom found in private collections, a representation is included here because of their unique position among the gastropods. Although the slit shells are moderately common on the sea bottom, where they feed upon algae and detritus that drifts down from coastal areas, they are not often captured except accidentally in fish traps. The large, beaded, colorful slit shells are readily recognized by the natural slit in the middle of the last whorl.

The internal anatomy of the pleurotomarids is considered primitive because of the existence of two long gill plumes. In most advanced snails the right gill is absent. Because waste water leaving the mantle cavity would otherwise pass over the two gills, a long, open slit in the shell serves as the point of exit for excreta and deoxygenated water. In other respects, the pleurotomarids resemble the trochid top snails (p. 68) in having a large, simple foot, a broad proboscis, and two delicate, long tentacles with an eye at the base of each. The radular teeth are of the rhipidoglossate type, with numerous fine teeth in each transverse row. This type of radula is adapted for rasping on delicate seaweeds, the main food of slit shells. The operculum, attached to the dorsal side of the posterior end of the foot, is round, thin, horny, and has many whorls. In most species, the operculum is quite small and incapable of sealing off the mouth of the shell. It serves as a protective pad upon which the shell rests when the snail has its soft foot fully extended. The sexes are separate in this family, and, although not as yet observed, it is probable that the eggs and sperm are shed freely into the water as in most trochids.

The identification of the slit shells is based upon the size and position of the open slit, the presence or absence of an umbilicus or opening on the underside of the shell, the nature of the beaded sculpturing, and the angle of the spire.

1. **LOVELY SLIT SHELL.** *Pleurotomaria amabilis* (F. M. Bayer). S. Fla.–Gulf of Mexico. The angle of the apex is about 65°; its sides slightly concave. The slit is about 1/5 the circumference of the body whorl. Exterior finely beaded. Rare; 300 to 700 ft.

Quoy's Slit Shell. *P. quoyana* Fischer and Bernardi (not illus.). From the Gulf of Mexico to the Antilles, is similar, 1.5 in., pale brownish, and with an apical angle of 80°. Deep water; rare.

2. **ADANSON'S SLIT SHELL.** *Pleurotomaria adansoniana* Crosse and Fischer. Bahamas–Lesser Antilles. Spire high, with somewhat convex sides, its angle about 85° to 90°. Slit narrow and very long, extending halfway around the body whorl. Exterior rough, cream with red splotches or maculations. Umbilicus large, round, and very deep. Lives on rough, steeply sloping areas at depths of 200 to 1,500 ft. Rare; sometimes found in fish traps set in deep water.

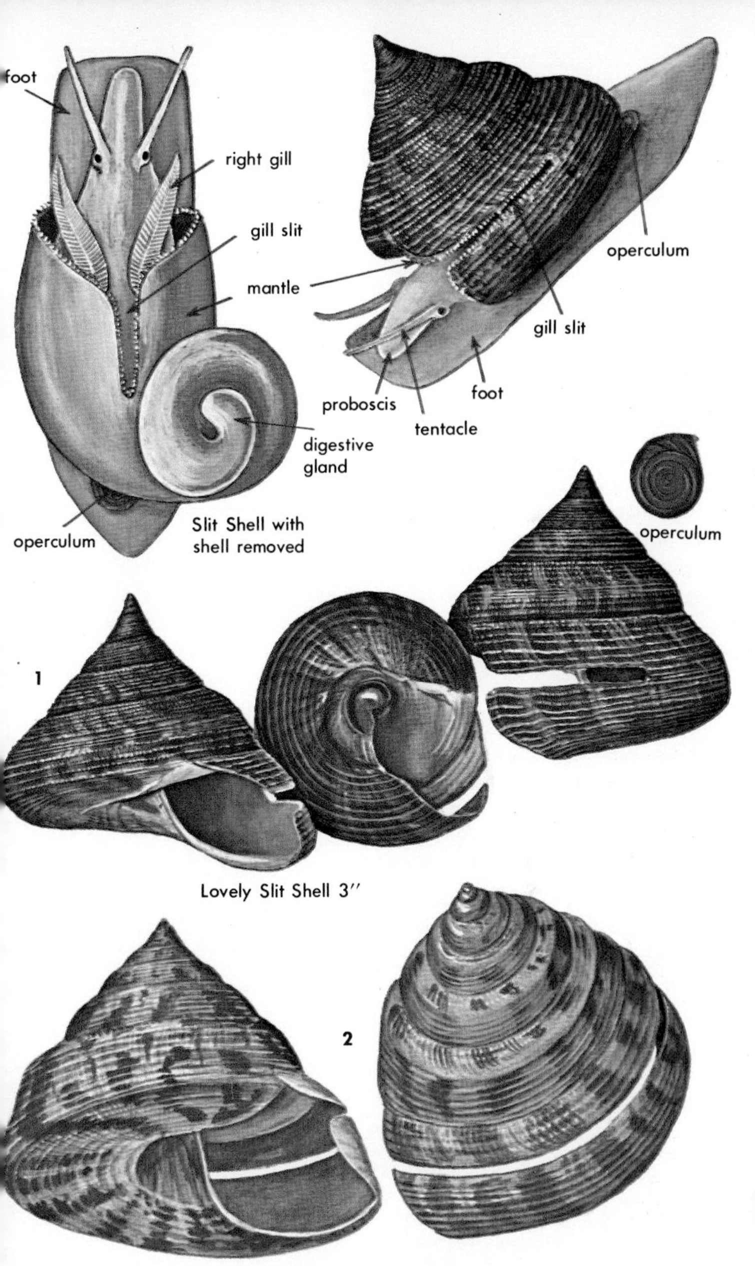

Lovely Slit Shell 3″

Adanson's Slit Shell 6″

ABALONES

Used by man since prehistoric times, abalones (family Haliotidae) are a favorite food and have decorative shells. These snails are characterized by their flat, dishlike shells, which are iridescent on the inside and have four to six natural holes through which the animal passes water and excrement. World-wide, there are over 100 species, 8 of which come from the Pacific coast of Mexico and California. Most abalones dwell on rocks in shallow water, but some may live as deep as 1,200 feet. They cling tenaciously to the substrate and can be removed only with the aid of a pry-bar. All are vegetarians, feeding mainly on algae, including the giant kelps. Adults usually browse in the same general area all their lives. The sexes are separate, with the gonads of the females being green and those of the male yellowish.

Californian abalones spawn in the spring and summer by ejecting sperm and eggs into the water. In 10 days the free-swimming larval veligers settle to the bottom and, within 2 months, develop into miniature adults. At an age of 1 year, the Red Abalone is 1-inch long, and within 4 years it reaches sexual maturity, at about 5 inches in length. Foot-long specimens are probably 15 to 20 years old.

Indians of the southwestern Pacific used abalones as ornaments and utensils. Numerous trade routes from California to southwestern and central United States have been traced by means of shell remains in ancient burial sites. Today, a flourishing fishery exists in southern California and Mexico, with an average annual catch of 3 million pounds of abalone meat. State laws strictly control abalone collecting, but divers gather legal-size shells at a rate of about 80,000 a year. Abalone steaks are delicious when properly prepared. Packing houses clean, trim, slice, pound, and market fresh and frozen fillets, with the bulk going to restaurants.

1. **PINK ABALONE.** *Haliotis corrugata* Wood. S. Calif.–Baja Calif. Exterior irregularly corrugated; scalloped edges. The 2–4 holes are open and have high rims. Interior pearly pink; muscle scar large. Moderately common; mostly from 20–80 ft.

Pourtales' Abalone 1″

2. **BLACK ABALONE.** *Haliotis cracherodi* Leach. Oregon–Baja Calif. Exterior smooth and bluish black, with 5–9 holes open (rarely none in the form *H. c. imperforata* Dall). Interior iridescent, whitish; muscle scar weak. Common; mostly in intertidal waters.

3. **POURTALES' ABALONE.** *Haliotis pourtalesi* Dall. N.C.—Fla. Keys. Small, low, elongate shell with 22–27 wavy spiral cords. Exterior yellow to light brown, rarely orange-red. Five holes open. Interior pearly white. A rare collector's item sometimes found in dredgings made at 200–1,200 ft.

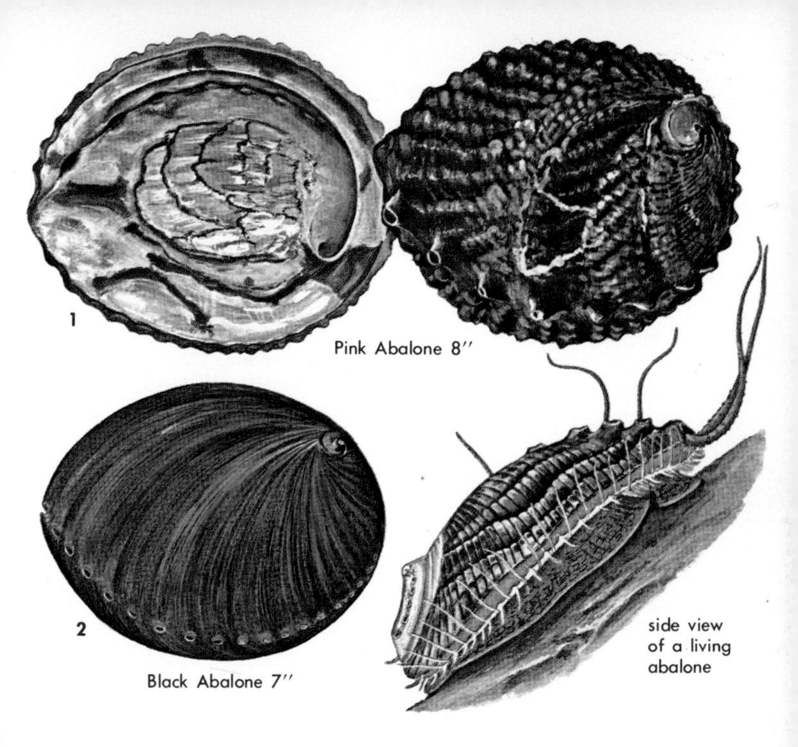

Pink Abalone 8″

Black Abalone 7″

Commercial abalone diver off California

ABALONES—continued

Large, sedentary mollusks, such as abalones, are likely to become covered with marine growths and serve as havens for other small creatures. The large, flat shell may sustain a small community of algae, sponges, barnacles, bryozoans, and hydroids. As many as 90 species of small living gastropods have been found on shells of the Red Abalone of California.

Probably feeding on these associated organisms is a small purple shrimp *(Betaeus),* which flits under the abalone shell and takes refuge in the mantle cavity in time of danger. Marine worms find protection in the miniature jungle of algae. Tiny, parasitic *Odostomia* snails crawl to the edge of the shell and plunge their proboscides into the fleshy mantle to suck the blood of the abalone.

The thick shells of abalones are sometimes riddled by various boring mussels *(Lithophaga)* and piddocks *(Penitella).* The abalone builds an interior reinforcement against these invaders by forming a shelly blister or, rarely, by creating an attached iridescent baroque pearl.

The predators of abalones, aside from man, include the rock crab, the octopus, bottom-feeding fish, and the sea otter. The last can smash the shell of a tightly clinging abalone with a rock, and scoop out the meat with its paw. An abalone can "shuffle" forward on its foot with surprising speed, especially when stimulated by a crab, starfish, or other natural enemy. As a last resort, the snail will withdraw and clamp fast to the substrate.

1. **PINTO ABALONE.** *Haliotis kamtschatkana* Jonas. S. Alaska–central Calif. Exterior rough, irregular, mottled green. Has 3–6 (usually 5) open holes with rims. Interior pearly white and quite irregular; muscle scar weak. Moderately common; intertidal to 50 ft. on black rocks.

2. **NORTHERN GREEN ABALONE.** *Haliotis walallensis* Stearns. B.C.–Calif. Elongate and flat; surface red and bluish and with fine threadlike lines. Has 4–8 (usually 6) open holes. Interior pearly; muscle scar very weak. Moderately common; subtidal to about 70 ft.

3. **GREEN ABALONE.** *Haliotis fulgens* Philippi. Oregon–Baja Calif. Shell oval and thick. Exterior olive to brownish; spirally grooved. Has 5–7 open holes. Interior bluish green; muscle scar bright iridescent. Formerly abundant; recently over-collected. Uncommon; 10–20 ft.

4. **WHITE ABALONE.** *Haliotis sorenseni* Bartsch. S. Calif.–Baja Calif. Deep oval shell; exterior reddish brown, with rough spiral ribs; 3–5 open holes with elevated rims. Interior pearly, pinkish; muscle scar obscure. An excellent tasting form. Uncommon; 80–100 ft.

5. **THREADED ABALONE.** *Haliotis assimilis* Dall. S. Calif.–Baja Calif. An oval, thin-shelled abalone, with exterior threaded. Greenish with red and white splotches. Has 4–6 open holes with raised rims. Interior pearly; muscle scar obscure. Rare; 70–100 ft.

6. **RED ABALONE.** *Haliotis rufescens* Swainson. Wash.–Calif. Exterior lumpy, brick-red, sometimes encrusted. Has 3 or 4 open holes. Interior pearly; muscle scar prominent, with a "smeared" look. Rim of shell is red. Fairly common and best-tasting abalone. On rocks; 20–50 ft.

1
Pinto Abalone 5″
2
Northern Green Abalone 4″
interior view
3
Green Abalone 7″
exterior view
4
White Abalone 7″
5
Threaded Abalone 5″
6
Red Abalone 10″

KEYHOLE LIMPETS

The keyhole limpets (family Fissurellidae), which have an anal hole at the top of the shell, belong to a major group of snails, the zeugobranchs, which have two well-developed gills. The hole in the shell is analogous to the series of holes found in the abalones, which also have two gills. Originally, in the primitive forms, the hole was a slit in the anterior edge of the shell. This slit is present in the adults of such a genus as *Emarginula*. In the embryonic shell of the keyhole limpets, such as *Fissurela* and *Diodora,* the slit is well developed, but as growth continues the slit is filled in at its front edge and becomes a hole. By the time adulthood is reached, the shell has grown so that the hole is located near the apex of the shell. Water is drawn in the underside of the shell, passed over the gills, and then passed out through the hole. Fecal matter and waste water are expelled through this keyhole also. Adults do not have an operculum. The interior of the shell may have a faint horseshoe-shaped scar opening anteriorly.

About 27 genera and over 500 species form the family Fissurellidae. Large species are eaten in some localities. A few other species are used extensively in the shellcraft industry. Members of the family are vegetarians, usually browsing at night on algae-covered rocks in the intertidal zone. The rhipidoglossate radula (see p. 25) is long and contains many rather hard teeth in each row. The continual rasping by many generations of keyhole limpets contributes to the gradual erosion of hard shoreline rocks.

1. **BRIDLE RIMULA.** *Rimula frenulata* Dall. N.C.–West Indies. In the genus *Rimula* the anal slit is long, arrow-shaped, and has a long gutter running toward the apex. This species is thin, delicate, elongate, and one third as high as it is long. Exterior finely cancellate. Color whitish to cream, sometimes with rusty stains near the hooked apex. Moderately common; found on sandy or on rocky bottoms at depths of from 30–1,000 ft.

2. **HOODED PUNCTURELLA.** *Puncturella cucullata* Gould. Alaska–Baja Calif. Apex of shell minute and hooked anteriorly. Behind it is a small, elongate slit. Internally the slit is separated from the apex by a strong, convex, shelly shelf. Exterior of shell has 15–23 major ribs, often with 2–4 smaller riblets between. Border of shell wavy. Moderately common; on rocky bottom, from 6–500 ft.

3. **PIGMY EMARGINULA.** *Emarginula pumila* (A. Adams). S.E. Fla.–Brazil. Shell shape variable. Slit at anterior end short and usually overlooked. Apex near center of shell and hooked backward. Exterior rough, with 17–20 radial ribs. Common; on rough bottom, 6–500 ft.

Ruffled Rimula. *E. phrixodes* Dall, (not illus.), is about 0.3 in. long. Found off Fla. at 200 ft.; white, with 20–24 ribs crossed by delicate threads. Uncommon.

4. **LINNÉ'S PUNCTURELLA.** *Puncturella noachina* (Linné). Circumpolar; south to Mass. and Alaska; also northern Europe. Shell delicate, compressed laterally, so that the base is elongated. Margin wavy. Slit very small just in front of apex, with a cup bordering it on the inside of the shell. Exterior gray-brown to white, with 21–26 main riblets. Common; under rocks, 10–200 ft. In Alaska, under intertidal rocks.

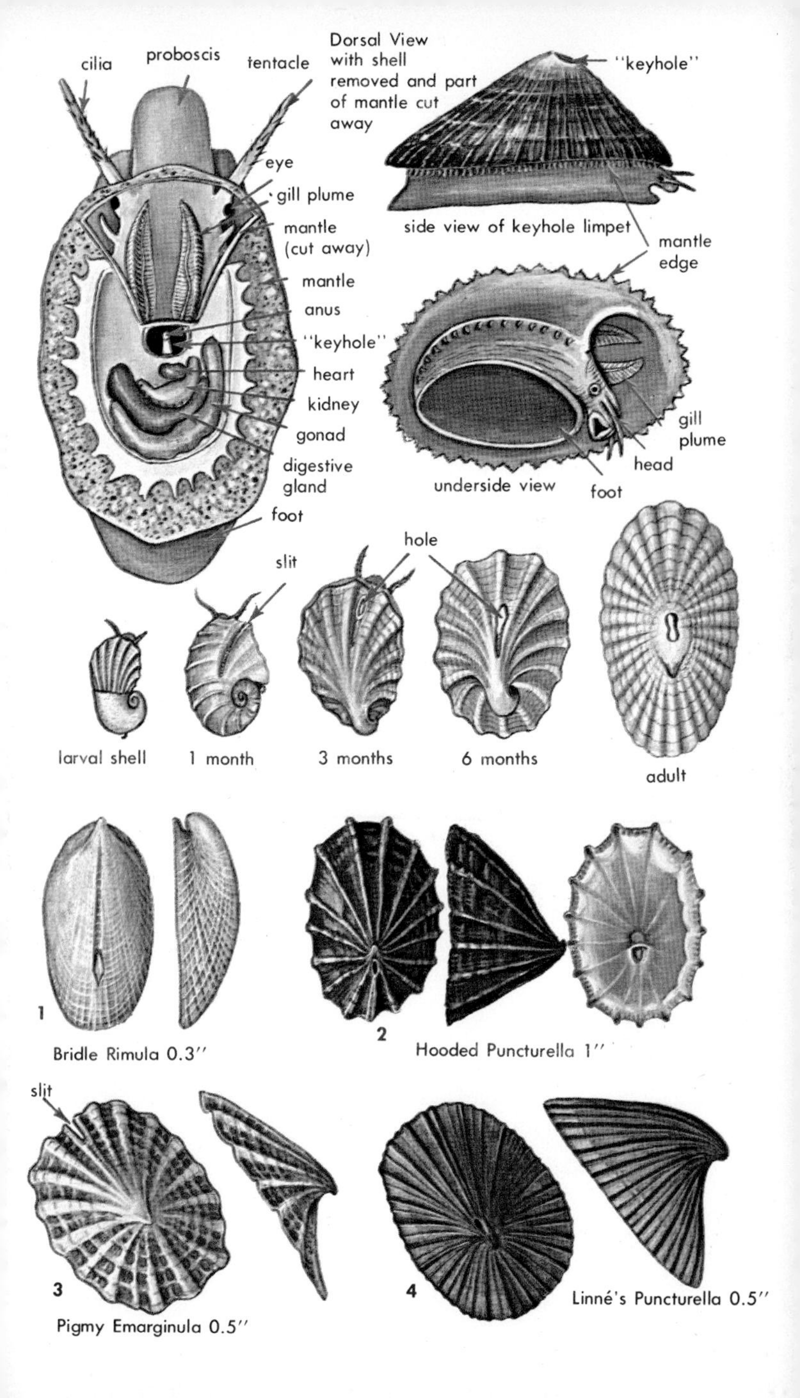

Bridle Rimula 0.3″

Hooded Puncturella 1″

Pigmy Emarginula 0.5″

Linné's Puncturella 0.5″

KEYHOLE LIMPETS—continued

The keyhole limpets on this page are species with a callus around the inside of the keyhole, as in the interior illustrations.

The mantle edges of *Diodora* species are usually festooned with numerous short filaments, which are sensitive to light changes. The animals are strictly marine and will not tolerate brackish water. They breed in the spring. The sticky, yellow eggs are exuded from the mantle cavity and adhere to nearby rocks. They hatch into crawling animals. This group does not have a free-swimming phase. In *Lucapina* the mantle extends over most of the exterior of the shell.

1. **LISTER'S KEYHOLE LIMPET.** *Diodora listeri* (Orbigny). Fla.–West Indies. Shell strong, with a rough exterior of strong radial ribs. Alternate ribs may be larger. Concentric threads strong, forming little squares as they cross the sometimes scaly radial ribs. Color cream to gray; rarely, with obscure bluish bands. Found on rocks at low-tide mark.

2. **CAYENNE KEYHOLE LIMPET.** *Diodora cayenensis* (Lamarck). N.J.–Brazil. Anal orifice (keyhole) is just in front of and lower than apex. Radial ribs numerous and irregular, every third or fourth one being larger. Interior gray or whitish. Exterior dark gray. Very common; on rocks, 3–30 ft.

3. **DYSON'S KEYHOLE LIMPET.** *Diodora dysoni* (Reeve). S. Fla.–Brazil. Shell small, elongate, with straight sides, and beautifully sculptured, with 17–19 strong ribs, between each of which are 2 or 3 small ones. Keyhole triangular in shape. Concentric threads strong. Interior creamy white with 8 solid or broken, narrow, blue-black rays. Moderately common under rocks near shore.

4. **DWARF KEYHOLE LIMPET.** *Diodora minuta* (Lamarck). S.E. Fla.–West Indies. Shell small, thin, low, and weakly sculptured, with many finely beaded ribs, most of which are pure black in color. The short front slope of the shell is slightly concave, while the longer posterior slope is slightly convex. Margin finely crenulate. Uncommon; 30–500 ft.

5. **CANCELLATE FLESHY LIMPET.** *Lucapina suffusa* (Reeve). S. Fla.–Brazil. Shell elongate with a low apex. The keyhole is stained bluish black outside and on its inner edges. Sculptured with alternately larger and smaller radial ribs. Interior color is delicate, with faint rays. Exterior gray to dirty white. Soft parts orange. Common; under rocks, 1–20 ft.

6. **SOWERBY'S FLESHY LIMPET.** *Lucapina sowerbii* (Sowerby). S.E. Fla.–Brazil. Shell oblong and low. Exterior has about 60 alternately large and small riblets. Concentric threads fairly strong. Color buff, with 7–9 splotched rays of pale brown. Outside of keyhole not stained. Soft parts cream with brown specklings. Uncommon; intertidally under rocks.

7. **ROUGH KEYHOLE LIMPET.** *Diodora aspera* (Eschscholtz). S. Alaska–Baja Calif. The keyhole is oval and its inner sides flat. It is located about one third of the way back from the narrower anterior end of the shell. Exterior purplish with darker radial bands and with many coarse radial riblets. Concentric threads fine. Common; on rocks from the low-tide line to 40 ft., rarely on stems of kelp.

Neat-Ribbed Keyhole Limpet. *Diodora arnoldi* J. McLean (not illus.). Central Calif.–Baja Calif. Shell is 0.7 in. long, similar to Rough Keyhole Limpet but longer and with much finer, neater cancellate sculpturing. The keyhole is nearer front end. Color white with broken, blackish rays. Uncommon; 3–70 ft. Formerly called *D. murina* Arnold.

1

Lister's Keyhole Limpet 1″

2

Cayenne Keyhole Limpet 0.7″

3

Dyson's Keyhole Limpet 0.5″

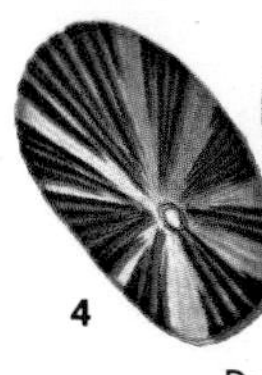

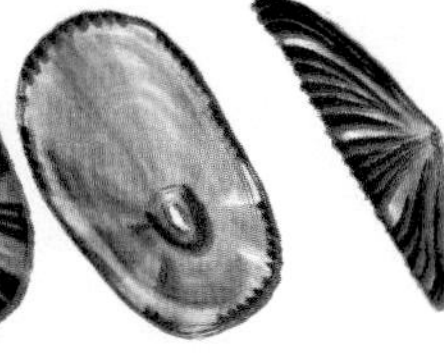

4

Dwarf Keyhole Limpet 0.5″

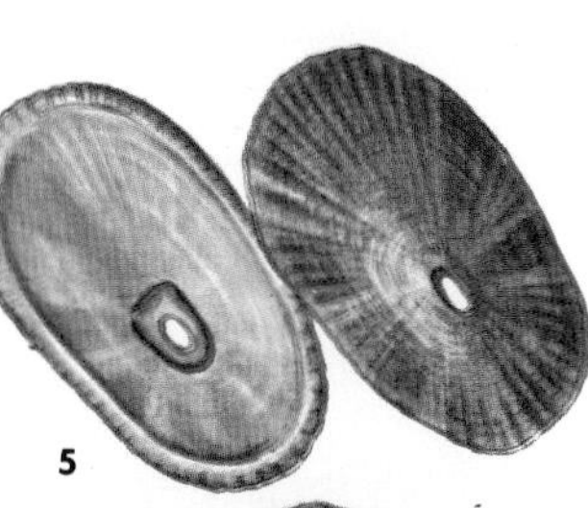

5

Cancellate Fleshy Limpet 0.7″

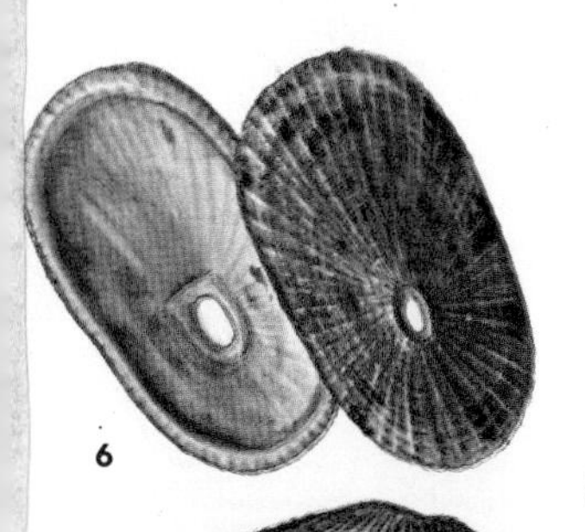

6

Sowerby's Fleshy Limpet 0.8″

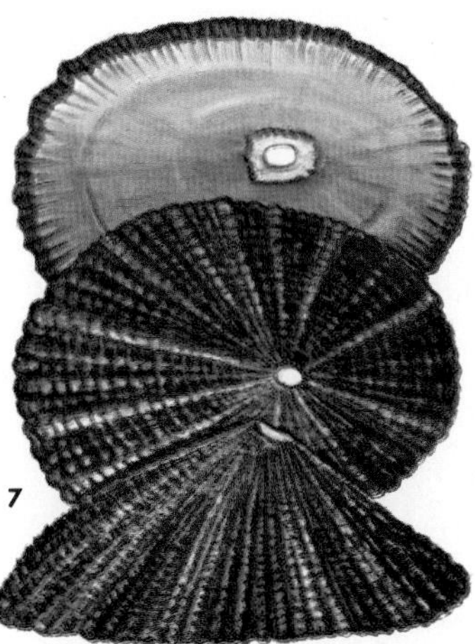

7

Rough Keyhole Limpet 1.2″

KEYHOLE LIMPETS—continued

One of the most inhospitable regions of the ocean is the lowest intertidal zone along a rocky coast. Despite the heavy pounding of the waves, a surprising number of plants and animals survive. Among the beds of *Modiolus* mussels and the mats of coralline algae, young trochids, chitons, and other mollusks live. Particularly adapted to this zone are the *Fissurella* keyhole limpets. One of the largest known members of the family, the Great Keyhole Limpet, is found on the Pacific coast of the United States. A large black mantle covers most of its shell.

1. **BARBADOS KEYHOLE LIMPET.** *Fissurella barbadensis* (Linné). S. Fla.–Brazil. Variable in shape and height. Characteristically, the interior is green with light, concentric streaks. Border of keyhole is green with a reddish brown line. Exterior has many irregular radial ribs, gray, pink, or purplish. Abundant; found intertidally.

2. **KNOBBY KEYHOLE LIMPET.** *Fissurella nodosa* (Born). Fla. Keys–West Indies. Interior pure opaque white. Keyhole oblong and not bordered by a red line. Exterior ash-gray or brownish with 20–22 strong radial ribs, which bear many rounded or pointed nodules. Margin of shell sharply crenulated. Lives on rocks just below low tide.

3. **VOLCANO LIMPET.** *Fissurella volcano* (Reeve). Pacific coast of U.S. Keyhole, elongate with flat inner sides, is located slightly nearer the narrower anterior end of the shell. Keyhole callus is bounded by a pink line. Exterior has very weak, broad radial ribs. Color grayish with wide mauve to dark rays. Interior white. Common; lives intertidally.

4. **WOBBLY KEYHOLE LIMPET.** *Fissurella fascicularis* Lamarck. S. Fla.–West Indies. Shell broad, flat, and rounded at ends. Both ends are turned up (so the shell wobbles if set on a flat surface). Exterior deep red-brown, with numerous round, radial ribs. The soft parts of the animal are larger than the shell. Uncommon; below low-tide mark on rocks.

5. **FILE FLESHY LIMPET.** *Lucapinella limatula* (Reeve). N.C.–West Indies. Shell oblong, flattish, and with a rather large keyhole located near the center of the shell. Ends of shell turned up slightly. Sides brownish, concave, with mauve blotches, and with 20–25 scaly, whitish ribs crossed by concentric ridges. Common on rocks or coral 30–400 ft.

6. **HARD-EDGED FLESHY LIMPET.** *Lucapinella callomarginata* (Dall). S. Calif.–Nicaragua. Shell flattish. Keyhole fairly large, narrowly elongate, with flat inner sides. Margin of shell has crenulations on the under edge. Exterior coarsely cancellate with scaly radial ribs. Color grayish with darker rays. Rather rare; under rocks in low-tide zone.

7. **TWO-SPOTTED KEYHOLE LIMPET.** *Megatebennus bimaculatus* (Dall). S. Alaska–Baja Calif. Keyhole very large: one third the length of shell. Shell flattish and with upturned ends. Exterior finely cancellate. Color gray to brown with a wide, darker blotch on either side. Common; under stones.

8. **GREAT KEYHOLE LIMPET.** *Megathura crenulata* (Sowerby). Central Calif.–Baja Calif. Animal is black and almost enveloping the shell, which may reach 6 inches. Keyhole one sixth length of shell, bordered with white, and with rounded edges. Exterior of shell finely beaded; mauve to brown. Common; in low-tide zone. This edible species is becoming increasingly scarce.

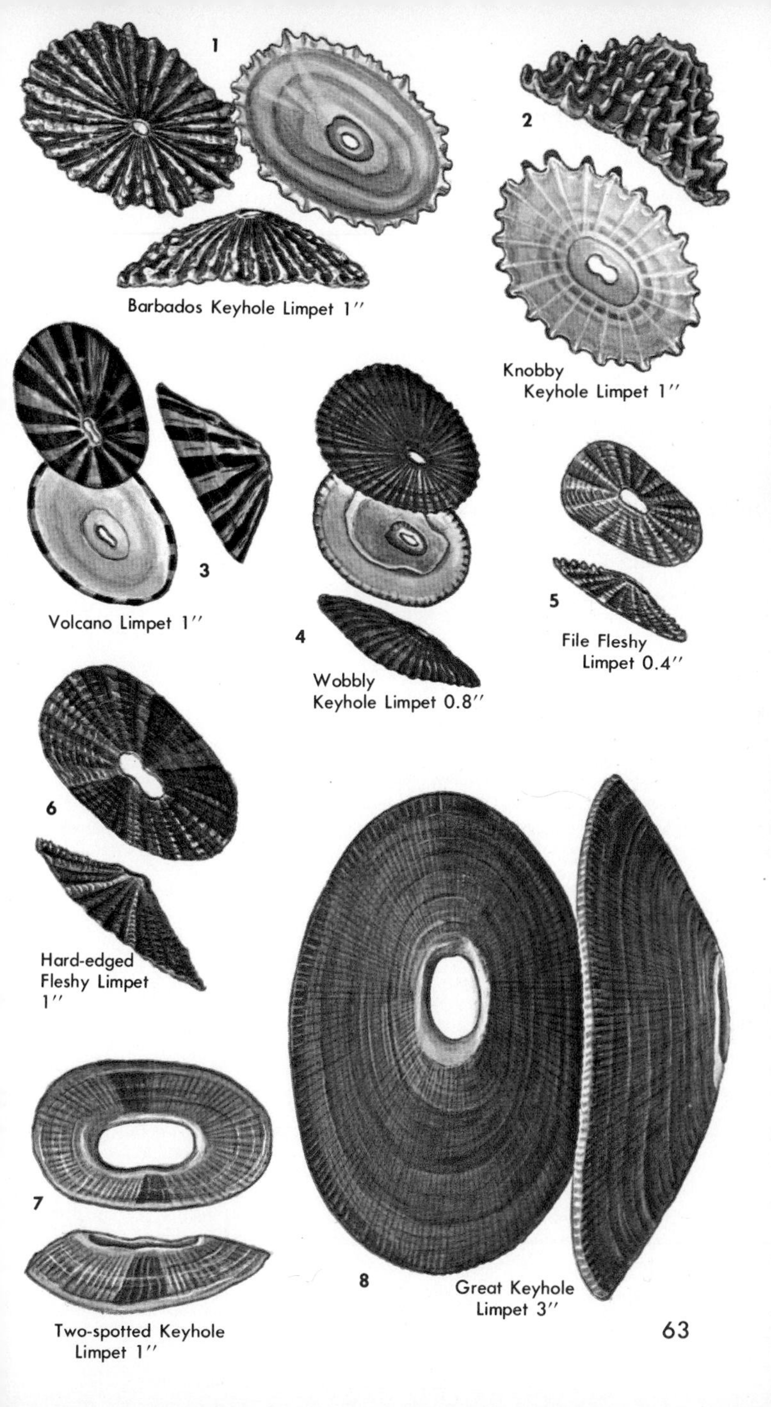

Barbados Keyhole Limpet 1″

Knobby Keyhole Limpet 1″

Volcano Limpet 1″

Wobbly Keyhole Limpet 0.8″

File Fleshy Limpet 0.4″

Hard-edged Fleshy Limpet 1″

Two-spotted Keyhole Limpet 1″

Great Keyhole Limpet 3″

TRUE LIMPETS

The rocky shores throughout the world are inhabited by many hundreds of species of limpets. They belong to several families, despite the fact that their simple cap-shaped shells look alike. All belong to the suborder Docoglossa, snails which have radulae of heavy, numerous, barlike teeth, a flat uncoiled shell, and simple open eyes that lack a lens. Members of the family Acmaeidae, including *Acmaea* and *Lottia,* have a single, feathery gill, but the family Patellidae, absent on our shores, has 2 lateral rows of gill leaflets.

The low, flat shell and the strong muscular foot permit limpets to adhere to shore rocks during the roughest weather. At night, limpets move about to graze on algae-covered rocks. The shapes of limpets are influenced by the type of bottom to which they attach and by the amount of wave action (p. 49). In exposed areas, the shells are low. The shells are elongate on species that live on narrow stems of seaweeds.

In many parts of the world, limpets are eaten raw or with a sauce or are made into soup. Too many should not be eaten at one meal, for the indigestible radulae sometimes become rolled into a small ball and may cause intestinal trouble.

Identification of the *Acmaea* limpets is usually difficult because of species variation and the possibility of hybridization between species. There are well over a dozen common species along our Pacific coast, but only three or four on the Atlantic. The only species in northern New England is the abundant Atlantic Plate Limpet.

The breeding season of limpets extends from April through September. The sexes are separate. The very small eggs are imbedded on the tops of rocks in a layer of thin mucus exuded by the sole of the foot. By October the young are at least a third of an inch long. The following spring they reach sexual maturity.

1. **GIANT OWL LIMPET.** *Lottia gigantea* (Sby). Calif.–Mexico. Apex near anterior end. Interior glossy, with a wide brown border and a rough bluish central area sometimes stained with brown. Foot is delicious when pounded, rolled in egg and flour, and fried. Common; on rocks at tide line.

2. **WHITE-CAP LIMPET.** *Acmaea mitra* Eschscholtz. Alaska–Mexico. Shell thick and dull-white, commonly covered with marine growths. Base round and apex usually very high. Lives on rocks below low-tide level in cold water to 200 ft. Moderately common; often washed ashore.

3. **SPOTTED LIMPET.** *Acmaea pustulata* (Helbling). S.E. Fla.–West Indies. Shell thick, with coarse radial ribs crossed by concentric threads. Sometimes flecked with reddish dots. A deep-water form of this species is thin, translucent, and red-flecked. Common; on intertidal coral rocks down to a depth of 30 ft.

4. **ATLANTIC PLATE LIMPET.** *Acmaea testudinalis* Müller. Arctic seas–N.Y.; Alaska. Interior color variable, in browns and blues. Exterior may be streaked with brown. The form living on eelgrass is long, thin, and mottled. A common intertidal rock dweller in cold water.

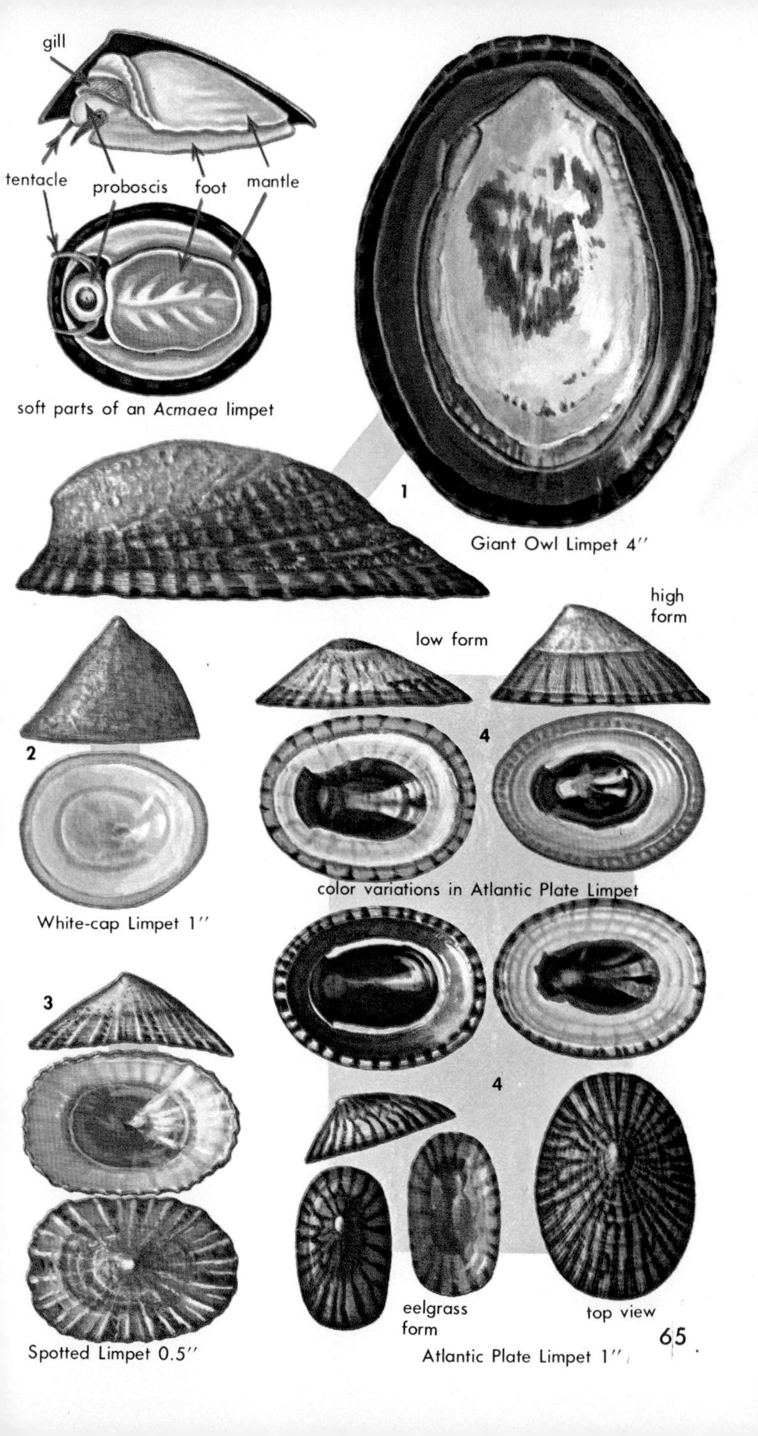
gill
tentacle
proboscis
foot
mantle
soft parts of an *Acmaea* limpet
1
Giant Owl Limpet 4″
2
White-cap Limpet 1″
low form
high form
4
color variations in Atlantic Plate Limpet
3
4
eelgrass form
top view
Spotted Limpet 0.5″
Atlantic Plate Limpet 1″

TRUE LIMPETS—(Pacific Coast)

The entire rocky Pacific Coast is abundantly supplied with *Acmaea* limpets, the commonest being the Fingered Limpet, *A. digitalis.* Other species are locally abundant. A few interesting brown species live on the leathery *Egregia* seaweeds. The bases of the shells are shaped to fit around the stalks. Some limpets excavate troughs in the stalks. Limpets are more active at night and when the tide is high.

1. **SHIELD LIMPET.** *Acmaea pelta* Eschscholtz. Bering Sea–Baja Calif. Shell elliptical; 20–27 weak radial ribs. Apex rather high and placed at the middle or one third of the way back from anterior end. Interior variable, whitish to bluish. A rough-ribbed form was formerly known as *A. cassis* Esch. Common; on rocks.

The similar Black Limpet. *A. asmi* Middendorff (not illus.). Alaska–Baja Calif. Only 0.3 in., all black; lives only on the snail *Tegula.*

The Fenestrate Limpet. *A. fenestrata* Reeve (not illus.), Alaska–Mex. (1–1.5 in.). Almost round in outline, high, smooth; exterior gray; interior brown with black edge. Common; on rocks.

2. **ROUGH LIMPET.** *Acmaea scabra* Gould. B.C.–Mexico. Elliptical; apex low and back from anterior end. With 15–25 strong ribs. Interior dull, rough. Common.

Test's Limpet. *A. conus* Test (not illus.). Calif.–Mexico. Similar but only 0.7 in., with a smooth, glossy interior. Common.

3. **FILE LIMPET.** *Acmaea limatula* Carpenter. Oregon–Mexico. Roundish in outline; apex low. Exterior greenish black and with numerous, narrow radial ridges of minute beads. Interior glossy cream, rarely with a brown patch. Border solid black-brown. Abundant; on rocks at tide mark.

4. **MASK LIMPET.** *Acmaea persona* Eschscholtz. Alaska–Calif. Elliptical, narrow at anterior end. Rather high hooked apex about one third of way back from anterior. Exterior smooth or with intertwining lines. Abundant; in wave-flushed crevices. Smaller (0.5 in.) in southern California.

5. **FINGERED LIMPET.** *Acmaea digitalis* Eschscholtz. Alaska–Baja Calif. Elliptical, narrowed at anterior end. Apex hooked forward, not centered. Weak, irregular, radial ribs produce a wavy edge. Exterior gray with white mottlings and black streaks. Abundant; on intertidal rocks.

6. **UNSTABLE LIMPET.** *Acmaea instabilis* Gould. Alaska–S. Calif. Shell solid, oblong, with a fairly high apex. Sides compressed. Exterior smoothish, dull, light-brown. Interior cream with brown stains; border a narrow band of brown. Lower edge of shell uneven. Lives on large brown seaweeds. Moderately common.

7. **PAINTED LIMPET.** *Acmaea depicta* Hinds. Calif.–Baja Calif. Shell small, very narrow, three times as long as wide. Sides compressed, cream to tan, with numerous fine, vertical brown stripes. Surface smoothish. Locally abundant; on eelgrass.

8. **SEAWEED LIMPET.** *Acmaea incessa* Hinds. S. Alaska–Baja Calif. Shell strong, rather high, with the apex hooked and placed near the anterior end. Outline elliptical. Exterior smooth and colored a uniform greasy light brown. Common; on brown seaweeds.

9. **CHAFFY LIMPET.** *Acmaea paleacea* Gould. This is the smallest of the Pacific-coast species—very thin, narrow, and fragile. Usually 3 or 4 times as long as wide. Sides straight, with fine, raised, radial threads. Brownish. Common; on eelgrass.

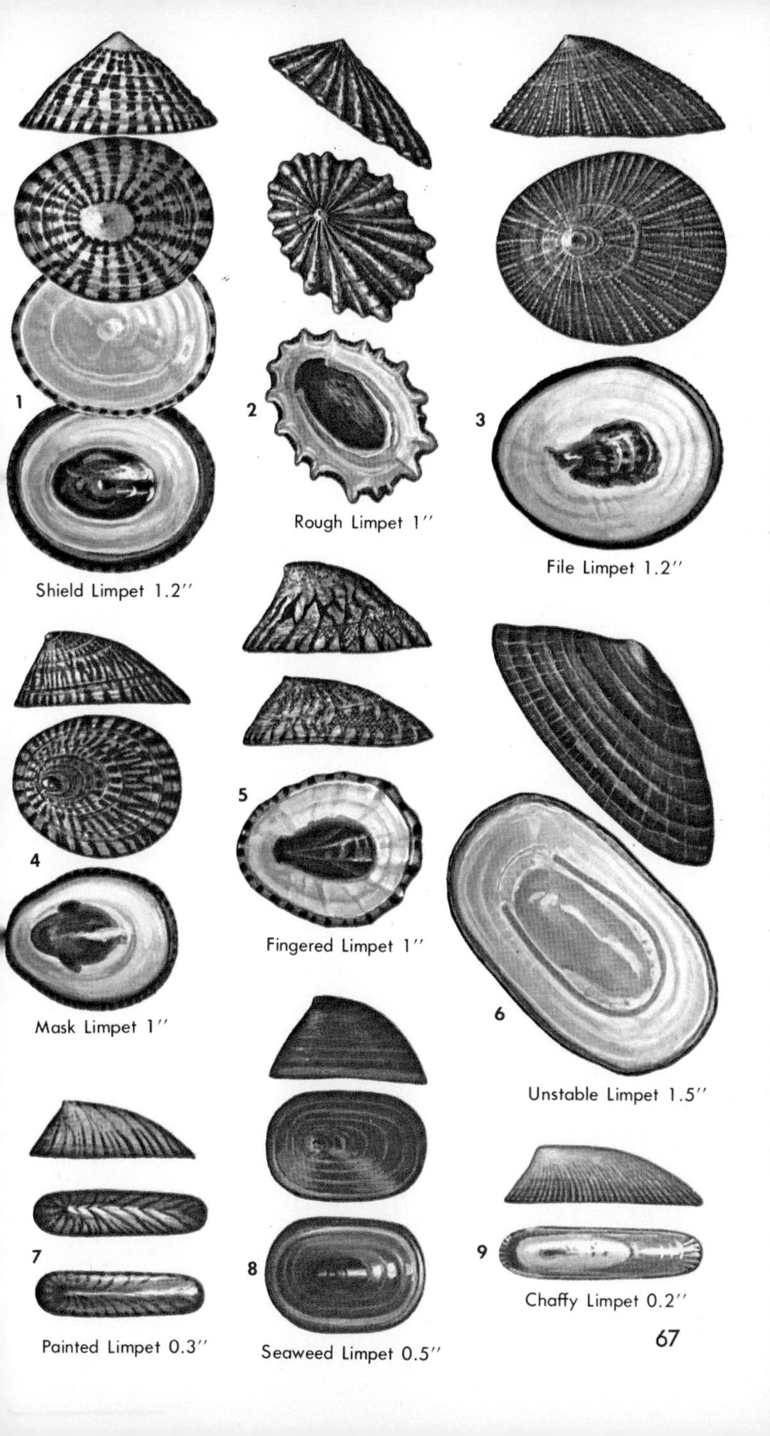
1
Shield Limpet 1.2″
2
Rough Limpet 1″
3
File Limpet 1.2″
4
Mask Limpet 1″
5
Fingered Limpet 1″
6
Unstable Limpet 1.5″
7
Painted Limpet 0.3″
8
Seaweed Limpet 0.5″
9
Chaffy Limpet 0.2″

MARGARITE TOP SHELLS

The family Trochidae contains dozens of genera and many hundreds of species, all of which have pearly shells and a round, many-whorled, horny operculum. Some top shells are found in shallow water, but many live at depths of several miles. The sides of the foot in the top shell bear delicate fleshy filaments that serve as sense organs. Compare top shells with the turban shells (p. 76), which are mostly from shallow tropical waters and which have an operculum of strong, shelly material. Top shells are all vegetarians, having numerous, delicate radular teeth of the rhipidoglossate type (p. 24). Sexes are separate. Females shed their eggs in the water or form a gelatinous egg mass.

The genus *Margarites* contains several dozen cold-water species with small, rather fragile shells, which are usually tinted with iridescent reds and greens. They are a main food for northern bottom-feeding fish. The shells of the females are larger than those of the males. Eggs are laid in jelly masses.

1. **NORTHERN ROSY MARGARITE.** *Margarites costalis* (Gould). Arctic Seas–Mass.; Alaska. Shell as wide as it is high; umbilicus whitish within, deep and narrow. Aperture pearly rose. Upper whorls with 5 or 6 spiral threads; last whorl with 10–12. Common; on sand bottoms, 60–200 ft. A common food for fish.

2. **UMBILICATE MARGARITE.** *Margarites umbilicatus* (Broderip and Sowerby). Arctic Seas–Mass. Shell wider than its height. Whorls rounded and smoothish. Umbilicus deep, smooth. Aperture pearly. Suture finely impressed. Exterior tan to cream and semiglossy. Common; on sand, 30–300 ft.

3. **PUPPET MARGARITE.** *Margarites pupillus* (Gould). Alaska–Calif. Shell higher than it is wide. Has 5 or 6 whorls and numerous spiral cords, between which are axial threads. Apex eroded. Umbilicus very minute. Exterior dull; interior iridescent. Common; on intertidal rocks and on pebble bottoms to 300 ft.

The similar Lirulate Margarite. *Margarites lirulatus* (Carpenter) (not illus.). S. Calif. Purple or whitish with purplish squares on the edge of the whorls; about 0.4 in. long. Base more rounded, umbilicus deep. Moderately common; 3–50 ft.

4. **VORTEX MARGARITE.** *Margarites vorticiferus* Dall. Arctic Seas–Alaska. Shell depressed; whorls rapidly expanding; umbilicus broad. Whorls have numerous spiral threads crossed by faint lines; edges sharp. Color purplish brown with rose tints. Common; offshore, in water 50–1,500 ft. deep.

5. **GREENLAND MARGARITE.** *Margarites groenlandicus* (Gmelin). Arctic Seas–Mass. Whorls rounded; spiral threads weak; umbilicus wide and deep. Apex smooth. Fine axial crenulations below suture. Exterior tan to rose. Moderately common; 30–700 ft.

6. **HELICINA MARGARITE.** *Margarites helicinus* (Phipps). Arctic Seas–Mass.; Alaska. Shell small and fragile, with a deep umbilicus. Aperture pearly and large. Exterior tan to brown. Surface smooth, very bright, and shining. Fairly common; 12–200 ft.

7. **OLIVE MARGARITE.** *Margarites argentatus* (Gould). Arctic Seas–Mass.; Alaska. Shell very small, globose, and smooth with fine spiral striae. There are 4 rounded whorls, with the suture deeply impressed. Umbilicus deep. Color translucent tan. Common; 20–300 ft.

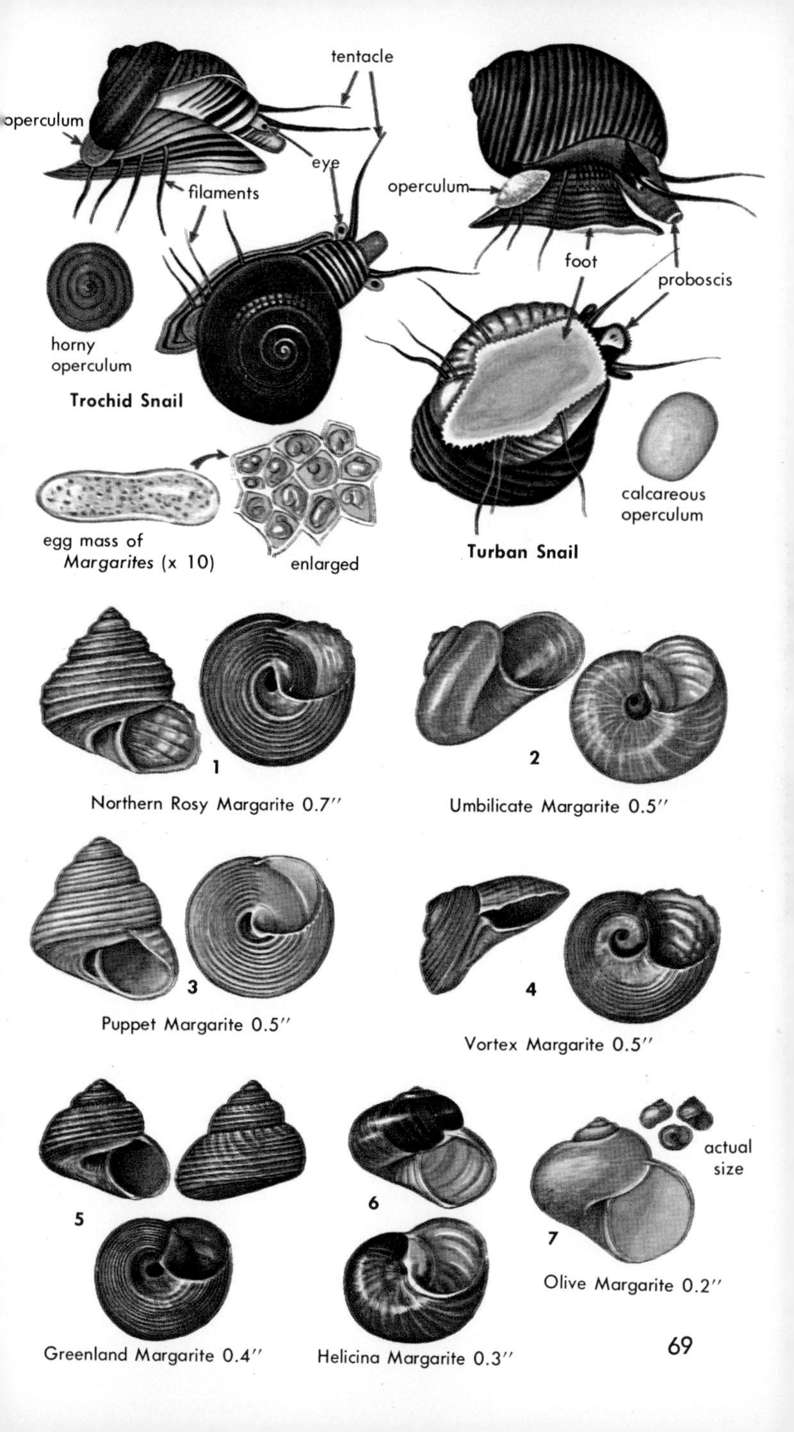
tentacle
operculum
eye
filaments
horny
operculum
Trochid Snail
operculum
foot
proboscis
calcareous
operculum
Turban Snail
egg mass of
Margarites (x 10)
enlarged
1
Northern Rosy Margarite 0.7″
2
Umbilicate Margarite 0.5″
3
Puppet Margarite 0.5″
4
Vortex Margarite 0.5″
5
Greenland Margarite 0.4″
6
Helicina Margarite 0.3″
actual
size
7
Olive Margarite 0.2″

CALLIOSTOMA TOP SHELLS

The handsome, finely sculptured shells of the genus *Calliostoma* contain about 100 species, with about 2 dozen in North American seas, mostly in fairly deep water. Some live in the kelp beds off California, while others are found on dead coral rubble off Florida. The translucent, round operculum has many whorls.

1. **BAIRD'S TOP SHELL.** *Calliostoma bairdi* Verrill and Smith. Mass.–Fla. Spire has straight sides; base flat and without umbilicus; whorls have 6 or 7 rows of neat beads, those just below the suture being largest. Uncommon; 100–1,200 ft.

2. **GEM TOP SHELL.** *Calliostoma gemmulatum* Carpenter. Calif.–Mexico. Spire high; periphery of whorls squarish. Apex brown; whorls gray-green and with 7 or 8 rows of neat beads. Common; under rocks and on wharf pilings, 1–200 ft.

3. **SCULPTURED TOP SHELL.** *Calliostoma euglyptum* A. Adams. N.C.–Texas. Variable in spire height. Whorls angular with 6 major, beaded cords. Color dull rose, splashed with cream or red. No umbilicus. Common; in sand, 3–200 ft.

4. **SAY'S TOP SHELL.** *Calliostoma sayanum* Dall. N.C.–Fla. Spire has flat sides; periphery has rose band and is well rounded. Umbilicus deep, round, and whitish. Whorls golden yellow with rows of whitish beads. Rare; 200–800 ft.

5. **CHOCOLATE-LINED TOP SHELL.** *Calliostoma javanicum* (Lamarck). Fla. Keys–West Indies. Base flat, with 5–7 fine, chocolate-brown lines. Umbilicus deep, white. Whorls have ten beaded threads. Rather rare; 2–30 ft. in coral sand. Formerly *C. zonamestum* A. Adams.

6. **JUJUBE TOP SHELL.** *Calliostoma jujubinum* (Gmelin). N.C.–West Indies. Shell has 9–10 well-beaded threads between the sutures. Brown to red, with white spots above suture. Umbilicus small, deep. Moderately common; in sand and weeds, 3–30 ft. Commonly cast on beaches.

7. **BEAUTIFUL ATLANTIC TOP SHELL.** *Calliostoma pulchrum* (C. B. Adams). S.E. United States. Spire high, somewhat flat. Whorls gray-green, with 6–7 beaded threads. Two above suture are larger, spotted. No umbilicus. Uncommon; 6–200 ft.

8. **PACIFIC RINGED TOP SHELL.** *Calliostoma annulatum* (Lightfoot). Alaska–S. Calif. Shell light; sides flattish. Whorls golden yellow with a mauve band. Has 5–9 rows of small, neat beads between sutures. No umbilicus. Lives on offshore kelps.

9. **CHANNELED TOP SHELL.** *Calliostoma canaliculatum* (Lightfoot). Alaska–S. Calif. Shell light; sides and base quite flat. Whorls yellowish tan, with about 9 sharp, weakly beaded, spiral cords between sutures. Common; on offshore kelps.

10. **WESTERN RIBBED TOP SHELL.** *Calliostoma ligatum* (Gould). Alaska–S. Calif. Shell solid. Whorls rounded dark brown, with 6–8 tan cords. No umbilicus. Aperture pearly white. Common; among stones and algae, in shallow water to 10 ft.

11. **TRICOLORED TOP SHELL.** *Calliostoma tricolor* Gabb. Central Calif.–Baja Calif. Shell solid. Whorls angular, with many spiral beaded threads. Brownish, with spiral lines of broken white. Rarely mottled or streaked. Commonly dredged from 30–200 ft.

12. **GRANULOSE TOP SHELL.** *Calliostoma supragranosum* Carpenter. Monterey, Calif.–Baja Calif. Shell solid, glossy. Whorls angular and with fine beaded cords; some spotted white. Interior bright and pearly. Moderately common; among intertidal rocks.

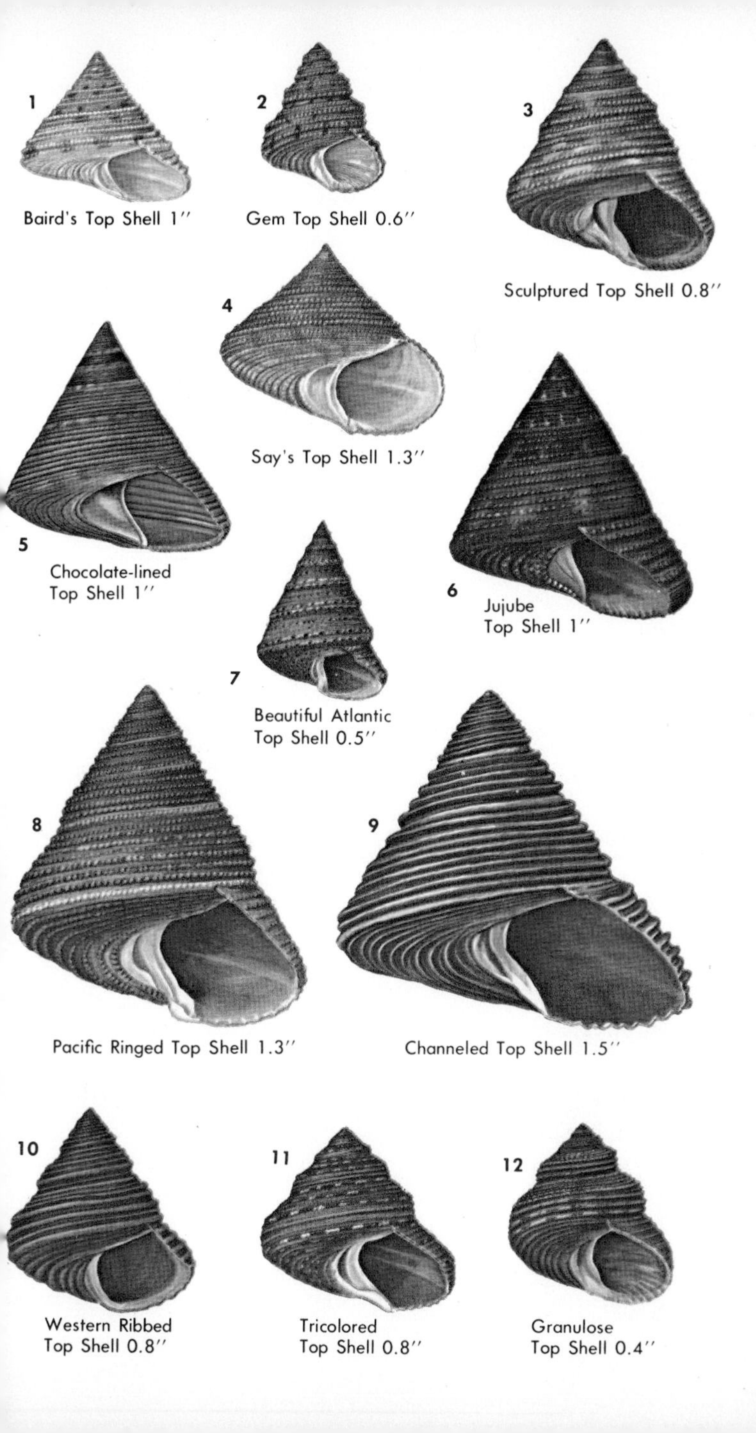
1
Baird's Top Shell 1″
2
Gem Top Shell 0.6″
3
Sculptured Top Shell 0.8″
4
Say's Top Shell 1.3″
5
Chocolate-lined
Top Shell 1″
6
Jujube
Top Shell 1″
7
Beautiful Atlantic
Top Shell 0.5″
8
Pacific Ringed Top Shell 1.3″
9
Channeled Top Shell 1.5″
10
Western Ribbed
Top Shell 0.8″
11
Tricolored
Top Shell 0.8″
12
Granulose
Top Shell 0.4″

VARIOUS TOP SHELLS

The range in kinds of trochids is very great, from the massive shells of the shallow-water *Norrisia* and *Cittarium* to the tiny, delicate *Solariella* from deep, offshore waters. The *Norrisia* is unusual in having a horny operculum with spiral rows of bristles.

1. **DALL'S DWARF GAZA.** *Microgaza rotella* Dall. N.C.–West Indies. Very small and glossy. Surface smooth, except for microscopic pimples below suture. Deep umbilicus bordered by row of minute creases. Whorls angular, pearly gray, with zigzag brown stripes on top. Common; dredged from sand bottoms at depths of from 80–600 ft.

2. **WEST INDIAN TOP SHELL.** *Cittarium pica* (Linné). S.E. Fla. (extinct)–West Indies (living). Shell heavy, rather rough. Purple-black with whitish mottlings. Umbilicus very deep. Interior of aperture pearly white. Operculum horny, circular, greenish-blue in life. Common; among intertidal rocks near surf in West Indies. Genus formerly *Livona*.

3. **NORRIS SHELL.** *Norrisia norrisi* Sowerby. Monterey, Calif.–Baja Calif. Shell heavy, solid, and smooth, with rounded, glossy blackish-brown whorls. Umbilicus deep and bordered on one side by the blue-green columella. Operculum horny, circular, with many whorls, and covered with spiral rows of bristles. Common; from shore to 40 ft. among kelps.

4. **SUPERB GAZA.** *Gaza superba* Dall. Gulf of Mex.–West Indies. Shell light but strong. Whorls rounded, light gray with a golden sheen. Early whorls purplish. Umbilicus deep but partially covered. Outer lip slightly thickened. Interior of round aperture is pearly rose. An uncommon collector's item; dredged from coarse sand in Gulf of Mexico, 200–600 ft.

5. **BAIRD'S SPINY TOP SHELL.** *Bathybembix bairdi* (Dall). Alaska–Mexico. Shell rather fragile, whitish, but covered with grayish or brownish periostracum, which flakes off when dry. Spiral cords may be beaded or weakly spined. No umbilicus. Interior of aperture pearly white. Formerly placed in the genus *Lischkeia*. A rare shell; dredged offshore from 300–1,800 ft.

6. **DIADEM SPINY TOP SHELL.** *Bathybembix cidaris* (Carpenter). Alaska–Baja Calif. Somewhat fragile. Whorls crossed by spiral and axial threads, which form distinct beads. Spire somewhat flat-sided. Sutures indented. Color gray to grayish green. No umbilicus. Interior of aperture greenish gray. Outer lip thin and scalloped. Moderately common; dredged from 60–1,400 ft.

7. **CHANNELED SOLARELLE.** *Solariella lacunella* Dall. Virginia–Key West. Shell thick, whitish. Whorls have 6 spiral cords, the upper 3 beaded. Suture channeled. Umbilicus deep, narrow. Aperture pearly. Common; in sand, 30–600 ft.

8. **LOVELY PACIFIC SOLARELLE.** *Solariella peramabilis* Carpenter. Alaska–Calif. Shell solid, semi-glossy, tan with mauve mottlings. Many weak spiral threads. Umbilicus deep. Interior iridescent. Common; in sand, 50–1,500 ft.

9. **LAMELLOSE SOLARELLE.** *Solariella lamellosa* Verrill and Smith. Mass.–West Indies. Suture broadly channeled and bordered below with numerous axial riblets. Umbilicus deep, edged with beaded thread. Color gray. Common on sand; 100–700 ft.

10. **OBSCURE SOLARELLE.** *Solariella obscura* (Couthouy). Labrador–Virginia. Whorls slightly angled by minutely beaded, spiral threads. Color gray to pink-tan. Axial slanting growth lines present. Deep umbilicus. Common; on sand, 20–1,500 ft.

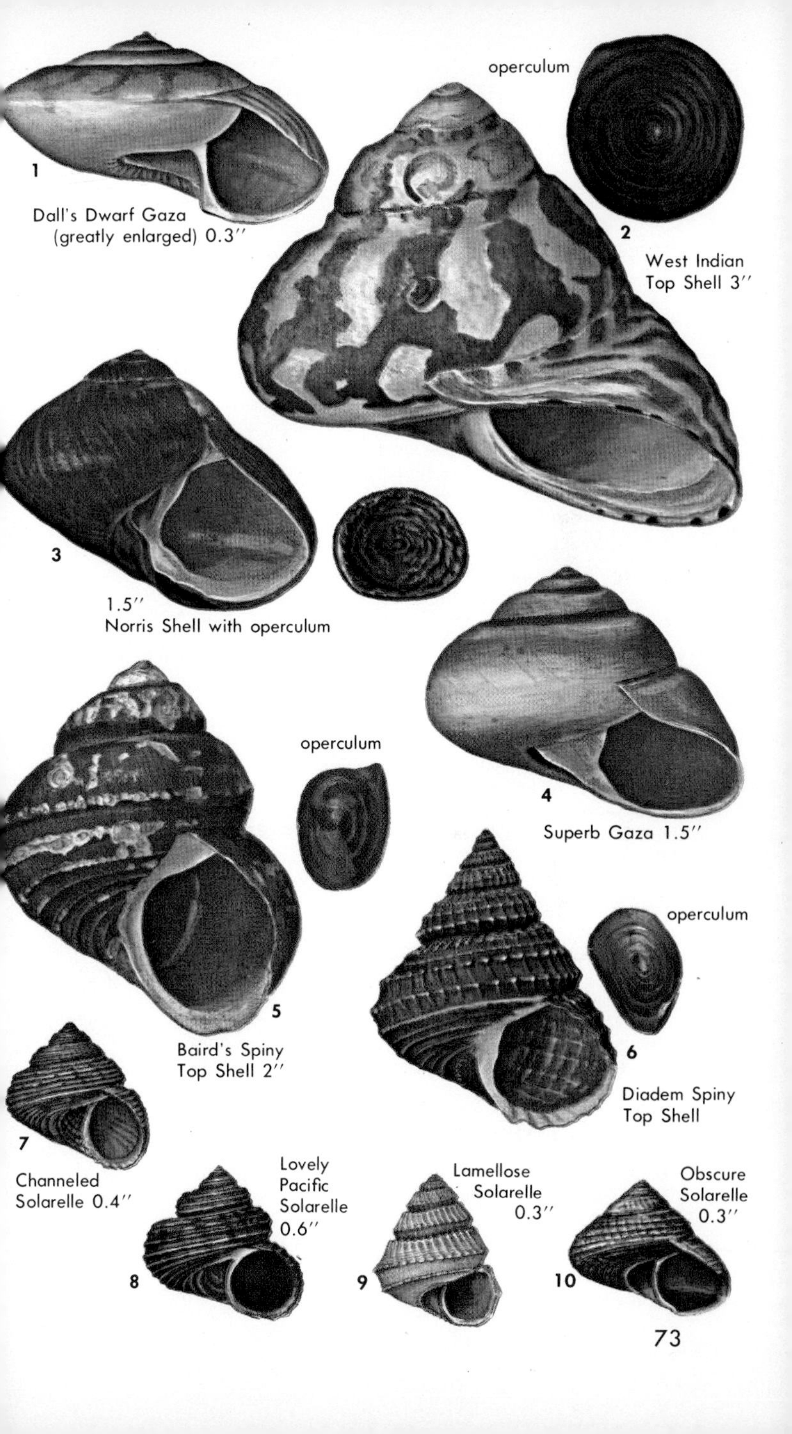
operculum
1
Dall's Dwarf Gaza
(greatly enlarged) 0.3″
2
West Indian
Top Shell 3″
3
1.5″
Norris Shell with operculum
operculum
4
Superb Gaza 1.5″
operculum
5
Baird's Spiny
Top Shell 2″
6
Diadem Spiny
Top Shell
7
Channeled
Solarelle 0.4″
Lovely
Pacific
Solarelle
0.6″
8
Lamellose
Solarelle
0.3″
9
Obscure
Solarelle
0.3″
10

TEGULA TOP SHELLS

The tegula top shells belong to a genus containing several hundred species of shallow-water, rock-dwelling snails, mainly from cool waters. They are vegetarians. Both eggs and free-swimming larvae are planktonic. About 30 species of tegulas inhabit North American waters, with the largest and most abundant species living along the rocky coast of California. Some species live with sea urchins and take shelter under the spines.

1. **SMOOTH ATLANTIC TEGULA.** *Tegula fasciata* (Born). S. Fla.–West Indies. Spire low; whorls rounded; umbilicus deep, round, with a white callus. Surface smooth, mottled with reds, browns, and black; often with spirals of red and white dots. Two tiny teeth are present on the base of the columella. Common; under shore rocks.

2. **DUSKY TEGULA.** *Tegula pulligo* (Gmelin). Alaska–N. Calif. Whorls smooth; umbilicus deep, rounded, and tan. Columella thin, usually without a tooth at base. Compare with Brown Tegula, which has no umbilicus. Moderately common, especially in the North; among subtidal rocks.

3. **WESTERN BANDED TEGULA.** *Tegula eiseni* (Jordan). Monterey, Calif.–Mexico. Surface grayish with black flecks, and with numerous beaded, spiral cords. Lower edge of outer lip has about 8 small nodules. Umbilicus smooth, deep, and narrow. Common; intertidal rocks.

4. **SPECKLED TEGULA.** *Tegula gallina* (Forbes). Calif.–Baja Calif. Shell heavy, its top usually eroded. Color grayish green, with narrow, zigzag, axial white stripes. Umbilicus closed. Compare with Black Tegula, which lacks stripes. Common; on intertidal rocks.

5. **BROWN TEGULA.** *Tegula brunnea* (Philippi). Ore.–Calif. Whorls rounded, light brown, with the base glossy tan. Usually encrusted with algae. One weak tooth at base of columella. Umbilicus closed. Compare with the Dusky Tegula. Common; among subtidal rocks.

6. **WEST INDIAN TEGULA.** *Tegula lividomaculata* (C. B. Adams). Fla. Keys–West Indies. Whorls have many fine spiral threads flecked with white. Umbilicus deep with 2 spiral cords, the upper one ending as a bead on base of columella. The periphery of the whorls is squarish. Among intertidal rocks; rare in Florida but common in the West Indies.

7. **BLACK TEGULA.** *Tegula funebralis* (A. Adams). W. Canada–Baja Calif. Shell heavy, dark purple-black. Surface smoothish, except for puckered cord below suture near the aperture. Umbilicus closed, but base of columella has two white nodules. Very common; among intertidal rocks.

8. **MONTEREY TEGULA.** *Tegula montereyi* (Kiener). Calif. Conical, with flat-sided whorls; base flat. Surface smooth, light gray-brown. Umbilicus very deep, lined with 1 or 2 spiral threads. Base of columella has one tooth. Rather rare; on giant kelps, 20–60 ft.

9. **QUEEN TEGULA.** *Tegula regina* Stearns. Catalina Is.–Baja Calif. Whorls and spire flat-sided with many slanting, axial ridges ending as small flutes. Dark purplish gray. Umbilical region bright golden yellow. An uncommon collector's item; found offshore, 20–80 ft.

10. **GILDED TEGULA.** *Tegula aureotincta* (Forbes). S. Calif.–Mexico. Shell heavy, grayish green, with strong spiral cords. Golden yellow stain inside deep, narrow umbilicus, which is sometimes bordered by a sky-blue band. Moderately common; among intertidal rocks.

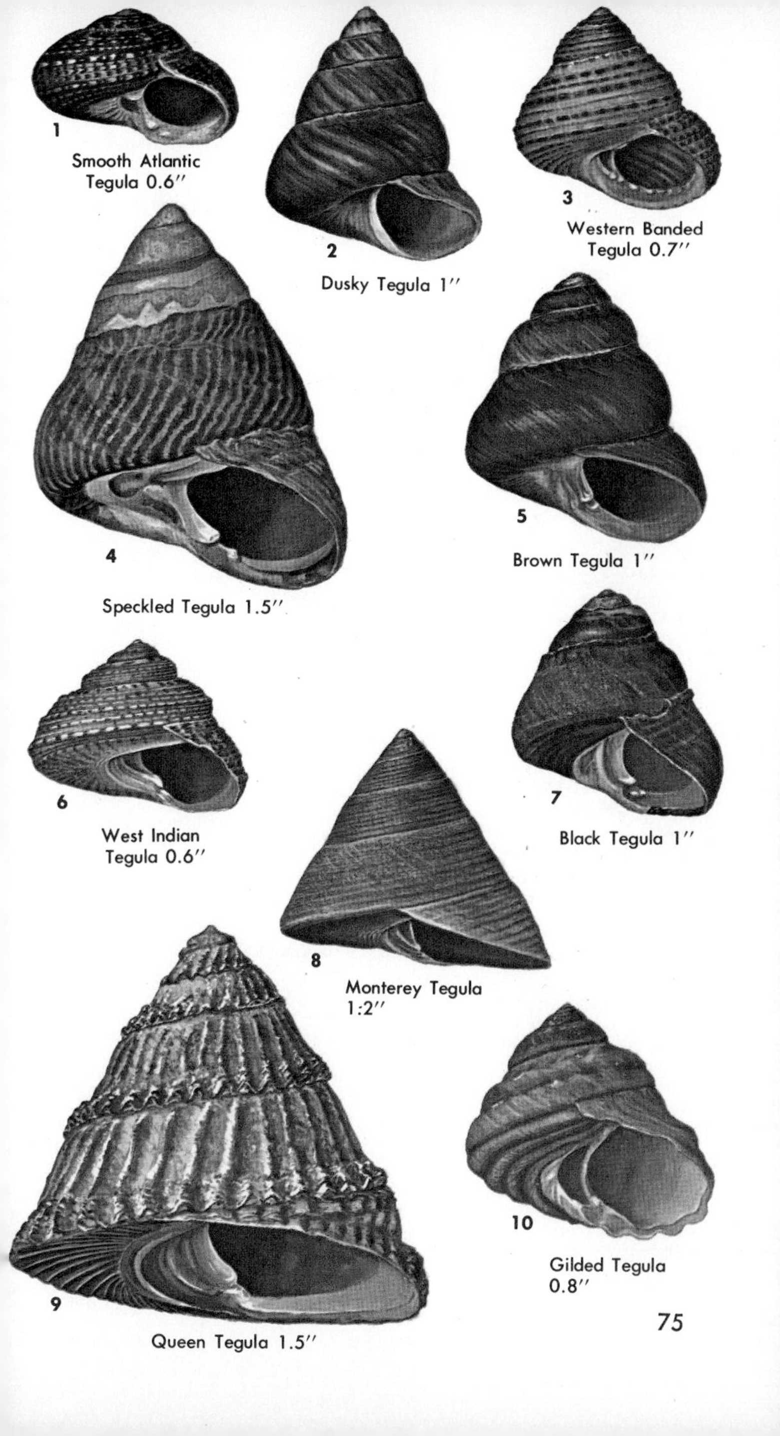

1 Smooth Atlantic Tegula 0.6″

2 Dusky Tegula 1″

3 Western Banded Tegula 0.7″

4 Speckled Tegula 1.5″

5 Brown Tegula 1″

6 West Indian Tegula 0.6″

7 Black Tegula 1″

8 Monterey Tegula 1:2″

9 Queen Tegula 1.5″

10 Gilded Tegula 0.8″

TURBAN AND PHEASANT SHELLS

Turban shells of the family Turbinidae have heavy shells and strong limy opercula by which most of the many species can be identified. Almost all turbans are tropical, shallow-water mollusks that feed exclusively on marine algae. Sexes are separate. The eggs are shed into the sea, and the larvae are free-swimming.

Pheasant shells of the family Phasianellidae are small, tropical, herbivorous snails. All have a limy operculum. There are about a dozen American members of the genus *Tricolia*.

1. **LONG-SPINED STAR SHELL.** *Astraea phoebia* Röding. S.E. Fla.–West Indies. Spire usually low, but some specimens are as high and as poorly spined as the Red Turban. Umbilicus round, rough, and usually open. Formerly called *A. longispina* Lamark. Common; found in grassy shallows.

2. **AMERICAN STAR SHELL.** *Astraea americana* (Gmelin). S.E. Fla. Spire always very high. Whorls whitish, with concave sides and axial riblets. Base has 5–8 fine, spiral threads. Operculum white, usually with a central depression. Common; in grassy shallows and under rocks at low tide.

3. **GREEN STAR SHELL.** *Astraea tuber* (Linné). S.E. Fla.–West Indies. Shell solid, whitish, with fine cross-hatch lines of green. Whorls have wide axial slanting ridges. Operculum has a thick, comma-like ridge. Common; 4–30 ft. in coral reefs.

4. **CARVED STAR SHELL.** *Astraea caelata* (Gmelin). S. Car.–West Indies. Top third of each whorl has slanting ridges; bottom part has 9 or 10 spiral cords bearing tiny flutings. Operculum whitish, oval, with a granular outer surface. Common; coral rubble or reefs.

5. **WAVY TURBAN.** *Astraea undosa* (Wood). S. Calif.–Baja Calif. Edge of whorls with heavy wavy cord. Outer covering brown and fuzzy. Base concave, with 3 small cords. Operculum has 3 strong, prickly ridges on the outer side. Common; among algae and rocks.

6. **CHESTNUT TURBAN.** *Turbo castanea* Gmelin. N.C.–Brazil. Surface beaded, rarely with tiny spines. Common; in shallow bays.

7. **CHANNELED TURBAN.** *Turbo canaliculatus* Hermann. Lower Fla. Keys—West Indies. Shell has 16–19 smooth, spiral cords. Suture channeled. Uncommon; 5–40 ft.

8. **RED TURBAN.** *Astraea gibberosa* (Dillwyn). W. Canada–S. Calif. Shell heavy, reddish brown. Lower edge of whorls has a wavy cord. Base has 5–7 distinct spiral cords. Operculum whitish, smooth, and somewhat elongate. Common; on rocks, 3–100 ft.

9. **STAR ARENE.** *Arene cruentata* (Mühlfeld). S.E. Fla.–West Indies. Periphery of whorl bears hollow, triangular spines. Suture deeply channeled. Umbilicus deep, bordered by 3 rows of beads. Operculum covered with circular rows of tiny, limy beads. Uncommon; under rocks.

10. **CALIFORNIAN BANDED PHEASANT.** *Tricolia compta* (Gould). N. Calif.–Baja Calif. Shell minute and smooth, with numerous green, red, or purple spiral threads and oblique lines. Abundant; on eelgrass in shallow bays; often washed ashore.

11. **SHOULDERED PHEASANT.** *Tricolia bella* (M. Smith). S.E. Fla.–Brazil. Whorls slightly carinate, with numerous spiral threads. Main thread at periphery may have white dots. Color variable. Operculum shelly, exterior half smooth, half ribbed. Common; on rubble bottom, 6–50 ft.

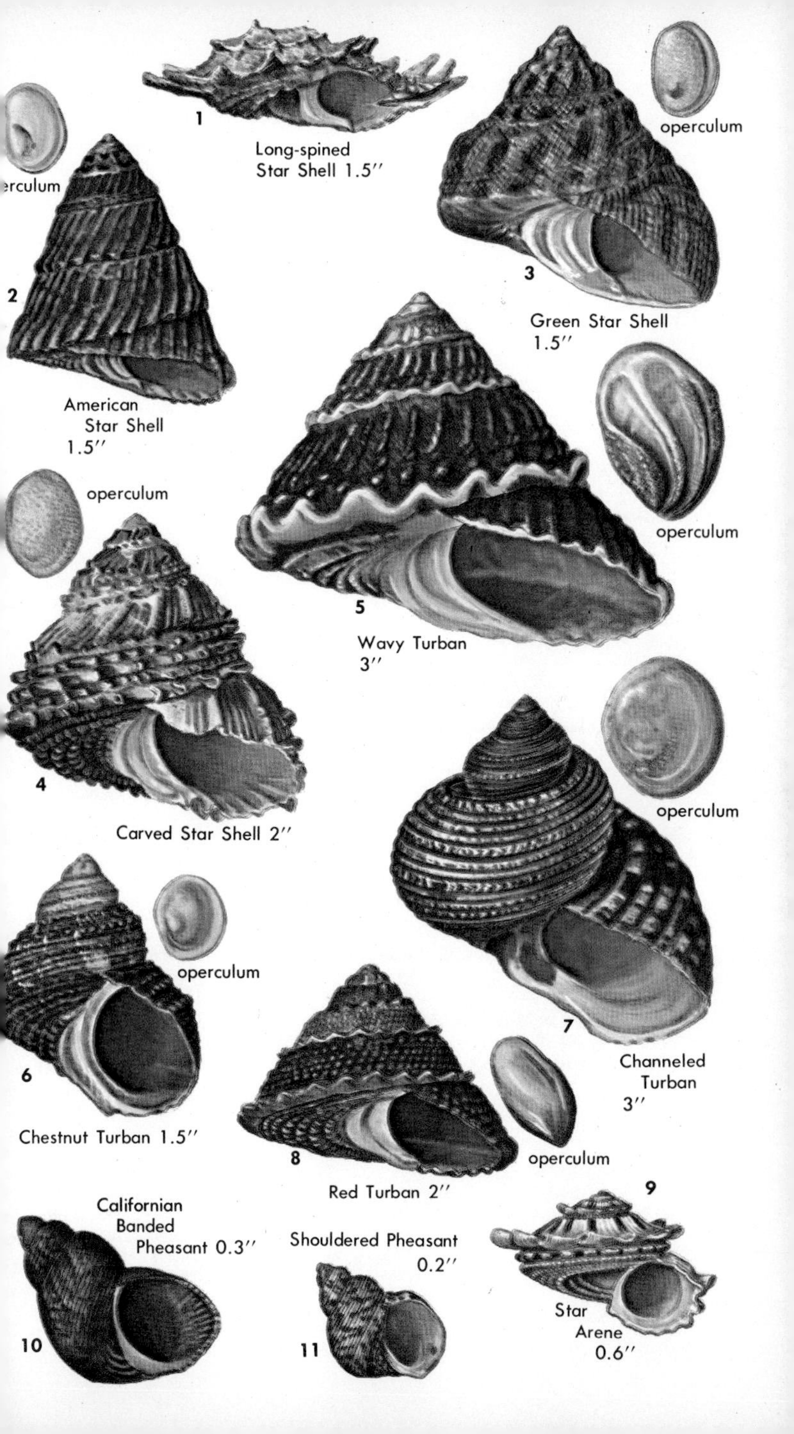
1
Long-spined
Star Shell 1.5″
operculum
3
Green Star Shell
1.5″
erculum
2
American
Star Shell
1.5″
operculum
operculum
5
Wavy Turban
3″
4
Carved Star Shell 2″
operculum
operculum
7
Channeled
Turban
3″
6
Chestnut Turban 1.5″
8
operculum
Red Turban 2″
9
Californian
Banded
Pheasant 0.3″
Shouldered Pheasant
0.2″
Star
Arene
0.6″
10
11

NERITES

The tropical family Neritidae has about 200 living marine species, although some live in brackish water. Nerites, solid and sturdy, are usually found clinging to wave-washed rocks. The operculum is limy, sometimes colorful, and often pimpled.

1. **BLEEDING TOOTH.** *Nerita peloronta* Linné. S.E. Fla.–Brazil. Inner lip has white teeth surrounded by orange blotch. Very common; on intertidal rocks, often above high water.

2. **FOUR-TOOTHED NERITE.** *Nerita versicolor* Gmelin. S. Fla.–Brazil. Shell dirty white with black and red markings. Inner lip usually has 4 strong, white teeth. Operculum gray-brown, pimpled on outer surface. Common; on intertidal rocks.

3. **ANTILLEAN NERITE.** *Nerita fulgurans* Gmelin. S.E. Fla.–Brazil. Shell has fine, spiral ridges. Inner lip pimpled and two-toothed. Operculum gray. Uncommon; found in brackish areas.

4. **TESSELLATE NERITE.** *Nerita tessellata* Gmelin. Fla.–Texas and Caribbean. Shell dirty white with small black dots. Rarely, all black. Operculum blackish. Common; intertidal.

5. **ZEBRA NERITE.** *Puperita pupa* (Linné). S.E. Fla.–Caribbean. Shell flat-white with numerous zebra-like, black stripes. Aperture yellow. Operculum smooth, pale yellow. Common; in intertidal rock pools.

6. **EMERALD NERITE.** *Smaragdia viridis* (Linné). S.E. Fla.–Brazil. Shell glossy, pea-green, with white (rarely purple) markings. Operculum green. Animal pea-green. Common; on eelgrass, 3–20 ft.

7. **OLIVE NERITE.** *Neritina reclivata* (Say). Fla.–Texas and Caribbean. Shell olive-green with numerous, very fine, blackish lines. Operculum black or brownish. Common; in brackish water.

8. **VIRGIN NERITE.** *Neritina virginea* (Linné). Fla.–Texas and Caribbean. Shell glossy, variably colored. Inner lip concave, with numerous small, irregular teeth. Operculum smooth. Very common; on brackish mud flats.

LACUNA PERIWINKLES

Lacuna periwinkles, family Lacunidae, are small, cold-water snails with horny opercula and shells with a chinklike umbilicus. The head and tentacles are brightly colored. Eggs are laid in jelly masses.

9. **COMMON NORTHERN LACUNA.** *Lacuna vincta* (Turton). Arctic Seas–Rhode Island; also to Calif. Whorls rounded; smoothish; apex purplish. Umbilicus chinklike. Common; on rocks and algae, 3–100 ft.

10. **CARINATE LACUNA.** *Lacuna carinata* Gould. Alaska–Calif. Body whorl carinate; umbilicus chinklike, long, and white. Periostracum thin, smooth, yellowish brown. Common; on kelp leaves, 6–100 ft.

11. **VARIEGATED LACUNA.** *Lacuna variegata* Carpenter. Wash.–Calif. Carina at periphery is very sharp. Umbilicus deep and bordered by sharp ridge. Yellow-tan, rarely mottled. Common; on eelgrass.

12. **ONE-BANDED LACUNA.** *Lacuna unifasciata* Carpenter. Calif.–Baja Calif. Weak peripheral carina has fine red or brown line, rarely with a series of streaks. Early whorls pink, rest yellow. Common; on kelp.

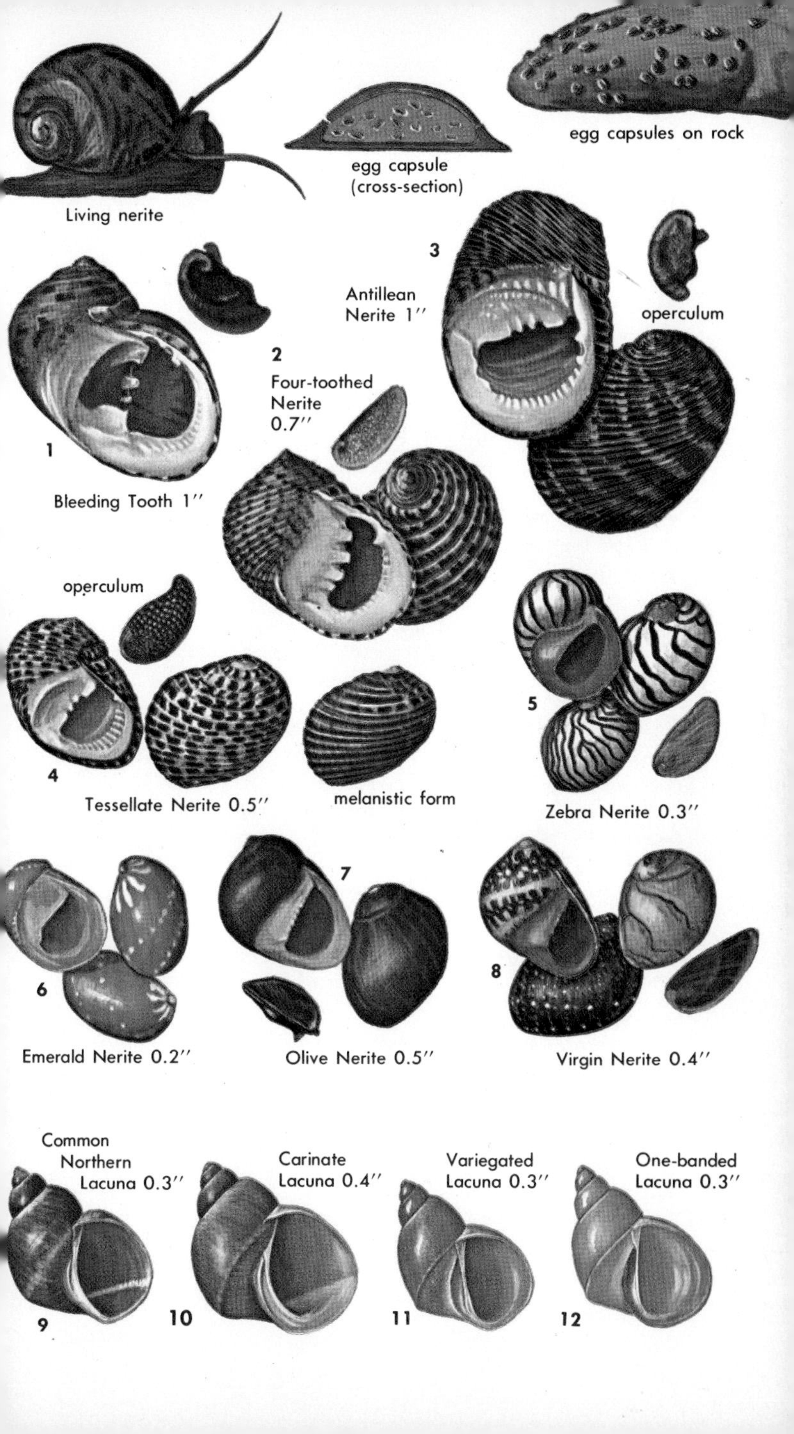
Living nerite
egg capsule
(cross-section)
egg capsules on rock
3
Antillean
Nerite 1″
operculum
2
Four-toothed
Nerite
0.7″
1
Bleeding Tooth 1″
operculum
4
Tessellate Nerite 0.5″
melanistic form
5
Zebra Nerite 0.3″
6
Emerald Nerite 0.2″
7
Olive Nerite 0.5″
8
Virgin Nerite 0.4″
Common
Northern
Lacuna 0.3″
Carinate
Lacuna 0.4″
Variegated
Lacuna 0.3″
One-banded
Lacuna 0.3″
9
10
11
12

PERIWINKLES

All of the true periwinkles (family Littorinidae) are strictly littoral snails, inhabiting a variety of habitats from rocks at the low-tide line to wharf pilings, grasses, and mangrove roots at and above the high-tide line. They feed on microscopic algae and diatoms, which they scrape from the surface with their long, coiled radula. Many species are able to survive for long periods out of water, and it is believed that a number of present-day land snails have evolved from ancient representatives of this family. Some species can tolerate periods of very low and quite high salinities. There are about 20 species in North America.

The shell is usually small, conical, and lacking an umbilicus. Most are rather solid, although a few tropical species are quite light in structure. The thin, horny operculum has only a few whorls. The sexes are separate. In some species, notably the angulate periwinkle (p. 83), the female is considerably larger than the male. The male possesses on the right side of the body a large, prominent penis, which varies in shape from species to species and is a good diagnostic character. The well-developed head bears a pair of long, delicate tentacles with an eye located at the base. The radula ribbon is exceptionally long and narrow, coiled behind the head like a watch spring, with 200 or 300 rows of minute, rasplike teeth. The operculum is chitinous and entirely fills the aperture when the snail has withdrawn.

A periwinkle leaves a trail of mucus exuded from a slit at the anterior end of the foot. The broad, rather square foot is muscularly divided longitudinally. The snail uses first one side of the foot and then the other as it glides forward slowly.

Some species lay eggs in capsules that float in the water for several weeks; others lay them in jelly-like masses that they attach to algae. *Littorina saxatilis* is ovoviviparous and gives birth to well-developed, shelled young. Free-swimming veligers of some species permit the populations to spread along the coasts to new areas.

1. **COMMON EUROPEAN PERIWINKLE.** *Littorina littorea* (Linné). Labrador–Md. Reputedly introduced from Europe (see p. 33). Shell usually smooth and thick; grayish brown to blackish, with numerous fine, dark, spiral bands. Outer lip brown-black within; columella white. Young, under 5 mm., are strongly corded. The floating egg capsules, about 0.03 in. in diameter, hatch in about a week. Very common; found on rocks and clinging to seaweeds in the intertidal zone. This abundant periwinkle is a favorite food in England and elsewhere in Europe.

2. **MARSH PERIWINKLE.** *Littorina irrorata* (Say). N.J.–central Fla. and Gulf States. Shell usually gray-cream, with red-brown streaks on the spiral cords. Columella orange-brown. Common; in brackish water marshes on reeds.

3. **CHECKERED PERIWINKLE.** *Littorina scutulata* Gould. Alaska–Calif. A moderately slender periwinkle; exterior smoothish; dark reddish brown, with spiral bands of irregular white or bluish spots. Aperture light brown; columella white. Common; on intertidal rocks in algae.

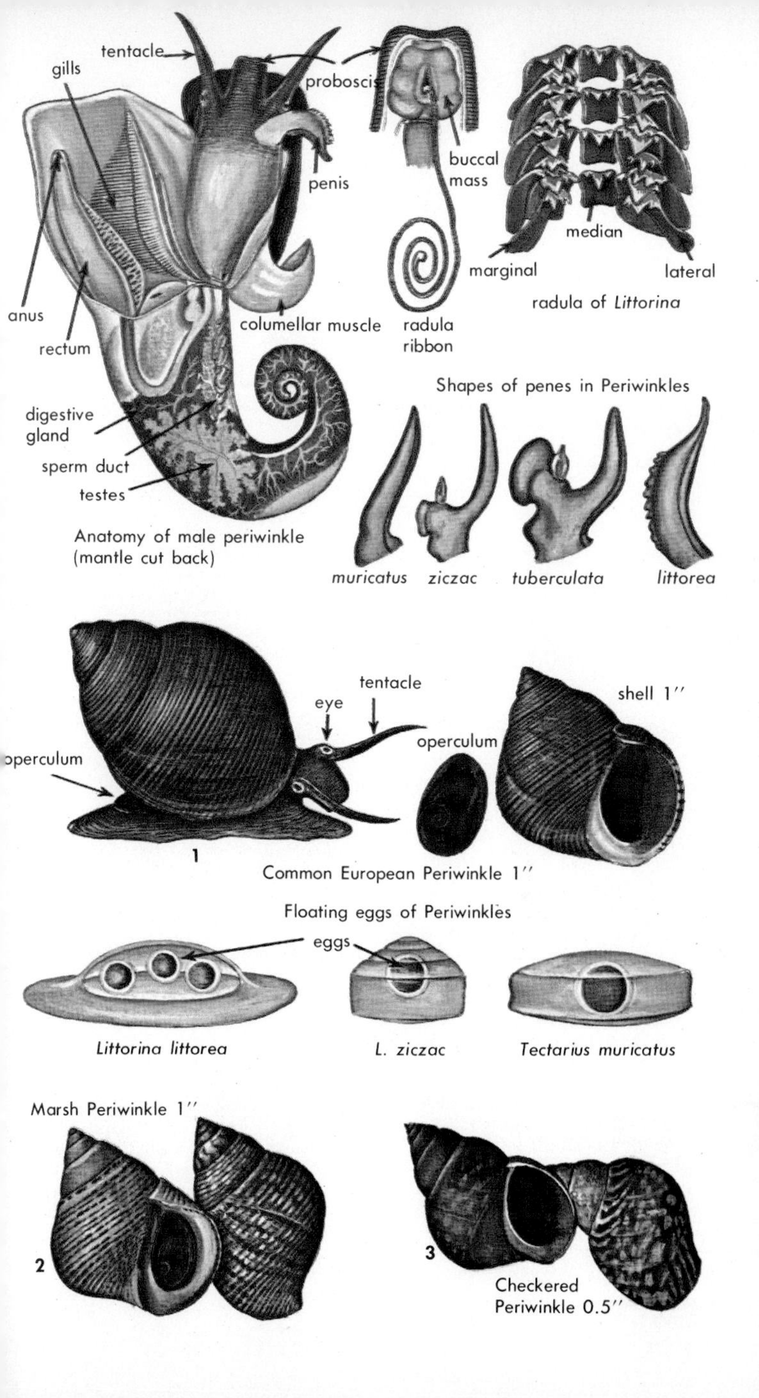
tentacle
gills
proboscis
buccal
mass
penis
median
marginal
lateral
radula of Littorina
anus
rectum
columellar muscle
radula
ribbon
Shapes of penes in Periwinkles
digestive
gland
sperm duct
testes
Anatomy of male periwinkle
(mantle cut back)
muricatus
ziczac
tuberculata
littorea
tentacle
eye
shell 1″
operculum
operculum
1
Common European Periwinkle 1″
Floating eggs of Periwinkles
eggs
Littorina littorea
L. ziczac
Tectarius muricatus
Marsh Periwinkle 1″
2
3
Checkered
Periwinkle 0.5″

PERIWINKLES—continued

Sibling species are those that live together and look similar, but which differ biologically and will not interbreed. Three sibling species, once thought to be only one, are found in the *floccosa-lineolata-ziczac* complex of periwinkles of the Gulf of Mexico and the Caribbean.

1. **FLOCCOSE PERIWINKLE.** *Littorina floccosa* (Mörch). Texas–Fla.; Caribbean. Aperture one half or more of the entire length of the shell. Upper whorls with 8–11 spiral lines. Operculum elongate, its nucleus small. The sides of the foot are a light yellowish gray. The sole of the foot is gray. Common; lives on rocks and on seaweeds in the intertidal zone.

2. **LINEOLATE PERIWINKLE.** *Littorina lineolata* (Orbigny). Texas–S. Fla.; Bermuda–Brazil. Shell elongate; aperture less than one half total length. Has 6–9 spiral lines on upper whorls. Sides of foot mottled black. Operculum almost round. Common; on rocks near high-tide mark.

3. **ZIGZAG PERIWINKLE.** *Littorina ziczac* (Gmelin). S.E. Fla.–Caribbean; Bermuda. Similar to Lineolate Periwinkle, but lighter; rarely, bluish gray; whorls more rounded. Has 20–26 fine spiral lines on upper whorls. Common; found in the intertidal zone in the crevices of rocks.

4. **ANGULATE PERIWINKLE.** *Littorina angulifera* (Lamarck). S. Fla.–Brazil; Bermuda. Shell thin but strong; smooth, but with numerous, fine spiral lines. Color variable, rarely orange or yellow. Aperture whitish; columella brownish purple. Common; on pilings and on mangrove roots in warm, shallow waters.

5. **NORTHERN YELLOW PERIWINKLE.** *Littorina obtusata* (Linné). Labrador–N.J. Shell small, solid, low-spired, and smooth. Color variable, usually a uniform orange-yellow or brownish yellow, sometimes banded. Columella whitish. Common; found on rockweed, where it also deposits its eggs.

6. **NORTHERN ROUGH PERIWINKLE.** *Littorina saxatilis* (Olivi). Arctic Seas–N.J.; Alaska–Puget Sound. Shell small, solid, with numerous, spiral ridges and fine growth lines. Color variable, gray to brown to almost black. Aperture brown within. Ovoviviparous. Common; on rocks, often above water.

7. **ERODED PERIWINKLE.** *Littorina planaxis* (Philippi). Puget Sound–Baja Calif. Shell rather smooth, grayish brown, with whitish spots and flecks. Always has an eroded, flattened area on the body whorl beside the whitish columella. Outer lip thin and sharp. Common; on rocks near the high-tide line.

8. **SITKA PERIWINKLE.** *Littorina sitkana* Philippi. Alaska–Puget Sound. Shell solid, grayish, with about 12 strong spiral threads. Common; on intertidal rocks.

9. **FALSE PRICKLY-WINKLE.** *Echininus nodulosus* (Pfeiffer). Fla.–Caribbean. Shell heavy, with square base. Whorls with 2 rows of sharp nodules. Common; above high tide.

10. **COMMON PRICKLY-WINKLE.** *Nodilittorina tuberculata* (Menke). S. Fla.–West Indies; Bermuda. Whorls with several rows of small nodules. Columella brown, flattened. Common; on tide-line rocks.

11. **BEADED PERIWINKLE.** *Tectarius muricatus* (Linné). Fla. Keys–West Indies; Bermuda. Pale bluish gray, with numerous spiral rows of neat, small, whitish beads. Columella creamy. Aperture interior reddish brown. Common; lives on rocks from high tide mark to several dozen feet above.

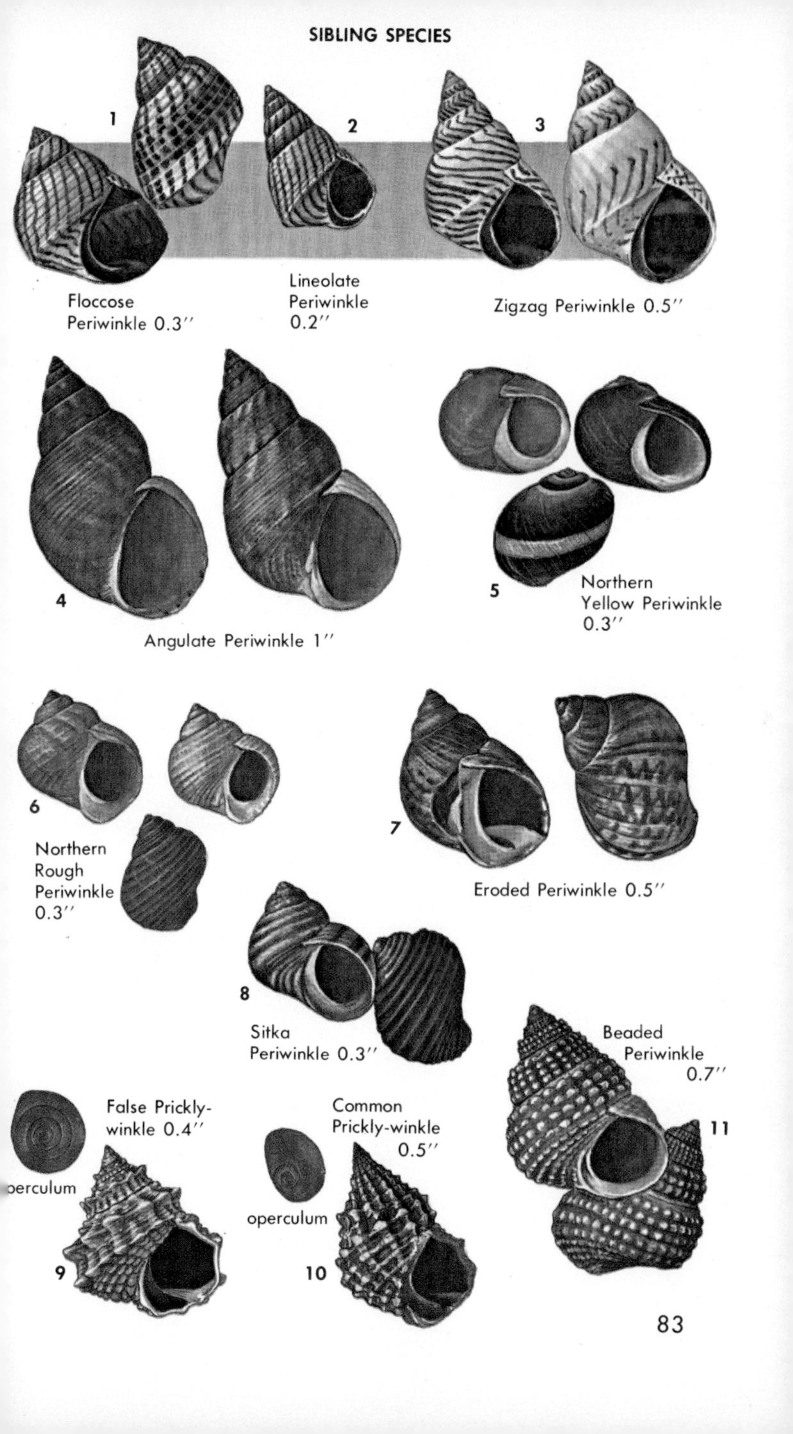
1
2
3
Floccose
Periwinkle 0.3″
Lineolate
Periwinkle
0.2″
Zigzag Periwinkle 0.5″
4
Angulate Periwinkle 1″
5
Northern
Yellow Periwinkle
0.3″
6
Northern
Rough
Periwinkle
0.3″
7
Eroded Periwinkle 0.5″
8
Sitka
Periwinkle 0.3″
Beaded
Periwinkle
0.7″
11
False Prickly-
winkle 0.4″
operculum
9
Common
Prickly-winkle
0.5″
operculum
10

TURRITELLAS AND WORM-SHELLS

The family Turritellidae contains the slender *Turritella,* herbivorous feeders whose shell is tightly coiled, and the *Vermicularia* whose shell is loosely coiled. The related family, Siliquariidae, contains the slit worm-shells, which live in sponges and have an open slit along the shell that allows water to enter the mantle cavity. The family Vermetidae, containing *Vermetus* and *Serpulorbis,* attach themselves to rocks. Similar calcareous worm tubes have only 2, not 3, shell layers and are chalky white, not shiny, inside the tube.

1. **EASTERN TURRITELLA.** *Turritella exoleta* (Linné). S.C.–Caribbean. Each whorl of this long, slender species bears a strong spiral cord below the suture and another at the periphery; the area between the cords is noticeably concave. Common; found at depths of 6–600 ft.

2. **COOPER'S TURRITELLA.** *Turritella cooperi* Carpenter. S. Calif.–Mexico. Similar to Eastern Turritella, but with numerous, much smaller spiral cords and threads. Aperture circular. Yellowish or light orange, with brownish axial patches. Fairly common; in sand below low tide.

3. **WESTERN SCALED WORM-SHELL.** *Serpulorbis squamigerus* (Carpenter). Alaska–Peru. Shell circular, minutely scaled or with rough, longitudinal cords. No operculum in this genus. Common; below tide line.

4. **SLIT WORM-SHELL.** *Siliquaria squamata* (Blainville). S.C.–West Indies. Early whorls smooth, later whorls spiny; characterized by a slit or row of small holes. Uncommon; in sponges, 20–200 ft.

5. **FARGO'S WORM-SHELL.** *Vermicularia fargoi* Olsson. Fla. west coast–Texas. Shell thick and sturdy; the early whorls or "turritella" stage 0.8–1 in. long. Later whorls with 3 strong, thick cords. Common; on mud flats.

The West Indian Worm-Shell, *V. spirata* Philippi (not illus.). S.E. Fla. and Caribbean. Lighter colored, and has a tiny 0.3 in. "turritella" stage. Later whorls with only 2 major spiral cords. Common.

6. **KNORR'S WORM-SHELL.** *Vermicularia knorri* Deshayes. N.C.–Fla.; Gulf of Mex. The 0.3 in. "turritella" stage is white; later detached whorls are yellow-brownish. Common; in sponges, 6–30 ft.

7. **BLACK WORM-SHELL.** *Petaloconchus nigricans* Dall. Fla. west coast. Shell rarely coiled. Gray to reddish brown with weak spiral and longitudinal sculpture. Each tube about 0.2 in. in diameter. Operculum present. Common; forms reefs at 3–15 ft.

RISSO SHELLS

The family Rissoidae contains many genera and species. There are about a dozen tiny *Rissoina* on both American coasts. The operculum is horny.

8. **CARIBBEAN RISSO.** *Rissoina bryerea* (Montagu). S.E. Fla.–Caribbean. Has 16–22 slanting ribs per whorl. Common; 1–10 ft.

9. **SMOOTH RISSO.** *Zebina browniana* (Orbigny). N.C.–Caribbean. Glistening white, rarely banded. Common; 1–10 ft.

VITRINELLA SHELLS

The family Vitrinellidae contains many genera and species of very small, low-spired shells, usually umbilicate. Operculum horny with many whorls. A few species are parasitic on worms

10. **PARASITIC VITRINELLA.** *Cochliolepis parasitica* Stimpson. N.C.–Caribbean. Spire low; whorls increase rapidly; umbilicus wide and deep. Translucent whitish. Lives under scales of polychaete worm. Common.

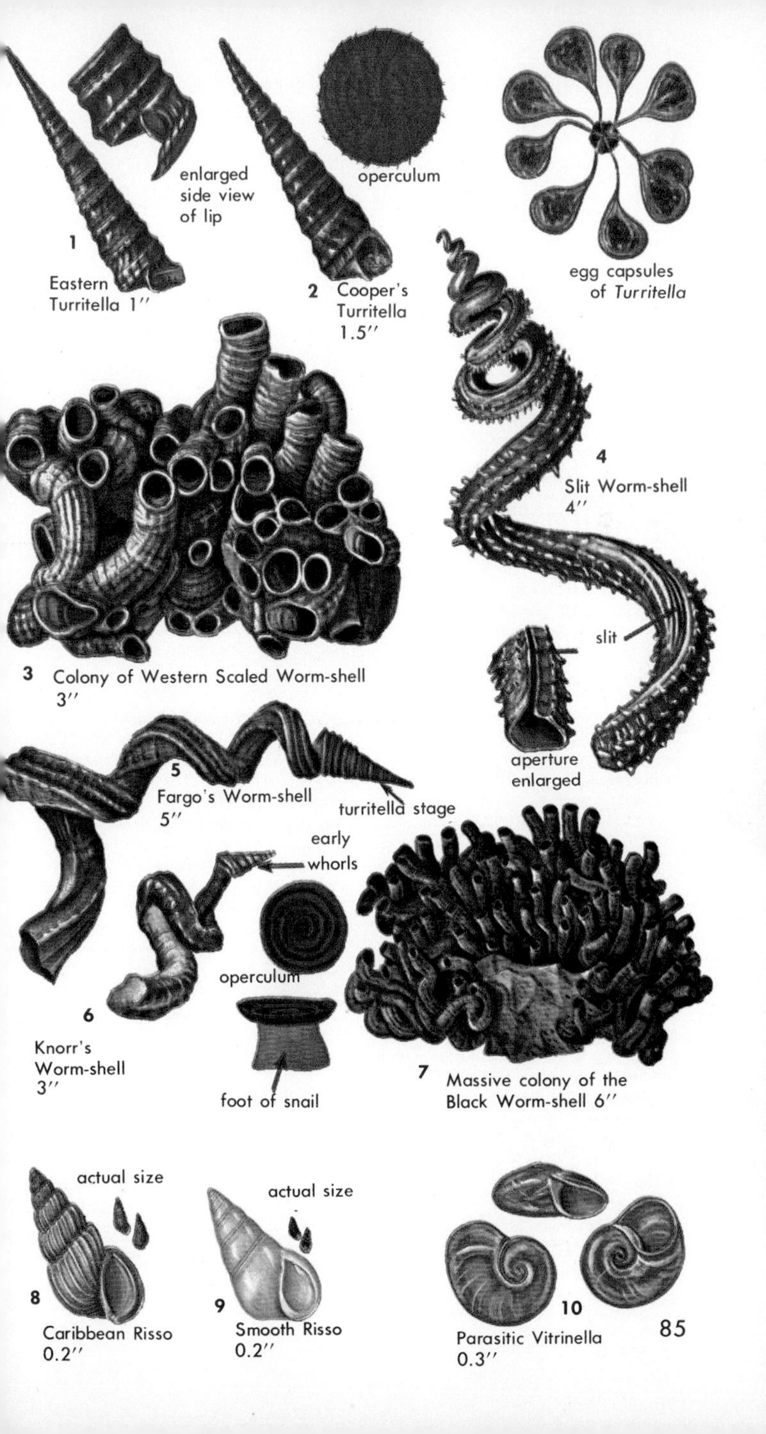
enlarged
side view
of lip
1
Eastern
Turritella 1″
operculum
2 Cooper's
Turritella
1.5″
egg capsules
of *Turritella*
3 Colony of Western Scaled Worm-shell
3″
4
Slit Worm-shell
4″
slit
aperture
enlarged
5
Fargo's Worm-shell
5″
turritella stage
early
whorls
operculum
6
Knorr's
Worm-shell
3″
foot of snail
7 Massive colony of the
Black Worm-shell 6″
actual size
8
Caribbean Risso
0.2″
actual size
9
Smooth Risso
0.2″
10
Parasitic Vitrinella
0.3″

CAECUMS, SUNDIALS, PLANAXIS

Caecums (family Caecidae) are very small, sand-dwelling gastropods with curved, tubular shells, open at one end. Nuclear whorls are lost and the hole is sealed by a septum or plug. Many of the species are extremely variable. They are allied to the Rissoidae (p. 84).

The shells of the family Modulidae resemble miniature top shells, but are entirely porcellaneous. Each bears a small toothlike spine at the base of the columella. There are two Atlantic species.

Sundials (family Architectonicidae) are characterized by widely coiled shells with the umbilicus open all the way to the apex. The nuclear whorls may be "sinistral." Most are associated with coelenterates. There are a dozen American species.

Planaxis snails (family Planaxidae) resemble some of the periwinkles (p. 80) but are heavier. The two Atlantic species live in the intertidal zone.

1. **FLORIDA CAECUM.** *Caecum floridanum* Stimpson. N.C.–Fla. Shell minute, opaque white, with 20–40 strongly developed axial rings; final ones quite large. Common; in muddy sand.

2. **LITTLE HORN CAECUM.** *Meioceras nitidum* (Stimpson). Fla.–Texas; Caribbean. Shell minute, glossy, translucent white, with irregular, opaque white mottlings. Has a swollen hump around middle or near aperture of the shell. Abundant; 1–30 ft.

3. **MANY-NAMED CAECUM.** *Caecum crebricinctum* Carpenter. S. Calif.–Mexico. Shell small, usually with about 100 very fine, crowded axial rings. Many forms have been unnecessarily named. Common; in sand, 6–200 ft.

4. **ATLANTIC MODULUS.** *Modulus modulus* (Linné). N.C.–Texas and Caribbean. Shell small and sturdy, low-spired, with numerous low, slanting, axial ribs that fade out at the periphery. Base with 5 strong spiral cords. Columella tooth purple-tinged. Animal greenish. Very common; on eelgrass, 2–30 ft.

The Angled Modulus, M. *carchedonius* (Lamarck), (not illus.), Fla. Keys—Caribbean, 0.5 in., is strongly angled at periphery. Columella tooth is uncolored; animal somewhat creamy. Uncommon.

5. **COMMON SUNDIAL.** *Architectonica nobilis* Röding. N.C.–Texas; Caribbean; also West Mex.–Peru. Shell heavy, with 4 or 5 somewhat beaded, spiral cords. Umbilicus deep, bordered by strongly beaded spiral cord. Fairly common; in shallow water among sea pansies.

6. **CYLINDER SUNDIAL.** *Heliacus cylindricus* (Gmelin). Fla. Keys–Caribbean. Small, high-spired, with numerous beaded spiral cords. Umbilicus round, very deep. The horny operculum is "top" shaped. Uncommon; at low-tide zone in clumps of soft corals.

7. **BLACK ATLANTIC PLANAXIS.** *Planaxis nucleus* (Bruguière). S.E. Fla.–Caribbean. Shell thick, polished, dark blackish brown, with numerous incised spiral lines strongest just below the periphery and at the base. Columella area scooped out, with a small pimple anteriorly. Aperture purplish, ridged within. Common; among intertidal rocks.

8. **DWARF ATLANTIC PLANAXIS.** *Planaxis lineatus* (da Costa). Fla. Keys–West Indies. Much smaller than the Black Atlantic Planaxis, but with numerous fine spiral cords at base and on early whorls. Usually cream-colored, with numerous neat, blackish brown spiral lines. Aperture white. Common; on rocks between tide lines.

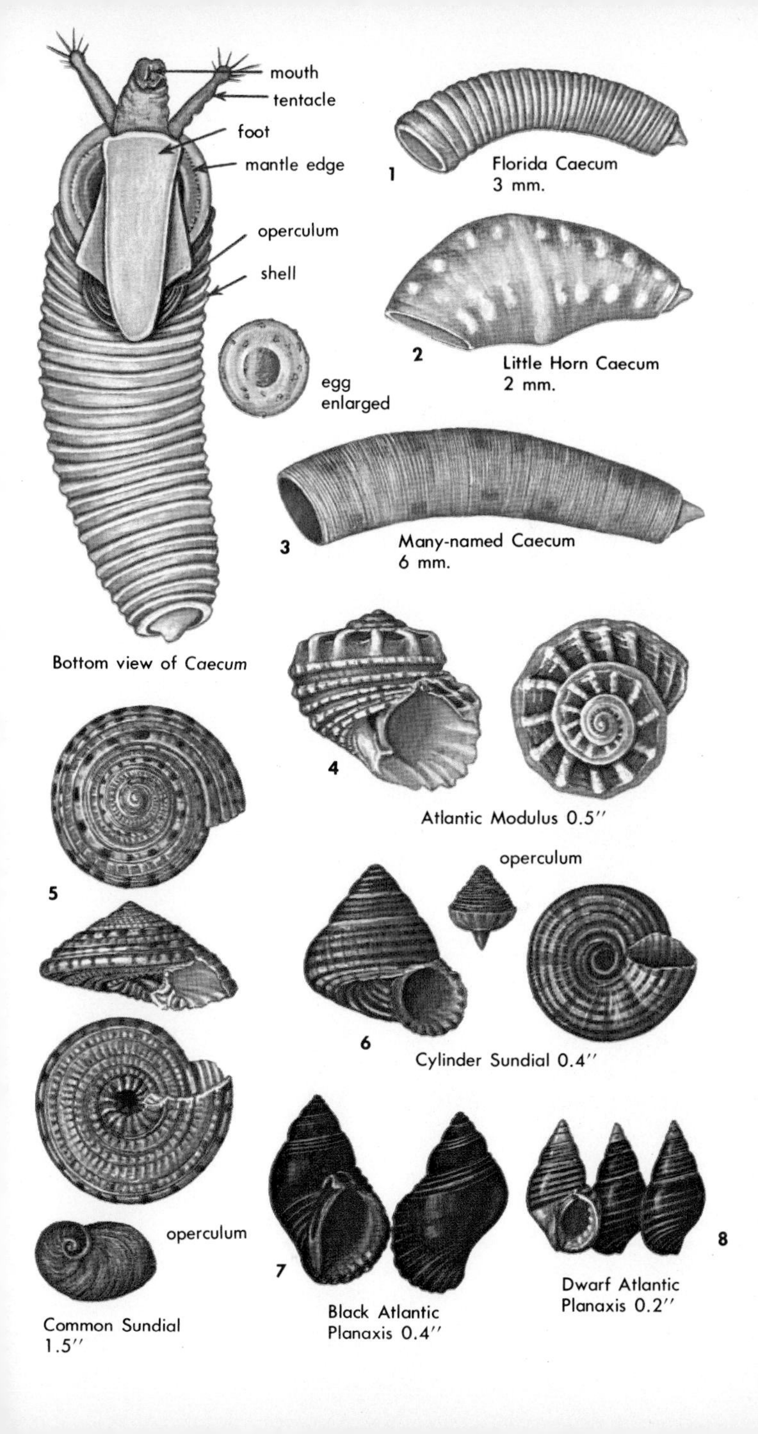
mouth
tentacle
foot
mantle edge
operculum
shell
egg
enlarged
1
Florida Caecum
3 mm.
2
Little Horn Caecum
2 mm.
3
Many-named Caecum
6 mm.
Bottom view of Caecum
4
Atlantic Modulus 0.5″
5
operculum
6
Cylinder Sundial 0.4″
operculum
Common Sundial
1.5″
7
Black Atlantic
Planaxis 0.4″
8
Dwarf Atlantic
Planaxis 0.2″

HORN SHELLS AND CERITHS

The small, usually dull-colored snails of these two families occur in large numbers in shallow, warm seas. Most are detritus feeders. Eggs are laid in jelly masses or strings, and the sexes are separate. The family Potamididae includes many species of brackish-water *Batillaria* and *Cerithidea*. These horn shells have a horny, round operculum with many whorls. The *Cerithium* snails, family Cerithiidae, live in shallow oceanic waters. The horny operculum has only a few whorls. They serve as a food for fish and waterfowl.

1. **FALSE CERITH.** *Batillaria minima* (Gmelin). S. Fla.–West Indies. Shell small, commonly deformed. Color variable, pale gray to black, usually with black or white spiral bands. Usually sculptured with nodules, swellings, and uneven spiral threads. Common; on mud in intertidal zone.

2. **COSTATE HORN SHELL.** *Cerithidea costata* (da Costa). Fla. west coast–West Indies. Shell strong, yellowish brown, with 25–30 axial ribs on the next to last whorl. Subspecies *C. c. turrita* Stearns (W. Florida) has 15–20 ribs. Common; on mud flats and in mangroves.

3. **PLICATE HORN SHELL.** *Cerithidea pliculosa* (Menke). La.–Texas; West Indies. Pale yellowish gray to brownish black, with numerous fine axial ribs crossed by spiral lines that give a somewhat beaded appearance. Several former varices present. Locally common, but not in Florida.

4. **LADDER HORN SHELL.** *Cerithidea scalariformis* (Say). S.C.–Caribbean. Shell pale reddish brown to violet-brown, with numerous whitish spiral bands. Axial ribs coarse. Base of shell with 6–8 well-defined spiral ridges. No old varices present. Common; on mud flats.

5. **CALIFORNIAN HORN SHELL.** *Cerithidea californica* (Haldeman). Central Calif.–Mexico. Shell dark brown to blackish, usually with 1 or 2 yellowish white former varices. Whorls with weak spiral threads and 12–18 strong axial ribs. Lip of aperture swollen and whitish in color. Very common; on mud flats.

6. **FLORIDA CERITH.** *Cerithium floridanum* Mörch. N.C.–Texas. Shell elongate, brownish, speckled with white. Siphonal canal well developed. Whorls with several neatly beaded, spiral cords between which are fine, granulated, spiral threads. Common; low tide to 20 ft.

7. **STOCKY CERITH.** *Cerithium litteratum* (Born). S.E. Fla.–Caribbean; Bermuda. Wide, stubby, whitish with spiral rows of small, dark squares. Sculptured with numerous coarse spiral threads and nodules along the suture. Common; on rocks below low tide. Often encrusted with lime.

8. **IVORY CERITH.** *Cerithium eburneum* Bruguière. S.E. Fla.–Caribbean. Shell white to cream in color, commonly with red-brown blotches. Whorls rounded, with 4–6 rows of 18–22 small, neat, raised beads. Very common; in shallow water and in grassy areas.

9. **FLY-SPECKED CERITH.** *Cerithium muscarum* Say. Fla.–Caribbean. Moderately elongate, with 8 or 9 noduled, axial ribs and several fine, spiral threads which bear many brown speckles. Siphonal canal long and twisted left. Common; on grass in shallow bays.

10. **MIDDLE-SPINED CERITH.** *Cerithium algicola* C. B. Adams. S. Fla.–Brazil. Similar in color and appearance to Ivory Cerith, but with a spiral row of 9–12 fairly large, pointed beads midway between sutures. The middle row bears the largest spires. Uncommon; in Florida, 6–60 ft.

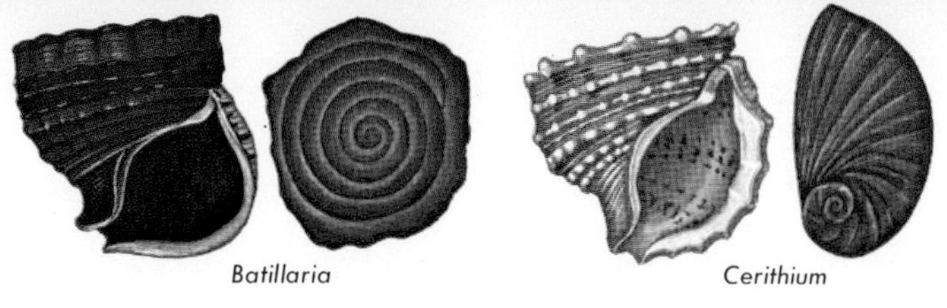

Enlargement showing apertures and opercula of False Cerith and Cerith

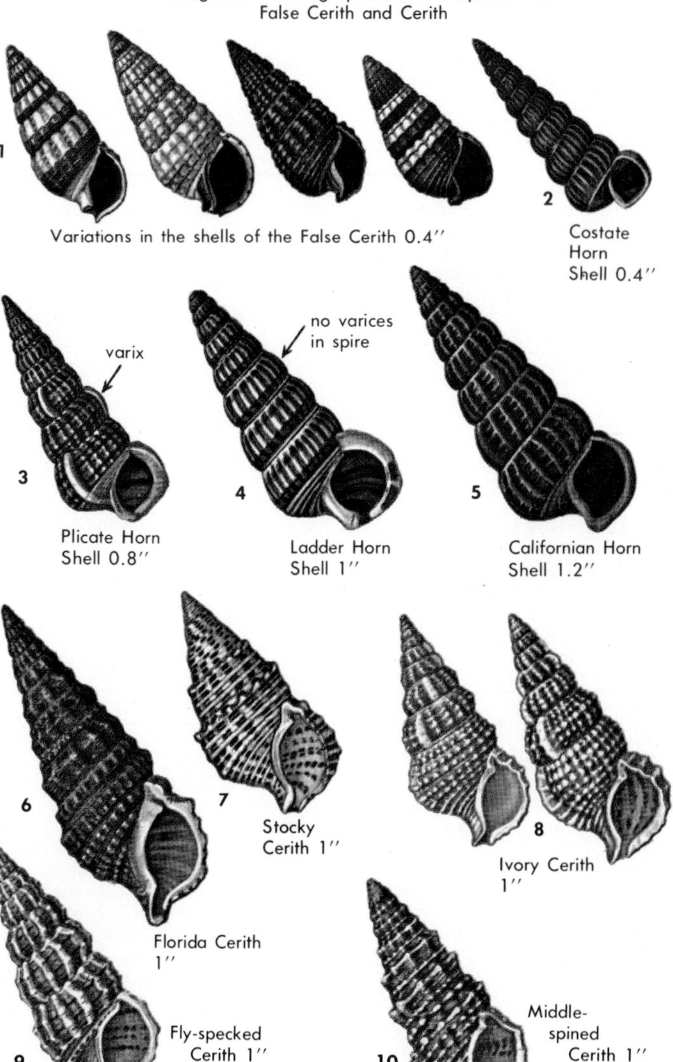

BITTIUMS AND TRIFORAS

Bittium, Cerithiopsis, and *Seila* (family Cerithiidae) are abundant, gregarious snails of shallow waters. They feed on diatoms, sponges, and detritus. The minute, sinistrally coiled Triphoridae (over 200 warm-water species) have a long proboscis for feeding on sponges. Because of the many species and their delicate, microscopic sculpturing, identification is difficult.

1. **ALTERNATE BITTIUM.** *Bittium alternatum* (Say). Gulf of St. Lawrence–Va. Very small, brownish; upper whorls cancellate or with 4 or 5 rows of beads. Columella short, twisted, brown-stained. Very common; from intertidal zone to 120 ft.

2. **VARIABLE BITTIUM.** *Bittium varium* (Pfeiffer). Md.–Fla. and Tex.; Mexico. Much smaller than Alternate Bittium, with a former, thickened varix and with fewer, less distinct axial ribs. Gregarious on eelgrass just below low tide.

3. **GIANT PACIFIC COAST BITTIUM.** *Bittium eschrichti* Middendorff. Alaska–N. Calif. Shell dirty white, with several wide, flattened, spiral cords. Subspecies *B. e. montereyense* Bartsch (to Baja Calif.) is glossy, somewhat shorter. Common.

4. **FOUR-THREADED BITTIUM.** *Bittium quadrifiliatum* Carpenter. S. Calif.–Mexico. Shell gray to reddish brown. Whorls rounded, with about 12 smooth axial ribs crossed by 4 or 5 small, raised, spiral threads. Very common; from the intertidal zone to 20 ft.

5. **ADAMS' MINIATURE CERITH.** *Seila adamsi* (H. C. Lea). Mass.–Texas and Caribbean. Shell small, with about 12 flat-sided, spirally corded whorls. Outer lip fragile. Sutures indistinct. Common; from low tide to 240 ft.

6. **AWL MINIATURE CERITH.** *Cerithiopsis subulata* (Montagu). Mass.–Caribbean. Slender, chocolate-brown, with about 14 flat-sided whorls bearing 3 rows of neat beads. Base concave, with fine axial growth lines. Common; 6–100 ft.

7. **GREEN'S MINIATURE CERITH.** *Cerithiopsis greeni* (C. B. Adams). Mass.–Caribbean. Very small, glossy brown, with 3 or 4 spiral rows of glassy beads connected by weak axial and spiral threads. Common; 6–60 ft.

8. **CARPENTER'S MINIATURE CERITH.** *Cerithiopsis carpenteri* Bartsch. Calif.–Mexico. Small, dark brown; whorls flat-sided, with 3 spiral rows of glassy beads. Base of shell with 2 distinct, smoothish, spiral cords. Abundant; 6–100 ft.

9. **MOTTLED TRIFORA.** *Triphora decorata* (C. B. Adams). S.E. Fla.–Caribbean. Sinistral; cream to gray, marked with red-brown. Whorls flat-sided, with 3 spiral rows of beads. Fairly common; 6–240 ft.

10. **BLACK-LINED TRIFORA.** *Triphora nigrocincta* (C. B. Adams). Mass.–Caribbean. Shell fusiform, sinistral; dark brown, with a blackish band just below the suture. Whorls bear 3 rows of glossy beads. Common; on seaweeds at low-tide line.

11. **SAN PEDRO TRIFORA.** *Triphora pedroana* Bartsch. Central Calif.–Mexico. Very small, glossy, yellow-brown. Whorls slightly convex, with 3 rows of whitish beads connected by axial threads. Moderately common; under stones at low tide.

12. **BROWN SARGASSUM SNAIL.** *Litiopa melanostoma* Rang (subfamily: Litiopinae). Pelagic in Atlantic Ocean; on sargassum weed. Shell thin and brownish. Surface smooth, with numerous fine spiral lines. Columella truncated below. Common. It hangs from seaweed by means of mucus strands.

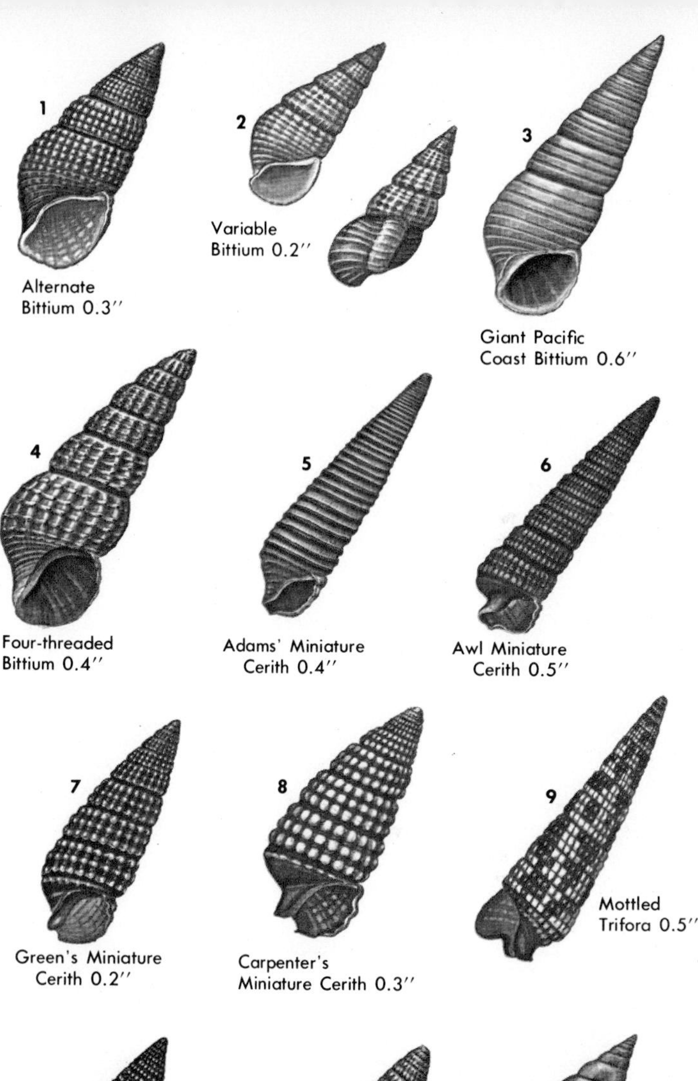

1 Alternate Bittium 0.3″

2 Variable Bittium 0.2″

3 Giant Pacific Coast Bittium 0.6″

4 Four-threaded Bittium 0.4″

5 Adams' Miniature Cerith 0.4″

6 Awl Miniature Cerith 0.5″

7 Green's Miniature Cerith 0.2″

8 Carpenter's Miniature Cerith 0.3″

9 Mottled Trifora 0.5″

10 Black-lined Trifora 0.2″

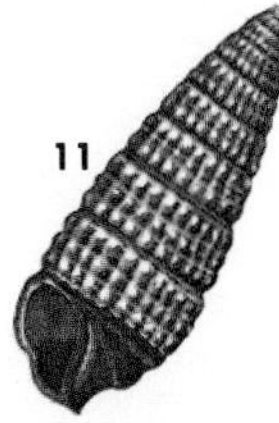

11 San Pedro Trifora 0.2″

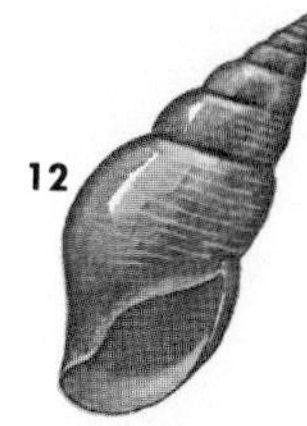

12 Brown Sargassum Snail 0.2″

JANTHINA SNAILS

The beautiful, pelagic janthina, or violet, snails (family Janthinidae) are very common and, fortunately for collectors, are sometimes blown ashore in great numbers. Even so, a perfect specimen is somewhat difficult to obtain, because the shells are extremely fragile. Violet snails trap bubbles of air in a gelatinous substance and attach them to the foot. A small operculum is present in the larval stage only.

Violet shells are hermaphrodites, with the male phase occurring first. There is no penis. Instead, a large, "feathery," mobile sperm carrier transports the sperm to the oviduct of a snail in the female stage. Some species, like the Elongate Janthina and the Pallid Janthina, attach egg capsules to the underside of the float. The Common Janthina is ovoviviparous. The young are born as free-swimming veligers. The *Recluzia* Snail attaches hundreds of yellowish egg capsules to the underside of its float.

If the float of a violet snail is in the way of the animal while it is feeding and crawling on a *Velella,* it may be discarded. Mucous secretions and the turbulence of the ocean water keep the snail near the surface. A new float is formed only if the anterior, bubble-blowing end of the foot can reach the surface film of the ocean.

Despite recent research, the speciation of this worldwide group of oceanic snails is imperfectly known. Environmental conditions appear to influence the coloration and shape of the shell. The so-called Dwarf Janthina may be the young of one of the three species found in American waters. The curious *Recluzia* Snail was unknown to the United States until 1953.

1. **COMMON JANTHINA.** *Janthina janthina* (Linné). Pelagic in warm waters; washed ashore on both coasts of the United States. Shell large, fragile, rather low-spired, with slightly angular whorls. Light purple above periphery, deep purple below. Outer lip slightly sinuate. Does not have eggs on float. Commonly found after storms in late spring.

2. **ELONGATE JANTHINA.** *Janthina globosa* (Swainson). Pelagic in warm waters; washed ashore on both coasts. Fragile, with rounded whorls and well-marked sutures. Aperture elongate. Outer lip slightly sinuate. Columella extends straight down. Coloring more uniformly violet and glossier than that of Common Janthina. Tentacles black with pale tips. Moderately common. *J. prolongata* Blainville is a synonym.

3. **PALLID JANTHINA.** *Janthina pallida* (Thompson). Pelagic; worldwide. Shell very globose; base of aperture rounded and without the slight projection seen in the Elongate Janthina. Sinus keel or scar can be seen in whorls of spire. Color light, whitish violet. Up to 400 elongate egg capsules are attached to float. Capsules about 4–7 mm. Tentacles pale. Moderately common; seasonally.

4. **RECLUZIA SNAIL.** *Recluzia rollandiana* Petit. Fla.–Texas–Brazil; pelagic. Shell high-spired, thin, and fairly strong. Suture well indented. Periostracum light brown. Body and tentacles sulphur-yellow. Float of brown bubbles is 2 or 3 in. long; brown egg capsules numerous on underside of float. Occasionally washed ashore on Texas beaches. Feeds on *Minyas* sea anemones.

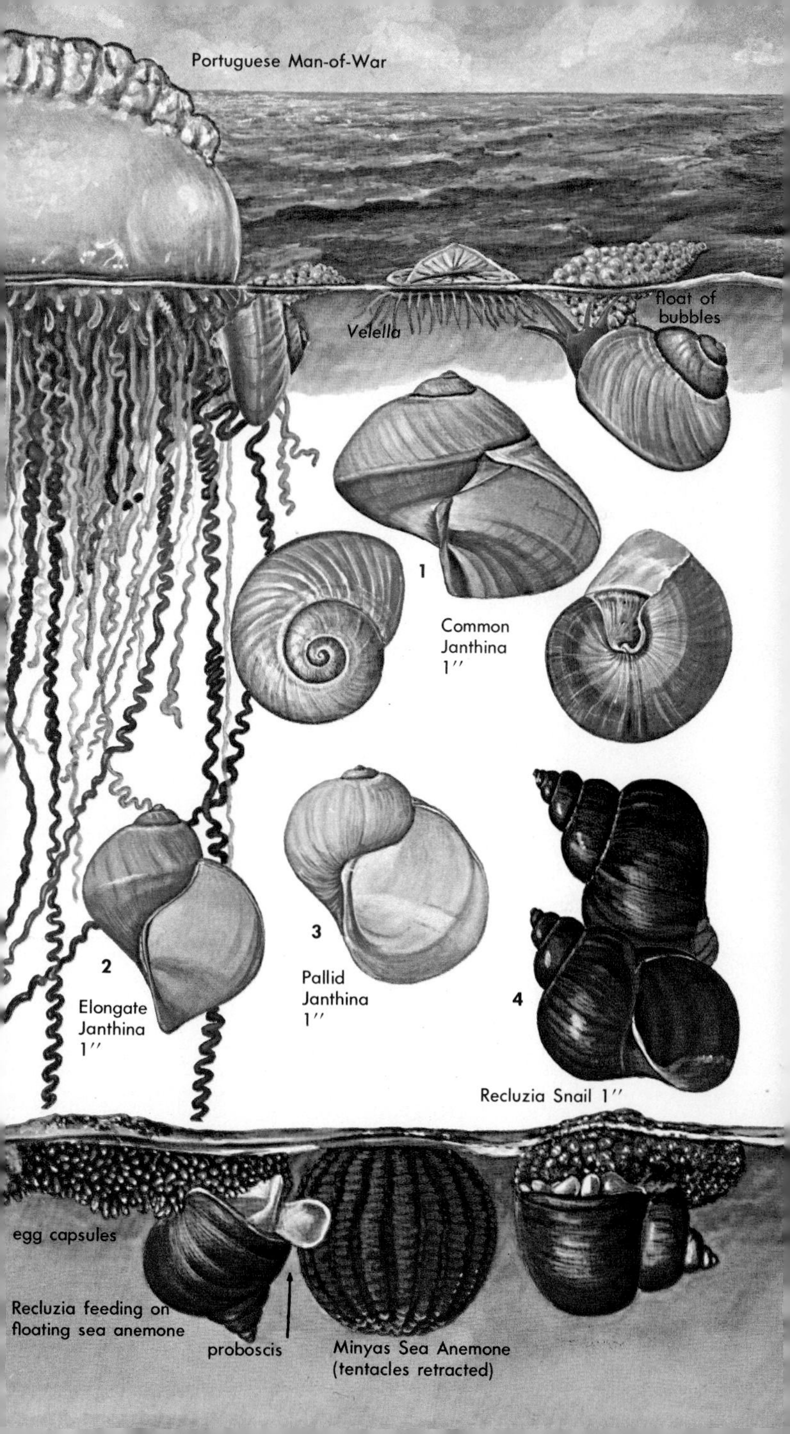

Portuguese Man-of-War
Velella
float of bubbles
1
Common Janthina 1″
2
Elongate Janthina 1″
3
Pallid Janthina 1″
4
Recluzia Snail 1″
egg capsules
Recluzia feeding on floating sea anemone
proboscis
Minyas Sea Anemone (tentacles retracted)

WENTLETRAPS

There are over 200 living species, and about as many fossil, in the assemblage of wentletraps (family Epitoniidae). At first glance, it seems impossible to separate them into species. Small, observable differences enable a careful student to identify most, if not all, the specimens he has collected, however. These significant differences include the number and shape of the axial ribs, or costae, the angle of the spire, and whether or not the shell has spiral sculpture or a basal ridge or cord. Except in a very few species, wentletrap shells are white.

The spawn of wentletraps consists of strings of angular capsules, often covered with mud or sand. These are joined to make a tough strand about 2 to 9 inches long. The larvae hatch in about nine days through apertures at the end of the capsules. Spawning may take place beneath the surface of the sand, as in *Natica*.

Several species of wentletraps have been observed sucking the juices of living sea anemones. From remains found in the croplike esophagus of other wentletraps, it seems possible others may feed on foraminifera. The radula consists of many rows of very small marginal teeth. These teeth show several characteristics that may eventually prove to be of generic significance.

The wentletrap's head bears a pair of tentacles with eyes set close to the base. When disturbed, many wentletraps exude a pink or purple dye, which may also serve as an anesthetic when it feeds on sea anemones.

The family has a wide distribution in all seas, from the low-tide line to depths well over 3,000 feet. Some are valued by collectors, such as the Precious Wentletrap, *Epitonium scalare* Linné, of the Indo-Pacific region, and the Caribbean Noble Wentletrap, *Sthenorytis pernobilis* (Fischer and Bernardi).

1. **SCALLOP-EDGED WENTLETRAP.** *Opalia insculpta* (Carpenter). S. Calif.–Mexico. Exterior chalky white, smooth, with numerous axial ribs that become very weak on the body whorl. Body whorl has a strong spiral ridge just below the periphery. Sutures somewhat pitted. Common; on rocks at low-tide line.

2. **WROBLEWSKI'S WENTLETRAP.** *Opalia wroblewskii* (Mörch). Alaska–Wash. Has 6–8 low, axial ribs, and with a smooth, spiral cord bounding the base of the body whorl. Common; in cold water, 3–60 ft.

The subspecies *O. w. chacei* (Strong) (not illus.), Ore.–Calif., is smaller, heavier, and broader. It is moderately common on rocky bottom near shore.

3. **NOBLE WENTLETRAP.** *Sthenorytis pernobilis* (Fischer and Bernardi). N.C.–Caribbean. This rare, variable species is considered a choice collector's item. Shell solid, pure white to grayish. Each whorl bears 12–14 very large, bladelike axial ribs, or costae, that taper to a thin margin. These costae, which are somewhat variable, are flattened in the area along the inner lip. Operculum black, horny, circular, with a central nucleus. Rare; usually dredged in 400 and 1,000 ft. of water.

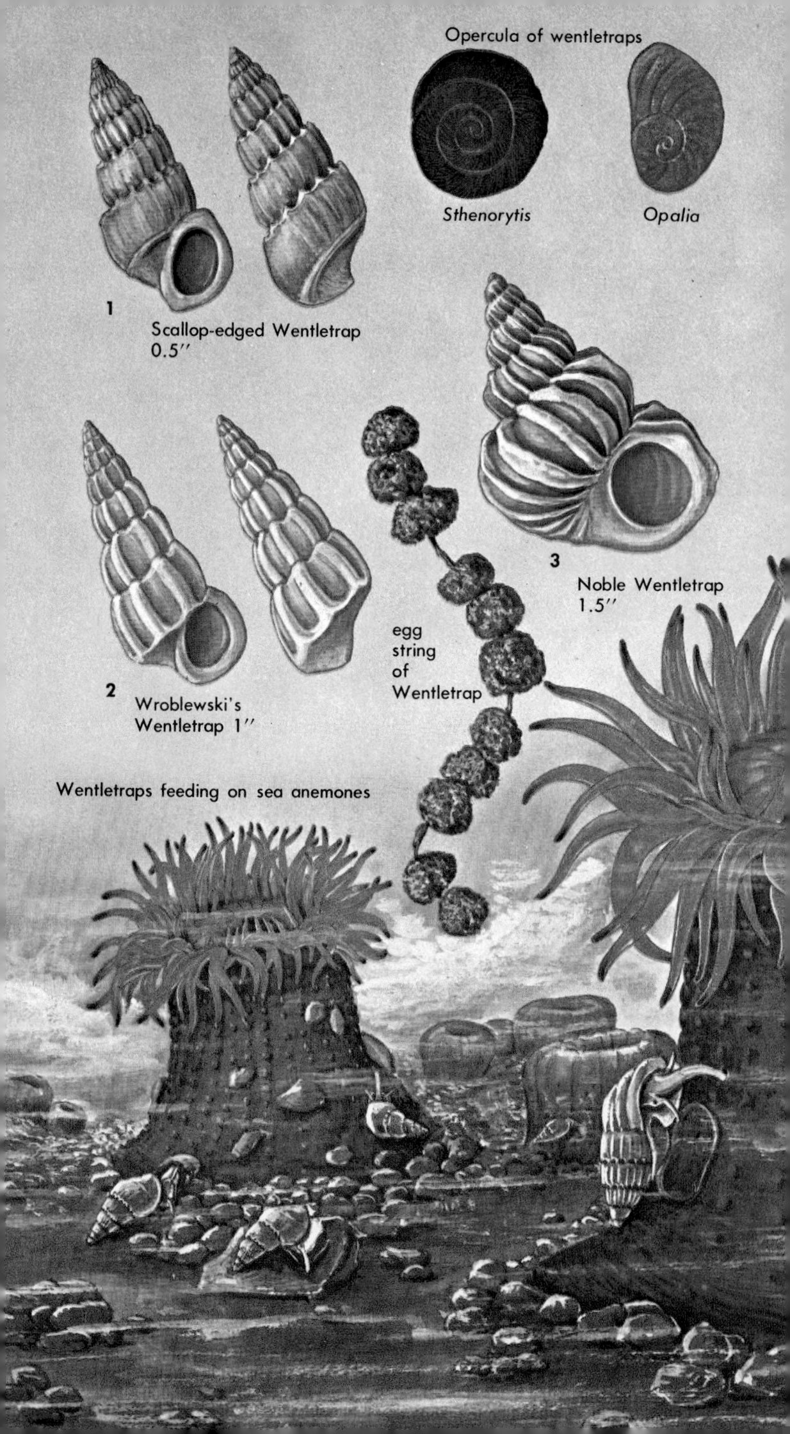
Opercula of wentletraps
Sthenorytis
Opalia
1
Scallop-edged Wentletrap
0.5″
3
Noble Wentletrap
1.5″
egg
string
of
Wentletrap
2
Wroblewski's
Wentletrap 1″
Wentletraps feeding on sea anemones

1. **MITCHELL'S WENTLETRAP.** *Amaea mitchelli* (Dall). Texas–Yucatan. Shell quite large, thin but strong; pale ivory with reddish brown bands; 2 bands on body whorl. Whorls rounded, with about 22 low axial ribs. Uncommon; in 3–30 ft. of water; occasionally washed ashore.

2. **DALL'S WENTLETRAP.** *Cirsotrema dalli* Rehder. N.C.–Brazil. Shell chalky, gray to tan, slender; surface pitted, with numerous fine, elevated ribs crossing the sutures and several moderately large varices at irregular intervals. Uncommon; found in 100–450 ft. of water.

3. **GREENLAND WENTLETRAP.** *Epitonium greenlandicum* (Perry). Arctic Seas–Long Island. Shell quite large, chalk-white to gray, with 9–12 variable ribs. Spiral cords numerous, flattened. Basal ridge may be strong. Uncommon; found in water 60–2,000 ft. deep.

4. **BROWN-BANDED WENTLETRAP.** *Epitonium rupicola* (Kurtz). Mass.–Fla. and Texas. Shell white to yellow, with broad brown bands, 2 on the body whorl. Rarely, shell is dark brown. Whorls globose, with 12–18 weak to strong ribs and a few varices. Basal ridge microscopic. Common; found from low tide to a depth of about 60 ft.

5. **WRINKLE-RIBBED WENTLETRAP.** *Epitonium foliaceicostum* (Orbigny) S.E. Fla.–Caribbean. Shell rather small, light, yellowish, moderately stout, with 7 or 8 thin, bladelike ribs per whorl, which are often acutely angled at the shoulder. Uncommon; down to 700 ft.

6. **LAMELLOSE WENTLETRAP.** *Epitonium lamellosum* (Lamarck). S.Fla.–Caribbean. Rather light, with numerous thin, white, bladelike ribs. Whorls usually with some brownish marking, stronger at suture. Basal ridges well developed. Fairly common; low tide to 100 ft.

7. **MONEY WENTLETRAP.** *Epitonium indianorum* (Carpenter). Alaska–Baja Calif. Shell slender, pure white, with 13 or 14 sharp, slightly recurved ribs on each whorl. Ribs somewhat pointed at the shoulder. Quite common; found in water 6–100 ft. deep.

8. **TINTED WENTLETRAP.** *Epitonium tinctum* (Carpenter). S. Calif.–Mexico. Shell delicate, white, sometimes tinged with brownish purple on the lower portions of the well-rounded whorls. With about 12 thin, sharp ribs. Uncommon; found in water 3–30 ft. in depth.

9. **HUMPHREY'S WENTLETRAP.** *Epitonium humphreysi* (Kiener). Mass.–Fla. and Texas. Shell rather thick and solid, with 8 or 9 well-developed ribs, heavy and rounded on body whorl. Ribs usually angled at the shoulder. Common in sand; from low tide to 300 ft.

10. **ANGULATE WENTLETRAP.** *Epitonium angulatum* (Say). N.Y.–Fla. and Texas. Shell glossy white, with 9 or 10 moderately thin, bladelike ribs, usually angled at the shoulder. No basal thread. Very common; low tide to 90 ft. Usually found in sand near sea anemones.

11. **MANY-RIBBED WENTLETRAP.** *Epitonium multistriatum* (Say). Mass.–Fla. and Texas. Shell rather small, fragile, and dull white. No basal thread. With up to 46 low, thin ribs on the early whorls and 16–19 on the body whorl of adult specimens. Spiral sculpture exceedingly fine. Rather uncommon; from low tide to 700 ft.

12. **KREBS' WENTLETRAP.** *Epitonium krebsii* (Mörch). N.C.–West Indies. Shell thin but strong; whorls globose, china-white, sometimes tinted with pinkish brown. Last whorl quite broad. Ribs bladelike, angled at the shoulder. Umbilicus very deep and wide. Rather rare; in coral sand, shore to 1,000 ft.

Mitchell's Wentletrap 2″

Dall's Wentletrap 1.5″

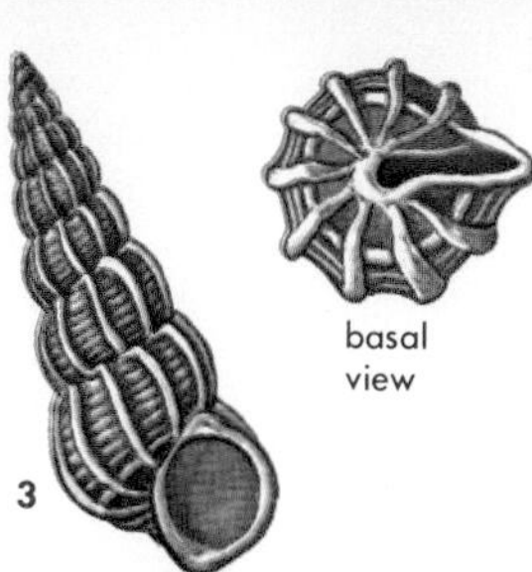

Greenland Wentletrap 1″

Brown-banded Wentletrap 0.5″

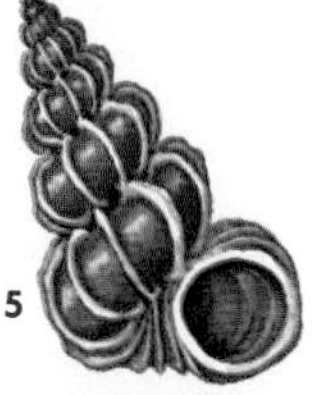

Wrinkle-ribbed Wentletrap 0.3″

Lamellose Wentletrap 1″

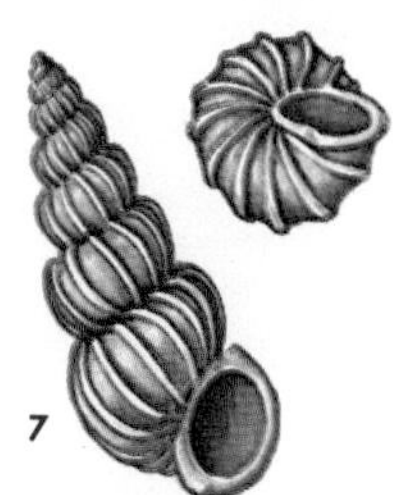

Money Wentletrap 1″

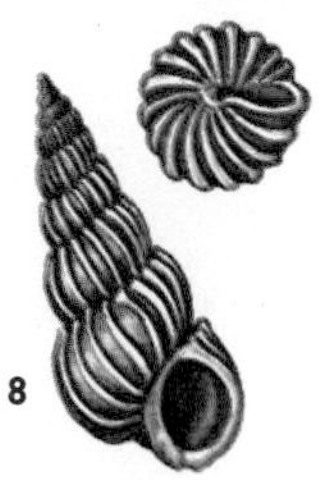

Tinted Wentletrap 0.5″

Humphrey's Wentletrap 0.6″

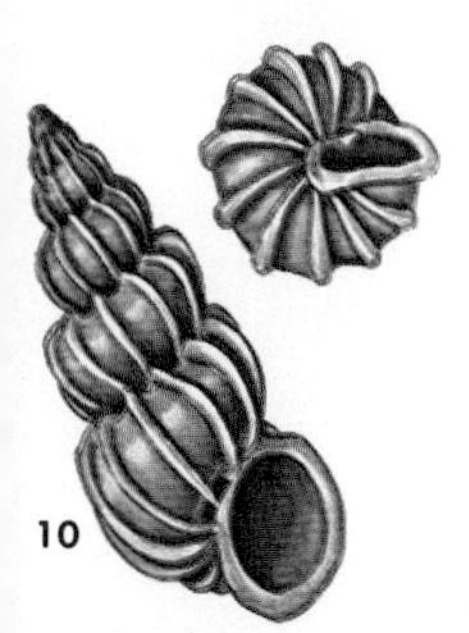

Angulate Wentletrap 0.9″

Many-ribbed Wentletrap 0.5″

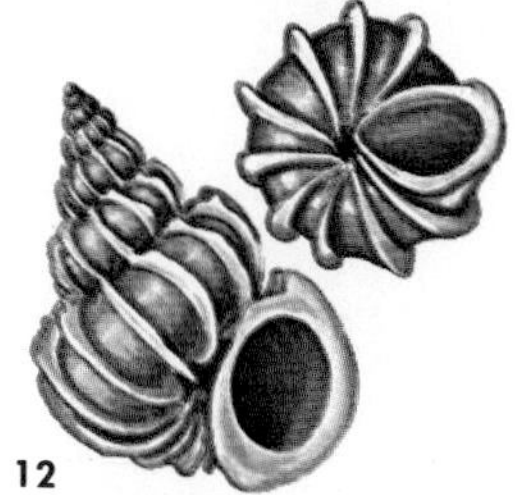

Krebs' Wentletrap 0.6″

HOOF AND HAIRY SHELLS

Hoof shells, of the family Hipponicidae, begin growing in the shape of typical coiled gastropods, but within a week of settling on a hard substrate they continue their growth in the cap-shaped form of limpets. In most species the foot secretes a shelly platform. The smaller males usually settle on the front side of the shell of the female or on rocks nearby. Feeding is limited to detritus of algae and animals, which is filtered by the gills from the ocean water. The saclike egg capsules, with 50 or more eggs, are brooded in the mantle cavity of the female. Most species are tropical.

Hairy shells, of the family Trichotropidae, are cold-water snails, mainly limited to boreal and arctic seas, with unusual periostracal bristles on the outside of the shell. Hairy shells may change sex; the smaller males later changing into females. When this happens, the penis gradually degenerates and the oviduct enlarges and becomes functional, passing eggs from the newly developed ovaries. There is no free-swimming larval stage. Hairy shells have a specially grooved extension of the proboscis, used to scoop up algae-laden mud.

1. **FALSE CUP-AND-SAUCER.** *Cheilea equestris* (Linné). S.E. Fla.–Brazil. Cap-shaped, lightweight, whitish, and with axial corrugations or, rarely, spines. Interior cup small, not complete, delicate, and white. Common; on subtidal rocks. The smaller males may be attached to the shell of the female or to rocks nearby.

2. **WHITE HOOF SHELL.** *Hipponix antiquatus* (Linné). S.E. Fla.–Brazil; Calif.–Peru. Shell solid, white, with concentric foliations. Periostracum thin, tan. Muscle scar inside shell is horseshoe-shaped. Common; lives in intertidal zone, attached to rocks.

3. **ORANGE HOOF SHELL.** *Hipponix subrufus* (Lamarck). S.E. Fla.–Caribbean; Calif.–Mexico. Whitish to orange; fine threads produce beaded effect. Periostracum sometimes heavy, tufted, and brown. Uncommon; on rocks, 6–100 ft.

4. **DALL'S HOOF SHELL.** *Hipponix benthophilus* (Dall). Fla.–Caribbean; deep water. Usually attached to the spines of large sea urchins. Shell sturdy, smoothish, except for crude growth lines. Color of shell and shelly platform is whitish.

5. **CANCELLATE HAIRY SHELL.** *Trichotropis cancellata* Hinds. Alaska–Oregon. Shell has netted surface with 5 or 6 spiral cords on upper whorls. Aperture small. Periostracum brown, thick, with curved bristles. Common; on rock bottom, 6–300 ft.

6. **BOREAL HAIRY SHELL.** *Trichotropis borealis* Broderip and Sowerby. Alaska–Wash.; Europe–Maine. Spire usually eroded. Numerous spiral cords covered with hairy, brown periostracum. Umbilicus narrow, deep. Common; 6–600 ft.

7. **TWO-KEELED HAIRY SHELL.** *Trichotropis bicarinata* Sowerby. Arctic Seas–W. Canada; Newfoundland. Shell fragile, with 2 spiral cords on body whorl and with a flattened columella. Periostracum with bristles on the cords. (When dry it flakes away.) Uncommon; 6–600 ft.

8. **GRAY HAIRY SHELL.** *Trichotropis insignis* (Middendorff). Bering Sea–Alaska. Shell heavy, with numerous uneven spiral cords. Aperture large and flaring. Periostracum rather thin and gray. Columella flattened. Uncommon; on rocky bottoms, 40–800 ft.

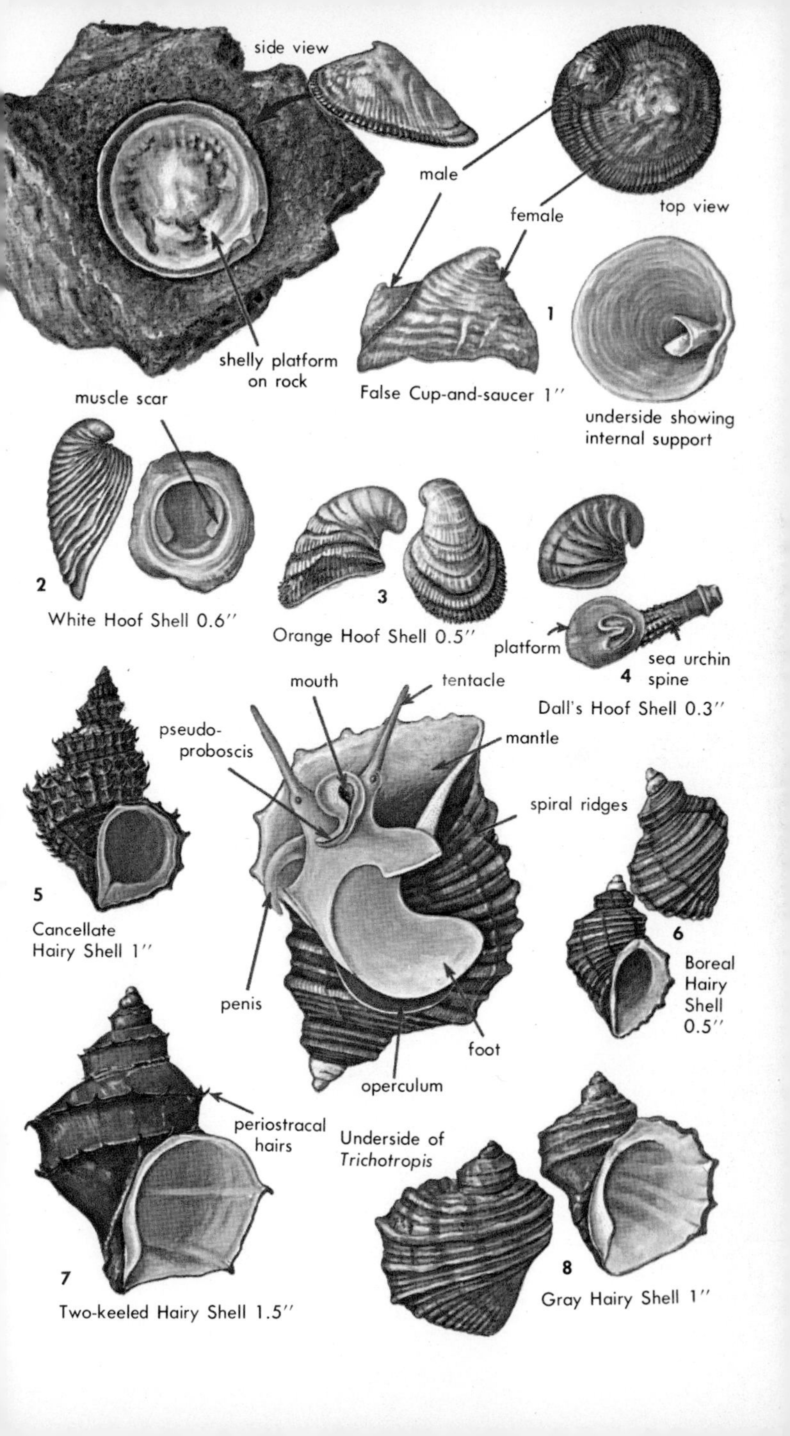
side view
male
top view
female
1
shelly platform
on rock
False Cup-and-saucer 1″
underside showing
internal support
muscle scar
2
White Hoof Shell 0.6″
3
Orange Hoof Shell 0.5″
platform
sea urchin
spine
4
Dall's Hoof Shell 0.3″
mouth
tentacle
pseudo-
proboscis
mantle
spiral ridges
5
Cancellate
Hairy Shell 1″
6
Boreal
Hairy
Shell
0.5″
penis
foot
operculum
periostracal
hairs
Underside of
Trichotropis
7
Two-keeled Hairy Shell 1.5″
8
Gray Hairy Shell 1″

CAP AND CUP-AND-SAUCER SHELLS

Chinese hats, family Capulidae, of which there are four American species, are cup-shaped snails that live in association with algae-eating scallops, snails, and sea urchins. Some species of *Capulus* that are attached to rocks obtain their food by sieving the algae-laden ocean water through their gills. Other species permanently attach themselves to the shells of scallops, so that they can steal the algal accumulations produced by the bivalves. A long, grooved proboscis permits the *Capulus* snail to reach within the scallop. The eggs are brooded in a thin-walled sac within the mantle cavity. An operculum is lacking in the adults.

Cup-and-saucer shells, family Calyptraeidae, including the genera *Crucibulum, Calyptraea, Crepipatella,* and (p. 102) *Crepidula,* is characterized by shells having a shelly "cup" inside that supports some of the soft parts. These genera, too, lack an operculum. There are about 26 American members of this family.

1. **CALIFORNIA CAP-SHELL.** *Capulus californicus* Dall. Calif.–Mexico. Cap-shaped shell is thin but strong. Obliquely ovate, with a small hooked apex. Periostracum soft, light brown. Interior glossy white, with a horseshoe-shaped scar. Uncommon; on the bottom valve of the San Diego Scallop (p. 207).

The Incurved Cap-shell (*C. incurvatus* Gmelin) (not illus.). N.C.–Fla. and West Indies. Similar, but 0.5 in. in diameter, with growth lines crossing spiral cords. Periostracum has rows of tiny tufts. Muscle scar is horseshoe-shaped with a swollen end just inside the columella. Rare; on rocks at low tide.

2. **SPINY CUP-AND-SAUCER.** *Crucibulum spinosum* Sowerby. S. Calif.–Chile. Exterior rough with radial rows of small prickles or erect tubular spines. Interior shiny brown with light rays. Complete interior cup is white and attached by one side. Common; on rocks, 1–90 ft.

3. **STRIATE CUP-AND-SAUCER.** *Crucibulum striatum* Say. Nova Scotia–S.C. Base circular; apex smooth, twisted, and near center. Exterior has small, wavy, radial cords. Interior glossy yellowish brown with a white shelly cup. Common; on rocks, 6–100 ft.

4. **WEST INDIAN CUP-AND-SAUCER.** *Crucibulum auricula* (Gmelin). N.C.–Caribbean. Exterior has coarse diagonal ribs that make the edges of the shell wavy. Apex hooked over. Edges of the shelly cup are free. Uncommon; attached to other shells, 6–50 ft.

5. **PACIFIC CHINESE HAT.** *Calyptraea fastigiata* Gould. Alaska–S. Calif. Base of shell circular with the apex at the center of the shell. Exterior smooth and white, covered by thin varnish-like periostracum. The interior, white, twisted "cup" is glossy and fragile. Common; 30–200 ft.

6. **CIRCULAR CUP-AND-SAUCER.** *Calyptraea centralis* Conrad. N.C.–Texas; West Indies. Shell small, grayish white, cup-shaped, with a circular base. Apex central and minutely coiled. Interior cup broad, with a thickened edge. Common; dredged from 20–200 ft.

7. **PACIFIC HALF-SLIPPER.** *Crepipatella lingulata* Gould. Alaska–Panama. Resembles the slipper shell, but the shallow interior deck is attached only along one side. Interior glossy tan to mauve. Exterior wrinkled and brownish. Common; on shells of living gastropods.

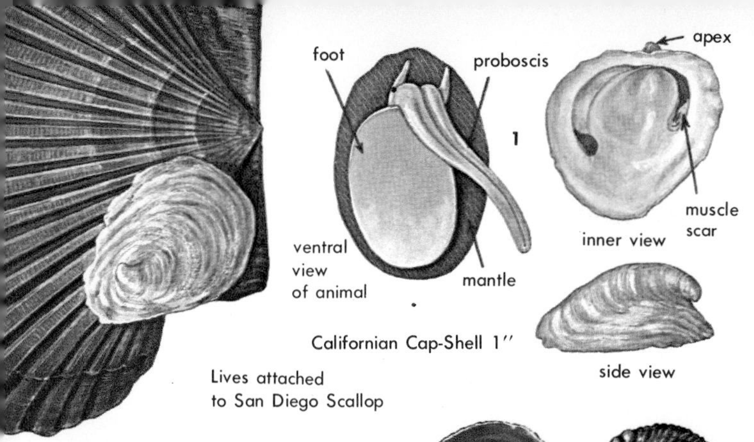

Californian Cap-Shell 1″

Lives attached
to San Diego Scallop

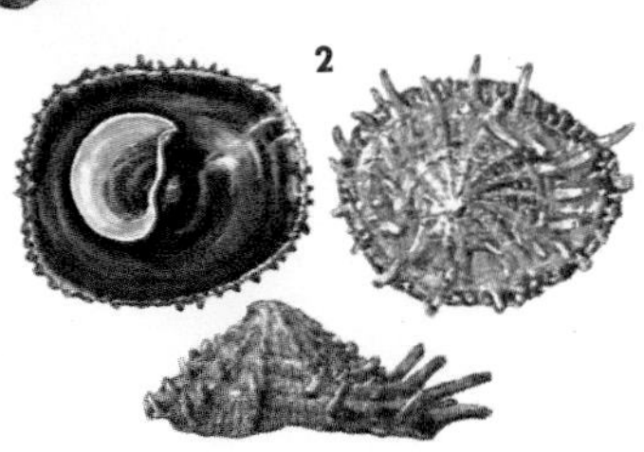

Spiny Cup-and-saucer 1″

Striate Cup-and-saucer 1″

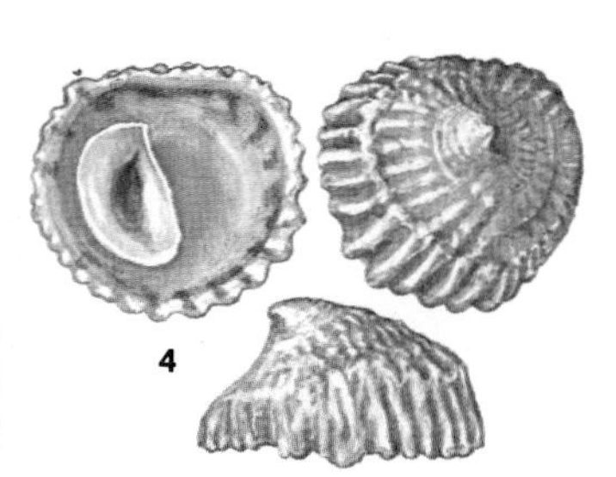

West Indian Cup-
and-saucer 1″

Pacific Chinese Hat 1″

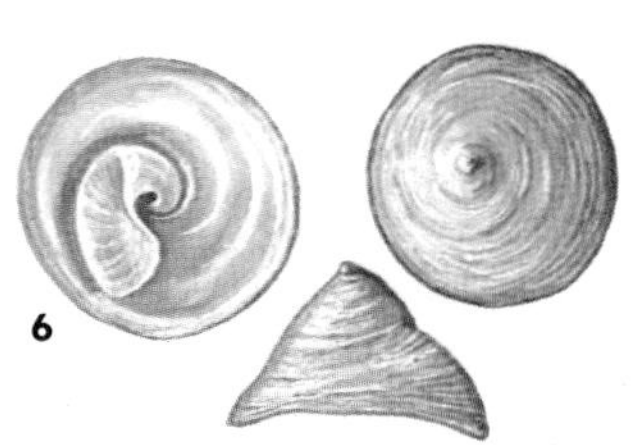

Circular Cup-and-saucer 0.5″

Pacific Half-slipper Shell 0.7″

SLIPPER SHELLS

Slipper shells of the genus *Crepidula* (family Calyptraeidae) are among the most abundant of shallow-water snails in most parts of the world. The cup-shaped shell is reinforced inside by a shelly platform that serves as a support for the soft digestive gland. Young slipper shells are fairly active in moving about, but two-year-old adults remain affixed in one place for life. In some species, such as *Crepidula fornicata* (Linné), individuals sit upon the shells of others, forming chains. The large older individuals die off and younger arrivals attach themselves to the top of the group (see p. 20).

Slipper shells cannot move much in search of food; hence, they have modified their manner of feeding to resemble that of algae-feeding oysters. Food is trapped by mucus in the mantle. *Crepidula* can change sex. The oldest animals are females, the youngest males. The degree to which female characteristics are developed is controlled by a hormone given off into the water by males. Females produce thin-walled capsules containing 70 to 100 eggs, which are attached to a rock. Then they brood the capsules until the free-swimming veligers hatch.

1. **EASTERN WHITE SLIPPER SHELL.** *Crepidula plana* Say. Nova Scotia–Texas. Shell white inside and out; exterior with thin, tan periostracum. Shape variable, usually concave from growing inside an empty snail shell. Males very small, found on top of larger females.

The Western White Slipper Shell, C. *nummaria* Gould (not illus.), one inch long. Almost identical. The internal, large deck has a weak, raised ridge. Lives on shells and rocks in shallow water from southern Alaska south to Panama.

2. **COMMON ATLANTIC SLIPPER SHELL.** *Crepidula fornicata* (Linné). Nova Scotia–Texas. Introduced to Puget Sound and Europe. Shell convex, lightly spotted. Interior deck buff. Commonly attached to each other in shallow water. Abundant; 1–50 ft.

3. **GIANT SLIPPER SHELL.** *Crepidula grandis* Middendorff. Alaska. Shell large, usually flat. Exterior chalky tan, with fuzzy, brown periostracum. Interior tan. Edge of shelf oblique. Common; along shore.

4. **ONYX SLIPPER SHELL.** *Crepidula onyx* Sowerby. Calif.–Peru. Interior glossy brown. Interior deck thin, white, and with wavy edge. Exterior usually coarse. Note ribbed form, which grows on scallop. Common; on rocks and on each other, 3–300 ft.

5. **CONVEX SLIPPER SHELL.** *Crepidula convexa* Say. Mass.–Texas; Caribbean. Shell small, reddish brown, strongly arched, rarely spotted. Brown apex strongly hooked. Common from 6–600 ft. The form C. *glauca* Say is 0.3 in. and lives on eelgrass.

6. **SPOTTED SLIPPER SHELL.** *Crepidula maculosa* Conrad. Gulf of Mexico. Very similar to Common Atlantic Slipper, but attractively spotted; with an oval muscle scar under one edge of the deck. Common; on dead pen shells, offshore.

7. **SPINY SLIPPER SHELL.** *Crepidula aculeata* (Gmelin). S.E. United States–West Indies. Exterior rough and spiny. Color white to orange-brown, rarely green-stained. Platform with central ridge. Common; on rocks, 1–30 ft.

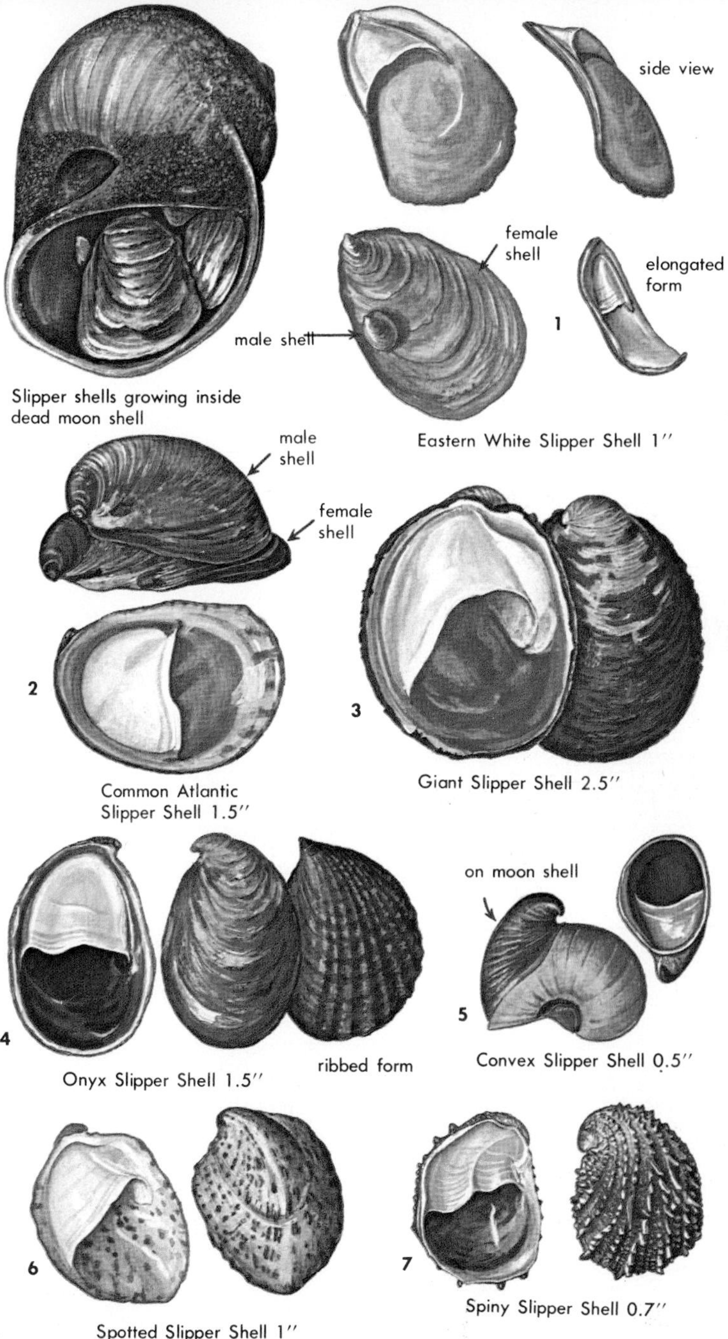

Slipper shells growing inside dead moon shell

Eastern White Slipper Shell 1″

Common Atlantic Slipper Shell 1.5″

Giant Slipper Shell 2.5″

Onyx Slipper Shell 1.5″

Convex Slipper Shell 0.5″

Spotted Slipper Shell 1″

Spiny Slipper Shell 0.7″

CONCHS AND CARRIER SHELLS

True conchs (family Strombidae) include about 50 shallow-water species, but only 11 in the Americas. They pry themselves forward with a narrow foot armed with a horny, sickle-shaped operculum. The colorful eyes are at the ends of short stalks that branch from the tentacles. Conchs feed on red algae, and the females lay long, jellylike strands of eggs. The lower part of the outer lip of the adult has a rounded "stromboid" notch. The pink conch rarely produces a pea-sized pink pearl. Conch meat makes tasty salads, steaks, and chowders.

Carrier shells (family Xenophoridae) cement stones, shells, and bits of coral to their shells as camouflage or reinforcement. In Florida there are one shallow-water and two deep-sea species. The operculum is thin, horny, and fan-shaped. The eggs and feeding habits are unknown.

Pelican foot shells (family Aporrhaiidae) live in cold waters and bury themselves just under the surface of soft, muddy sand. A sand trough made at the front permits water to enter the snail's siphon and allows the extended proboscis to search for plant and animal detritus. The genus *Aporrhais* is limited to the North Atlantic and the Mediterranean, with two species off New England. The flaring lip marks the mature animal. The operculum is small, horny, and narrow. The small, attractive Mediterranean species, *A. pespelicani* (Linné), is used in shell jewelry.

1. **PINK CONCH** (also Queen Conch). *Strombus gigas* Linné. S. Fla.–Brazil. Common; 2–30 ft., on sand and eelgrass. The young conchs, called "rollers," do not have flaring lips on shells.

The Milk Conch (*S. costatus* Gmelin) (not illus.). Fla. Keys–West Indies. Similar but has cream-white lip and no knobs on spire. 4–6 in. long. Common.

2. **FLORIDA FIGHTING CONCH.** *Strombus alatus* Gmelin. N.C.–Yucatan. Note down-sloping outer lip and short spines on spire. Albinos rare. Usually mottled, or banded. Common; 5–25 ft., on sandy mud.

3. **HAWK-WING CONCH.** *Strombus raninus* Gmelin. S.E. Fla.–West Indies. Exterior mottled with browns and purples. Interior of aperture reddish. Fairly common; lives in grassy shallow water.

The Rooster-tail Conch (*S. gallus* Linné) (not illus.). Fla.–West Indies. Similar. Has long extension of upper lip. Body whorl has 2 triangular spines. Rare.

4. **WEST INDIAN FIGHTING CONCH.** *Strombus pugilis* Linné. S.E. Fla.–Brazil. Note upward-sloping outer lip, long spines on next-to-last whorl. Compare with Florida Fighting Conch. Common; in shallow water.

5. **ATLANTIC CARRIER SHELL.** *Xenophora conchyliophora* (Born). S. Fla.–Brazil. Animal is red; operculum horny, amber. Uncommon; 3–20 ft., near coral reefs.

6. **AMERICAN PELICAN'S FOOT.** *Aporrhais occidentalis* (Beck). Labrador–N.C. Shell has 15–25 axial ribs per whorl. Operculum clawlike. Common; offshore, 60–1,200 ft., on sandy mud.

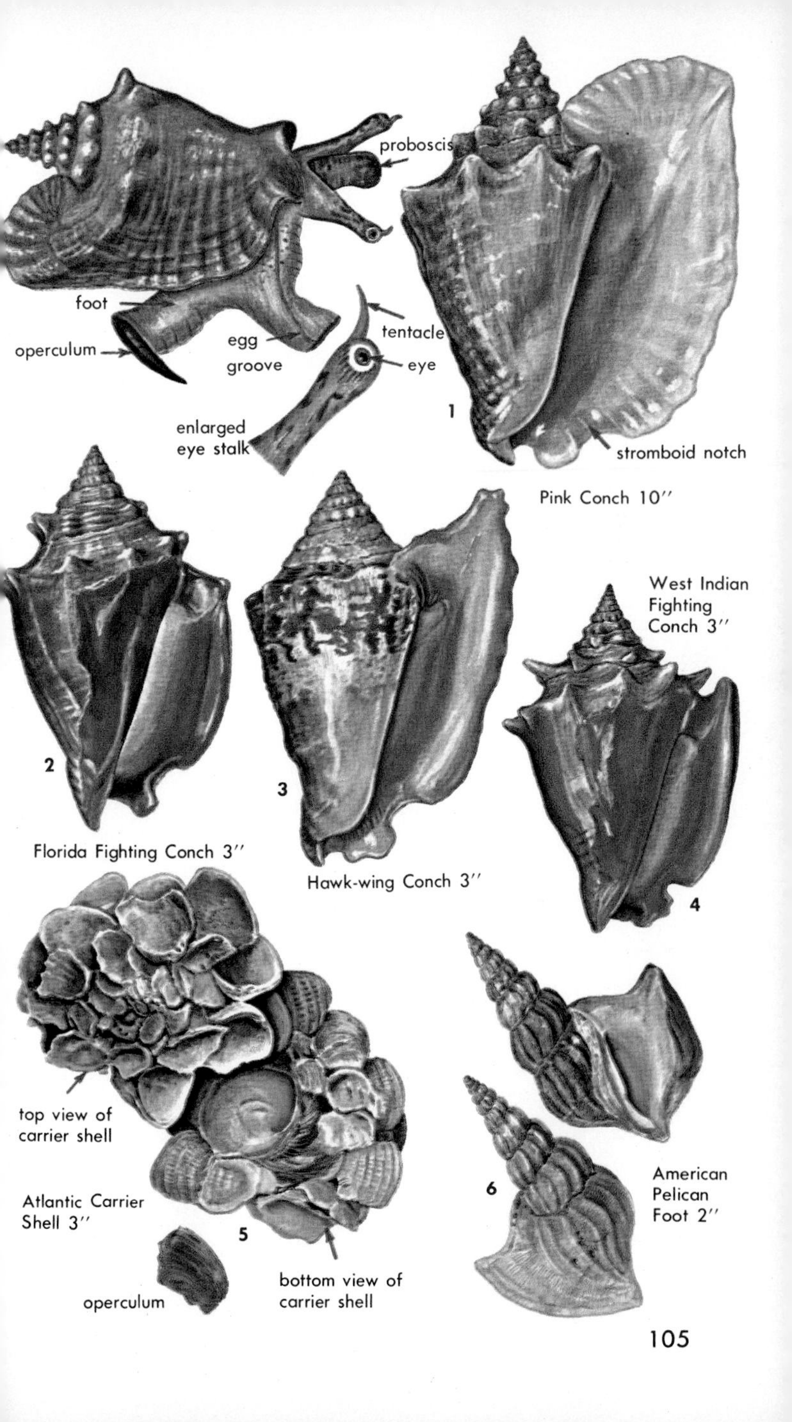
proboscis
foot
operculum
egg
groove
tentacle
eye
enlarged
eye stalk
1
stromboid notch
Pink Conch 10″
West Indian
Fighting
Conch 3″
2
3
Florida Fighting Conch 3″
Hawk-wing Conch 3″
4
top view of
carrier shell
Atlantic Carrier
Shell 3″
5
operculum
bottom view of
carrier shell
6
American
Pelican
Foot 2″

TRIVIAS

Trivias (family Eratoidae) resemble the cowries (p. 110), but live closely associated with tunicates in shallow water. There are about 100 worldwide species, 20 in the Americas. The female trivia feeds upon and places the cup-shaped horny egg capsule deep into the flesh of the tunicate host. *Erato,* which has a very short intestine, feeds on the zooids of the tunicate *Botryllus*. The mantles of *Trivia* are highly ornamented with minute fronds and extend up over the shell as in true cowries.

1. **FOUR-SPOTTED TRIVIA.** *Trivia quadripunctata* (Gray). S.E. Fla.–West Indies. Bright pink with 2–4 very small, red-brown dots on the center line of the back; 19–24 riblets cross the outer lip. Common; in shallow water.

2. **CALIFORNIAN TRIVIA.** *Trivia californiana* (Gray). Calif.–Mexico. Mauve with a whitish crease on the back. Outer lip has 15 riblets. Common; 6–250 ft. *T. sanguinea* (Sowerby) is all purple and has 20 riblets. It occurs in Baja Calif. south.

3. **COFFEE-BEAN TRIVIA.** *Trivia pediculus* (Linné). Fla.–West Indies. Tan to pinkish brown with 3 pairs of irregular, dark brown splotches on the back. Outer lip has about 16–19 ribs. Common; on coral reefs, 3–50 ft. This attractive species is used extensively in the shell jewelry business in the West Indies.

4. **SUFFUSE TRIVIA.** *Trivia suffusa* (Gray). S.E. Fla.–West Indies. Siphonal canal has a weak brownish blotch on each side. Riblets on top are beaded. Outer lip white and crossed by 18–23 riblets. Common; in shallow water.

5. **ANTILLEAN TRIVIA.** *Trivia antillarum* Schilder. N.C.–West Indies. Small, purple-brown. Riblets smooth, 18–22 on outer lip. Uncommon; 20–600 ft.

6. **SOLANDER'S TRIVIA.** *Trivia solandri* (Sowerby). S. Calif.–Peru. Purplish brown with cream nodules on each side of top groove. Common; in littoral zone.

7. **MALTBIE'S TRIVIA.** *Trivia maltbiana* Schwengel and McGinty. N.C.–West Indies. Shell globose, but flattened above, with 24–28 riblets across outer lip. Area between ribs microscopically granular. Dorsal groove slight. Common; 200–300 ft.

8. **WHITE GLOBE TRIVIA.** *Trivia nix* Schilder. S.E. Fla.–West Indies. Groove on back with 15–25 riblets. Uncommon; intertidal.

9. **LITTLE WHITE TRIVIA.** *Trivia candidula* (Gaskoin). N.C.–West Indies. All white, with 17–24 riblets; no groove on back. Common.

10. **PUSTULATE TRIVIA.** *Jenneria pustulata* (Lightfoot). Baja Calif.–Ecuador. Top covered with bright-orange pustules, each of which has a dark ring. Teeth on base lighter. Interior violet. Moderately common; in shallow water.

11. **MAUGER'S ERATO.** *Erato maugeriae* (Gray). N.C.–West Indies. Glossy tan or yellowish. Thickened outer lip has about 10–15 tiny teeth. Apex bulbous. Common; 12–300 ft.

12. **COLUMBELLE ERATO.** *Erato columbella* (Menke). Central Calif.–Panama. Thickened outer lip with about 12–15 teeth. Apex brown. Siphonal canal purple-brown. Common; to 300 ft.

13. **APPLESEED ERATO.** *Erato vitellina* (Hinds). S. Calif.–Mexico. Columella has 5–8 whitish teeth; outer lip has 7–10 small teeth. Body whorl blotched with purple. Apex glazed. Common; in shallow water.

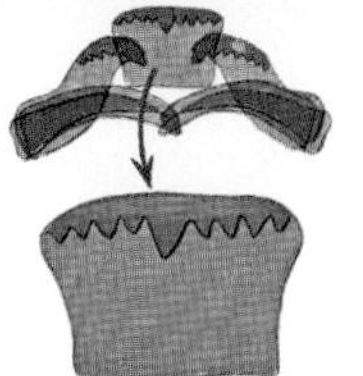

radula of *Trivia*

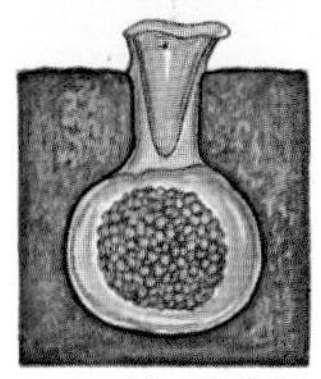

egg capsule of *Trivia* embedded in compound ascidian.

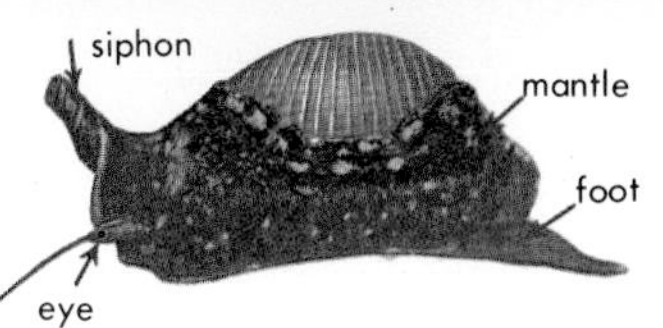

Living animal of *Trivia*

Four-spotted Trivia 0.3″

Californian Trivia 0.4″

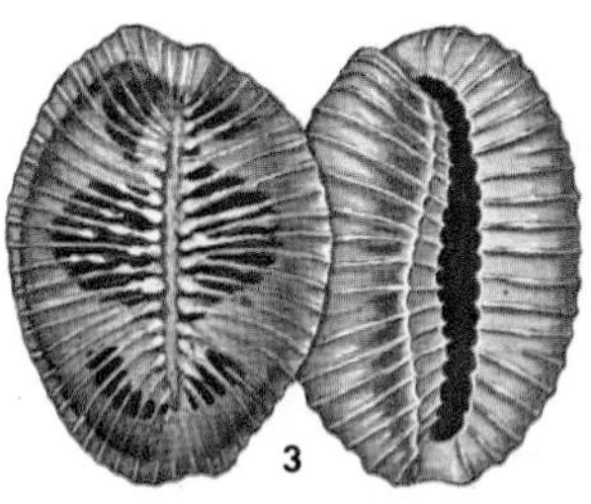

Coffee-Bean Trivia 0.5″

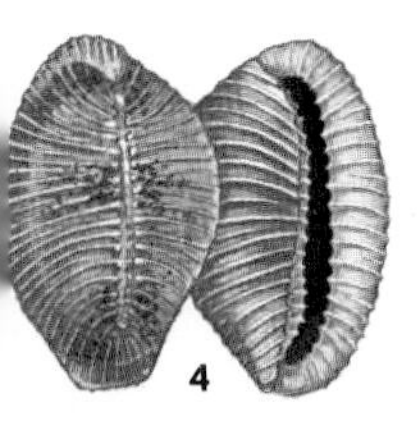

Suffuse Trivia 0.4″

Antillean Trivia 0.2″

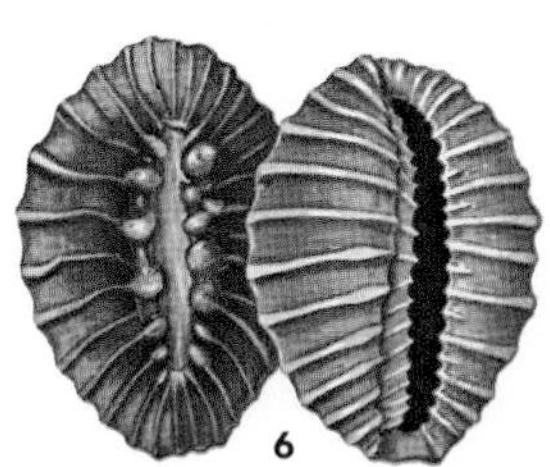

Solander's Trivia 0.6″

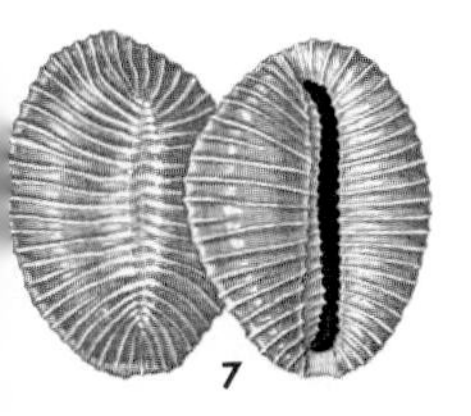

Maltbie's Trivia 0.4″

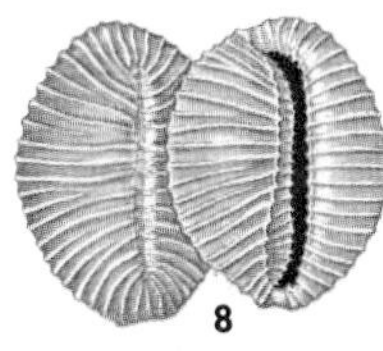

White Globe Trivia 0.4″

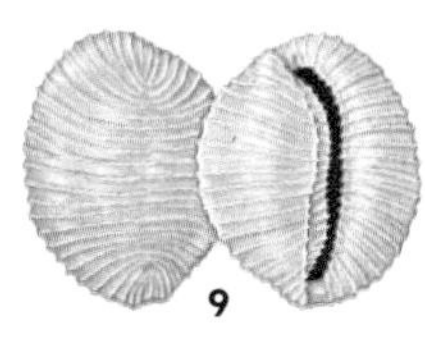

Little White Trivia 0.2″

Pustulate Trivia 0.4″

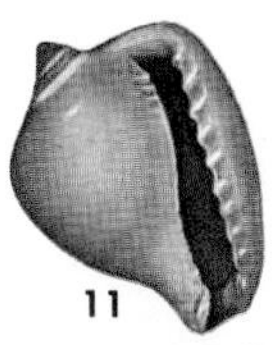

Mauger's Erato 3 mm.

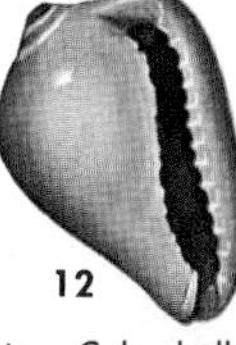

Columbelle Erato 5 mm.

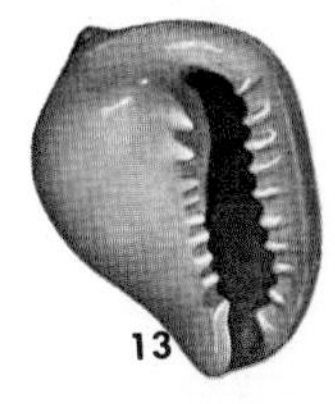

Appleseed Erato 15 mm.

CYPHOMA AND SIMNIA SHELLS

Cyphoma and Simnia snails (family Ovulidae) live in tropical waters on sea whips and sea fans, browsing on the living polyps. The glossy shells are protected by an expansible fleshy mantle. No members of this family produce an operculum. Eggs are laid on the branches in small, oblong clumps of crowded capsules, each capsule containing several hundred eggs, which hatch into free-swimming veligers.

The mantle of each species of *Cyphoma* has a characteristic color pattern. Simnias live on sea fans. The yellow and purple shells are believed to be related to the color of the sea fan upon which the snail feeds. The closely related *Pedicularia* lives and feeds on hard hydrocorallines.

Because these marine snails live on such a specialized and restricted habitat they have evolved a competitive way of life that is seldom seen in mollusks. The male controls a major part of the sea fan or sea whip upon which he lives, and will ward off other males that encroach upon his territory. These species are so brightly colored that divers are reducing their numbers very rapidly.

1. **FLAMINGO TONGUE.** *Cyphoma gibbosum* (Linné). S.E. United States–West Indies. Shell stout, glossy, with a swollen ridge running over the back. Exterior cream with orange edges. Interior of aperture whitish. Common; 4–20 ft. This bright species is used in shell jewelry. It is the commonest *Cyphoma* in the Atlantic.

2. **McGINTY'S CYPHOMA.** *Cyphoma mcgintyi* Pilsbry. Lower Fla. Keys–Bahamas. Shell like Flamingo Tongue's but longer; whitish with lilac tints, Aperture whitish pink. Mantle has solid spots and bars of black-brown. Rather uncommon; 4–25 ft.

3. **FINGERPRINT CYPHOMA.** *Cyphoma signatum* Pilsbry and McGinty. Lower Fla. Keys–West Indies. Similar to others, but ridge on back is weaker, and interior of aperture is tan. Mantle is pale yellow with numerous, crowded, thin, black lines. Rather rare; 6–20 ft.

4. **VIDLER'S SIMNIA.** *Neosimnia vidleri* (Sowerby). Calif.–Mexico. Shell somewhat stout, wine-red, with lavender outer lip. Exterior with microscopic spiral scratches. Uncommon; on purple gorgonia, 20–100 ft.

5. **COMMON WEST INDIAN SIMNIA.** *Neosimnia acicularis* (Lamarck). S.E. United States—West Indies. Columellar area flattened or dished, and bounded by a white ridge on either side. Color of shell may be purple, whitish, or yellow. Common; on sea fans and sea whips, 6–30 ft.

6. **SINGLE-TOOTHED SIMNIA.** *Neosimnia uniplicata* (Sowerby). S.E. United States–West Indies. Similar to Common Simnia, but with a short, twisted ridge at end of columella. White ridge borders the columella. Has both purple and yellow phases. Common; 6–30 ft.

7. **CALIFORNIAN PEDICULARIA.** *Pedicularia californica* (Newcomb). South half of Calif. Exterior uneven, rarely with spiral threads. Aperture very large and glossy rose. Nuclear whorls reticulate. Uncommon; 20–60 ft., attached to a red hydrocoral.

8. **DECUSSATE PEDICULARIA.** *Pedicularia decussata* (Gould). Ga.–West Indies. Thick, whitish gray. Aperture very large and flaring, often distorted. Exterior with fine crisscross threads. Uncommon; on dead coral branches, 50–1,000 ft.

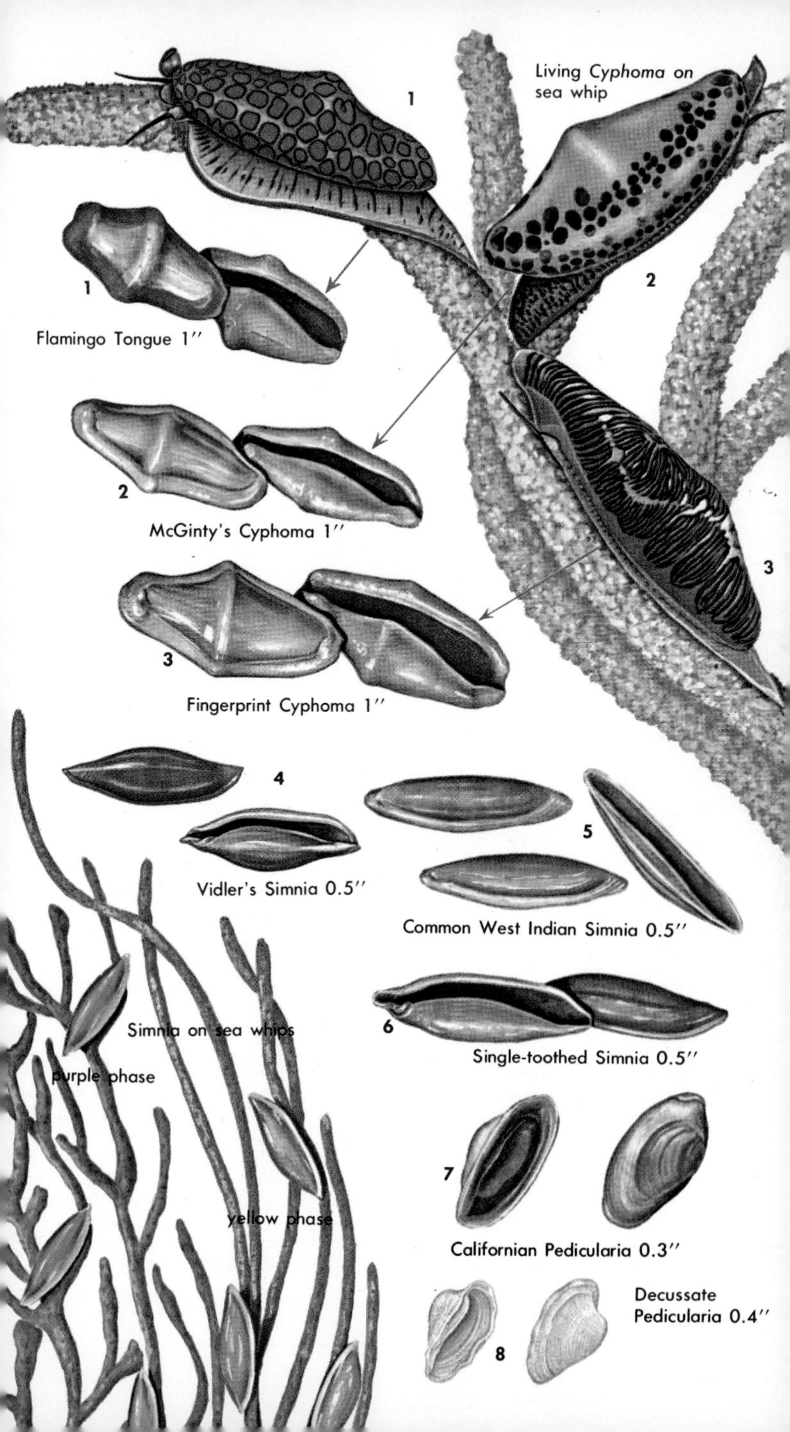
Living *Cyphoma* on sea whip
1
2
3
1
Flamingo Tongue 1″
2
McGinty's Cyphoma 1″
3
Fingerprint Cyphoma 1″
4
Vidler's Simnia 0.5″
5
Common West Indian Simnia 0.5″
Simnia on sea whips
purple phase
yellow phase
6
Single-toothed Simnia 0.5″
7
Californian Pedicularia 0.3″
Decussate Pedicularia 0.4″
8

COWRIES

Members of the family Cypraeidae are well-known for their smooth, glossy, colorful shells. Most of the 200 or so species live in shallow tropical seas, but only 4 kinds are found in southeastern United States and 1 in California. The sexes are separate. The females sit on the clumps of egg capsules, which are laid in protective rock crevices. The free-swimming young hatch in about three weeks and are carried to new areas by ocean currents. The shell of the immature cowries (opposite, upper right) is fragile and thin-lipped, but later it becomes thickened, with a curled-in outer lip. It is banded at first. At maturity, teeth form along either side of the aperture, and the spots develop. Adulthood is reached in about eight months. The mantle, which is usually adorned with fleshy filaments and warts, secretes the shell material. Cowries presumably feed on colonial invertebrates, such as tunicates and hydroids. They have no operculum.

The size of adults, especially in such species as the Measled Cowrie, varies considerably, and may range from 1.5 to 3 inches. The males are usually smaller. The ultimate size of an adult depends on the maximum size to which the immature shell grows. Sex, racial inheritance, and nutritional conditions modify the adult shell size.

Cowries were used by early man as a form of exchange. Most frequently used was the inch-long, yellow Money Cowrie of the Indian and western Pacific oceans. It has been found in the burial grounds of North American Indians, evidently brought to this continent by earlier fur traders as a trade item. The most valuable cowries come from the tropical waters of the southwestern Pacific.

1. **MEASLED COWRIE.** *Cypraea zebra* Linné. S.E. Fla.–Brazil. Dark brown with small whitish dots, those near the sides having brown centers. Apertural teeth brown. Moderately common near protective rocks in shallow water; rare near mangroves and shady rock areas.

2. **ATLANTIC YELLOW COWRIE.** *Cypraea spurca acicularis* Gmelin. S.C.–Brazil. Shell flattened; speckled orange-brown or whitish. Base and apertural teeth ivory-white; small indentations around base of the top half of shell. Uncommon; in coral rubble.

3. **ATLANTIC DEER COWRIE.** *Cypraea cervus* Linné. S. Fla.–Yucatan. Like Measled Cowrie, but larger, more inflated; smaller spots on top have no brown centers. Apertural teeth brown. Moderately common; from low tide mark to 30 ft. May grow to 7 inches.

4. **CHESTNUT COWRIE.** *Cypraea spadicea* Swainson. Monterey, Calif.–Baja Calif. Back light chestnut-brown with bluish undertone and bordered by a dark black-brown wavy band. Base white; with about 20–25 whitish teeth on each side of the narrow aperture. This is the only true cowrie in western U.S. Moderately common; low-tide zone to 120 ft. Usually found among rocks near sea urchins.

5. **ATLANTIC GRAY COWRIE.** *Cypraea cinerea* Gmelin. Fla. Keys–Brazil. Round; back brownish mauve; base cream with mauve-brown between some teeth; sides flecked with black-brown. Moderately common among coral on shallow reefs.

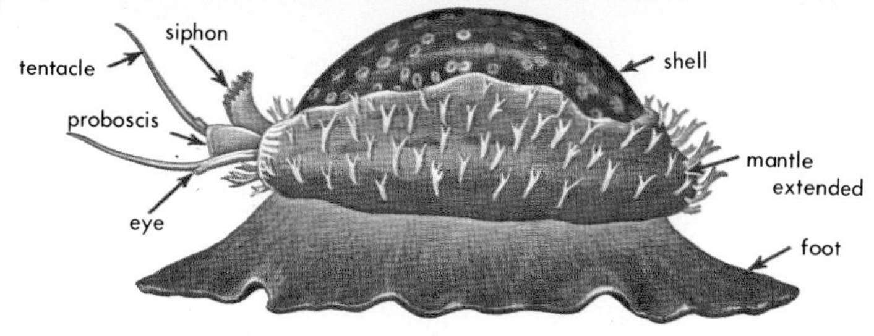

side view of living Measled Cowrie

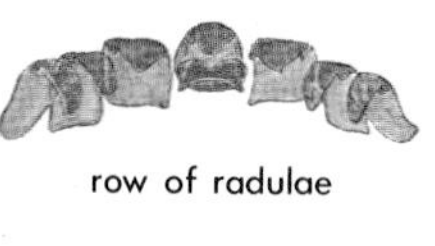

row of radulae

egg cluster is brooded by female

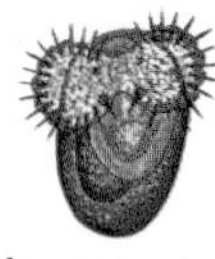

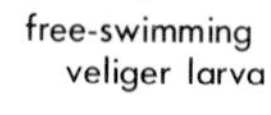

free-swimming veliger larva

young stage at 3 months

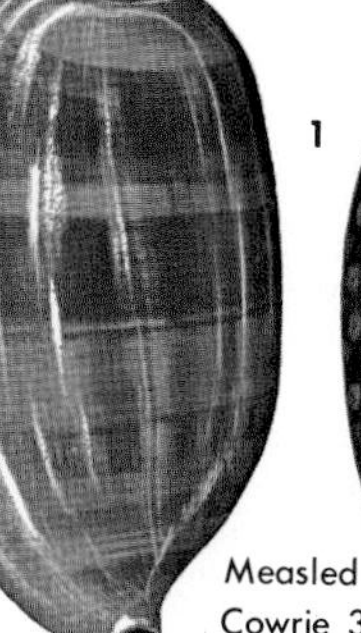

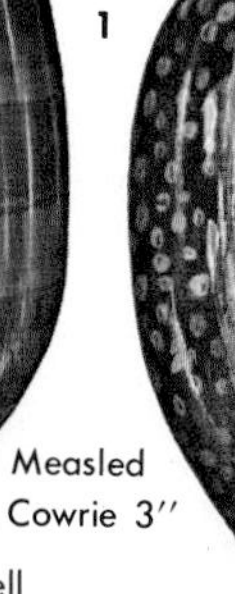

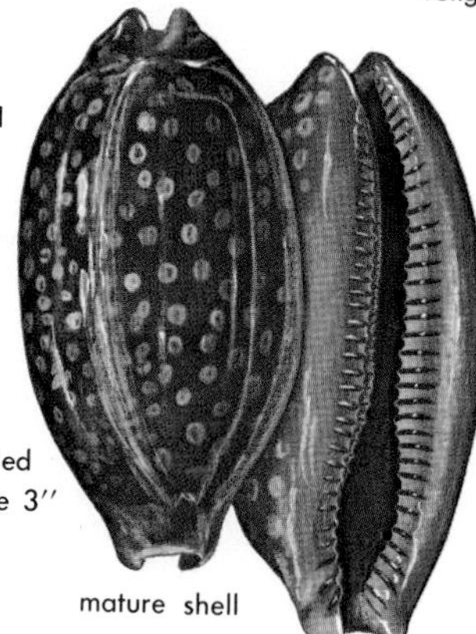

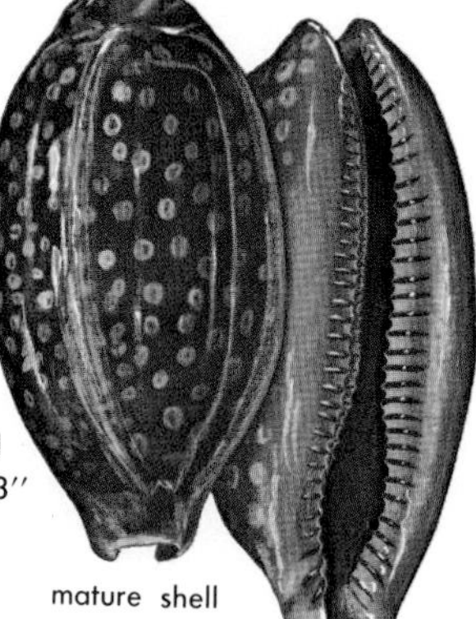

1

Measled Cowrie 3″

immature shell

mature shell

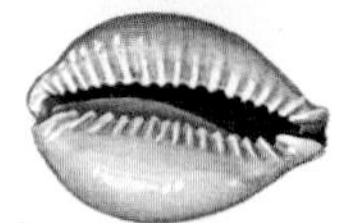

2

Atlantic Yellow Cowrie 1″

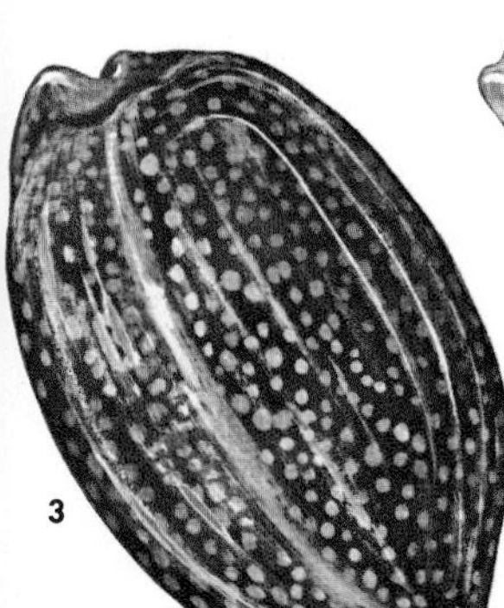

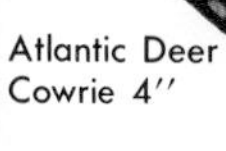

3

Atlantic Deer Cowrie 4″

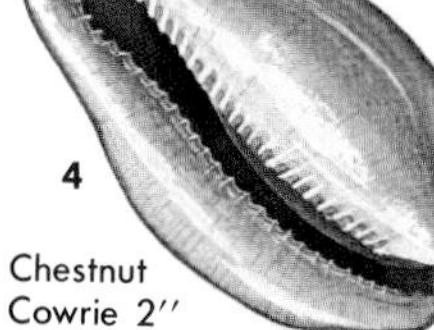

4

Chestnut Cowrie 2″

5

Atlantic Gray Cowrie 1″

MOON SNAILS

The moon snails of the family Naticidae are among the most active of the gastropod carnivores. They feed on bivalves, some consuming three or four small clams a day. Moon snails dig through the sand, uprooting the clam, which is then held by the strong foot and is pierced by a neat round hole made by a combination of rasping by the radula teeth and corrosion by a gland-produced acid (see p. 24). At low tide these snails leave a broad meandering trail over sand bars.

Moon snails include several genera and many hundreds of species; all are sand dwellers and most live in temperate seas. Tropical species are colorful; deep-sea and northern species are drab. Identification of moon snails depends largely on the structure of the umbilicus and of the umbilical callus. The latter may not be fully developed in young specimens. Members of the genera *Polinices, Lunatia, Sigatica,* and *Amauropsis* have a pliable horny operculum. Those in the genus *Natica* are shelly and hard. *Sinum* has none. The egg mass of the typical moon snail is gelatinous and is surrounded by a protective "sand collar," formed while the eggs are deposited. This sand collar tends to have a characteristic shape for each species, and its form is determined by the shape and size of the female's shell. When dried the "sand collar" becomes very fragile.

1. **ATLANTIC MOON SNAIL.** *Polinices duplicatus* (Say). Cape Cod, Mass.–Texas. Shell large, gray to tan, smooth; base lighter in color. Shell wider than high in most cases, but some have a high spire. Umbilicus deep, but almost completely covered by a purple-brown callus. Operculum horny, light brown. Common; on shallow-water sand bottoms.

2. **RECLUZ'S MOON SNAIL.** *Polinices reclusianus* (Deshayes). Calif.–Mexico. Shell heavy, not very glossy, grayish with rusty-brown stains. Spire height varies. Brown or whitish callus has a small scar and may or may not fill the umbilicus. Operculum is thin, horny, and bright reddish brown. Common; from 1–150 ft.

3. **DRAKE'S MOON SNAIL.** *Polinices draconis* Dall. Alaska–Mexico. Shell rounded, globe-shaped, moderately thick, and grayish brown in color. Umbilicus deep and fairly wide. Umbilical callus absent or very small. A moderately common species; found offshore in 60–150 ft.

4. **MILK MOON SNAIL.** *Polinices lacteus* (Guilding). N.C.–West Indies. Shell thick, white, and glossy. Periostracum very thin. A heavy callus extends along the aperture wall and borders the deep umbilicus. Operculum horny, yellow or amber-red. The egg collar is covered with whitish sand and has wavy edges. Common; on sand.

5. **IMMACULATE MOON SNAIL.** *Polinices immaculatus* (Totten). Gulf of St. Lawrence–N.C. Shell small, glossy white, but with a thin greenish yellow periostracum. Narrow ivory-white callus does not cover the deep, round umbilicus. Common; in 20–600 ft. of cold water, on sandy bottom.

6. **BROWN MOON SNAIL.** *Polinices hepaticus* (Röding). Fla.–West Indies. Shell thick, glossy, and colored either purple-tan or orange-brown. The white callus extends as a low spiral ridge deep into the white umbilicus. Operculum horny, light brown to amber. Uncommon; from shore to 50 ft.

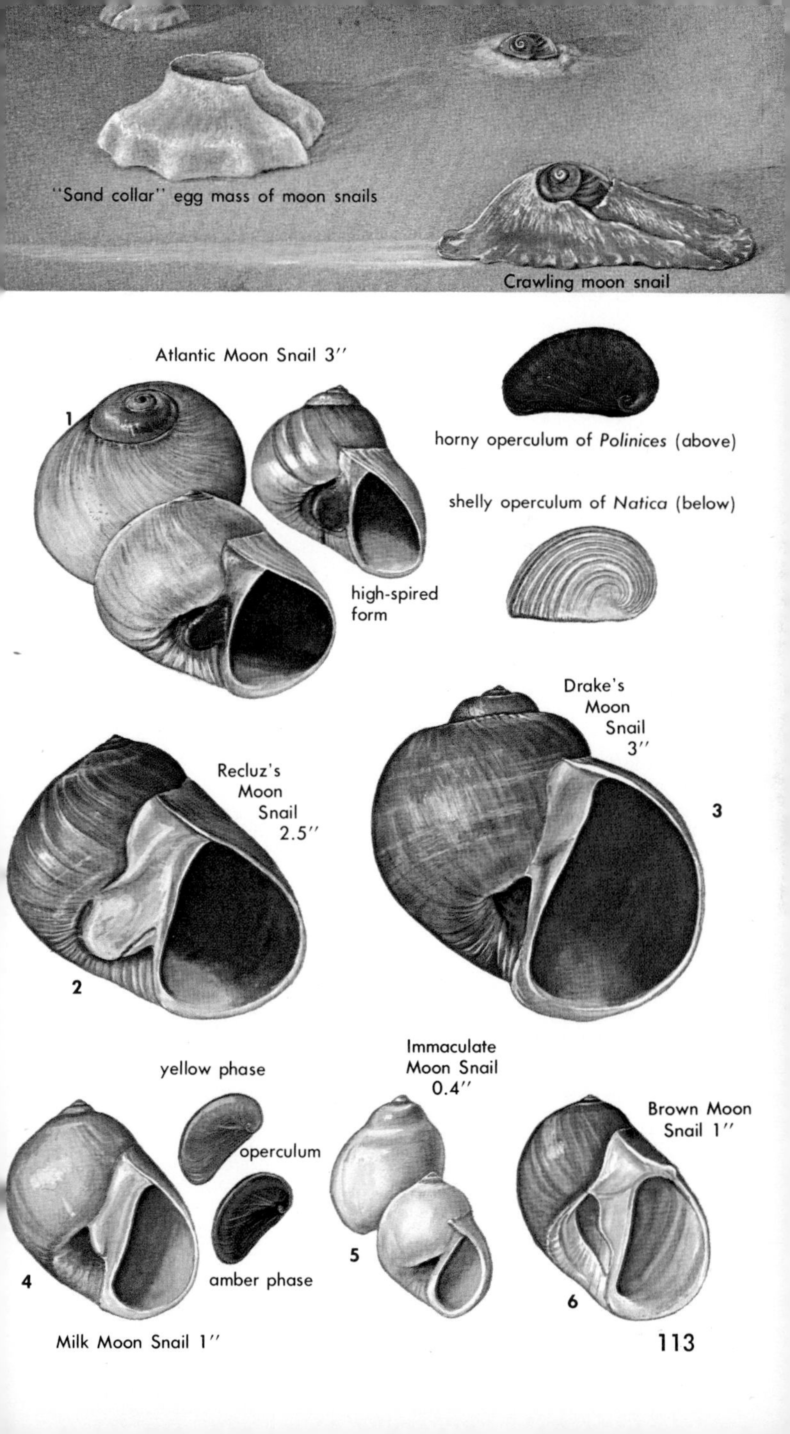
"Sand collar" egg mass of moon snails
Crawling moon snail
Atlantic Moon Snail 3″
1
high-spired form
horny operculum of *Polinices* (above)
shelly operculum of *Natica* (below)
Recluz's Moon Snail 2.5″
2
Drake's Moon Snail 3″
3
yellow phase
operculum
amber phase
4
Milk Moon Snail 1″
Immaculate Moon Snail 0.4″
5
Brown Moon Snail 1″
6

1. **COMMON NORTHERN MOON SNAIL.** *Lunatia heros* (Say). Canada–off N.C. Gray to brownish. Periostracum thin and yellow-brown. Umbilicus small, very deep, and only slightly covered by the edge of the columellar wall. Common; in shallow water in the north; offshore in the south.

2. **LEWIS' MOON SNAIL.** *Lunatia lewisi* (Gould). W. Canada–Baja Calif. Shell rounded, heavy. Whorls slightly shouldered. Umbilicus deep, round, narrow, with a brown-stained, button-like callus. Operculum horny and brown. Common; from 1–150 ft. in sandy areas.

3. **SPOTTED NORTHERN MOON SNAIL.** *Lunatia triseriata* (Say). Gulf of St. Lawrence–off N.C. Shell colorful, thick, and small and similar in form to Common Northern Moon Snail. Last whorl with 2 or 3 indistinct spiral rows of bluish or brown squares. Common; on sand from 6–400 ft.

4. **PALE NORTHERN MOON SNAIL.** *Lunatia pallida* (Broderip and Sowerby). Arctic Seas–N.C.; also to Calif. The rather thin, white or gray shell is longer than wide. Periostracum a light tan. Umbilicus almost closed. Common; offshore in cold water.

5. **COLORFUL ATLANTIC NATICA.** *Natica canrena* (Linné). N.C.–West Indies. Attractive, smooth shell with spiral rows of brown spots. Umbilicus and internal callus white. Exterior of limy operculum has about 10 strong grooves (p. 113). Moderately common; in sand 3–50 ft.

6. **SOUTHERN MINIATURE NATICA.** *Natica pusilla* Say. Cape Cod, Mass.–West Indies. Shell very small and glossy; similar to Arctic Natica. The umbilicus is almost sealed by a whitish callus. Nucleus of shelly operculum often stained brown. Common; 1–100 ft.

7. **LIVID NATICA.** *Natica livida* Pfeiffer. Fla.–West Indies. Shell gray to tan, small, glossy, with a brown aperture and columella. Dark brown callus almost fills the umbilicus. Shelly operculum is light tan. Moderately common; on sandy flats to about 10 ft. Do not confuse with *Polinices duplicatus* (p. 113, no. 1) which has a chitinous operculum.

8. **ARCTIC NATICA.** *Natica clausa* Broderip and Sowerby. Arctic Seas–N.C.; also to Calif. Shell fairly thin, yellowish white, smooth, covered with brownish periostracum. White callus seals umbilicus. Shelly operculum. Common; offshore.

9. **COMMON BABY'S EAR.** *Sinum perspectivum* (Say). Va.–West Indies. Shell very flat with a characteristic large, white aperture, and a curved columella. Top of whorls have many fine, spiral threads. Periostracum very thin and tan. Huge foot. A common sand-burrowing species.

10. **WESTERN BABY'S EAR.** *Sinum scopulosum* (Gould). Calif.–Mexico. Shell thin, chalky white but covered with a thin, yellowish periostracum. Spire elevated; aperture large. Top of last whorl has many shallow spiral grooves. Fairly common; on beaches and sand bars.

11. **CAROLINA MOON SNAIL.** *Sigatica carolinensis* (Dall). S.C.–West Indies. A nearly white shell with sutures well channeled. Body whorl has about 20 fine, cut, spiral lines. Umbilicus deep, round, and without a callus. Operculum tan. Uncommon; from 30–600 ft.

12. **ICELAND MOON SNAIL.** *Amauropsis islandica* (Gmelin). Arctic Seas–Va. Shell smooth, yellowish tan, thin but quite strong. Suture narrowly channeled. Umbilicus narrow and slitlike. Operculum horny with microscopic spiral lines. Common; in cold waters down to 400 ft.

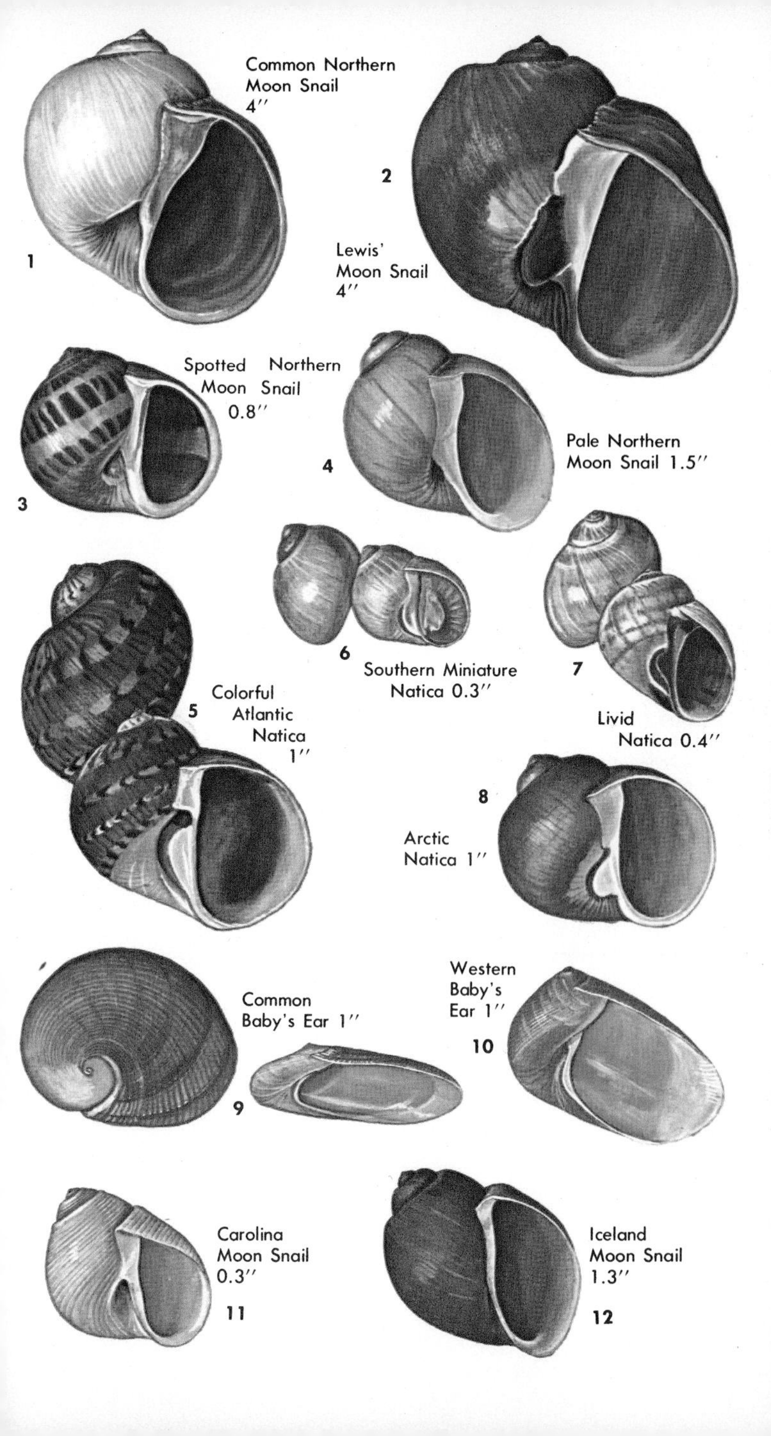
Common Northern
Moon Snail
4″
2
1
Lewis'
Moon Snail
4″
Spotted Northern
Moon Snail
0.8″
Pale Northern
Moon Snail 1.5″
4
3
6
Southern Miniature
Natica 0.3″
7
Colorful
5 Atlantic
Natica
1″
Livid
Natica 0.4″
8
Arctic
Natica 1″
Western
Baby's
Ear 1″
Common
Baby's Ear 1″
10
9
Carolina
Moon Snail
0.3″
11
Iceland
Moon Snail
1.3″
12

BONNET AND HELMET SHELLS

The family Cassidae contains about 70 living species, mainly in shallow, tropical waters. The shells are large, colorful, and have a broad, heavy parietal shield (i.e., the area adjoining the inner margin of the aperture). Every few whorls a thickened lip, or varix, is formed. The former ones can be seen in the spire of the shell. The operculum is horny, brown, and oblong in *Cassis,* fan-shaped in *Phalium,* and very small and oval or, rarely, absent in *Cypraecassis.* Helmets feed on sea urchins and sand dollars. The shells of the females are usually larger than the male's. Horny egg capsules are laid in the shape of a round tower or in irregular clumps. The newly hatched veliger is believed to be free-swimming for several weeks.

The genus *Cassis* has three eastern American species, all with large, heavy shells from which cameos can be carved. The most abundant American member of this family is the Scotch Bonnet, sometimes cast ashore in large numbers on the beaches of western Florida and the Carolinas. Largest is the foot-long Emperor Helmet of southern Florida, while the rarest is the deep-water Royal Bonnet of the Gulf of Mexico.

1. **SCOTCH BONNET.** *Phalium granulatum* (Born). N.C.–Brazil. Whorls with square spots are spirally grooved. Columella shield has numerous fine pustules. The spiral cords vary considerably in width, strength, and amount of beading. Rarely with axial riblets. Common; in shallow water on sand bottoms.

2. **SMOOTH SCOTCH BONNET.** *Phalium cicatricosum* (Gmelin). Fla.–Brazil. Surface smooth; columellar shield with pustules. May be only a variety of Scotch Bonnet. Fairly common; in shallow water.

3. **RETICULATED COWRIE–HELMET.** *Cypraecassis testiculus* (Linné). N.C.–Texas–Brazil. Shell heavy, reddish; surface crisscrossed; columellar shield orange, heavy, smooth. Operculum small or absent. Not uncommon; near coral reefs.

4. **ROYAL BONNET.** *Sconsia striata* (Lamarck). Gulf of Mex. and Caribbean. Shell thick, whitish, with brown spots and numerous spiral cut lines. Two former varices can be seen in spire. Rare; 150–500 ft.

5. **ATLANTIC WOOD-LOUSE.** *Morum oniscus* (Linné). Lower Fla. Keys and Caribbean, under old coral. Shell heavy, rough; parietal wall has tiny white pustules. Periostracum gray, velvety. Common; below low-tide mark.

6. **FLAME HELMET.** *Cassis flammea* (Linné). Fla. Keys–Brazil. Oval parietal shield. No brown-black between teeth; last whorl is without spiral threads. Opercula in all *Cassis* are horny, small, brown, and oblong. Fairly common in shallow water.

7. **EMPEROR HELMET.** (Queen Helmet). *Cassis madagascariensis* Lamarck. S.E. United States and West Indies. Largest *Cassis;* up to 14 in. Knobs on shoulder vary in number and size. Black between teeth on parietal shield. Fairly common; on sand, 10–60 ft.

8. **KING HELMET.** *Cassis tuberosa* (Linné). N.C.–Brazil. Parietal shield triangular. Black between teeth. Surface netted; with 7 or 8 brown stripes over outer lip. Commonest Atlantic helmet; in shallow water, almost buried in the sand.

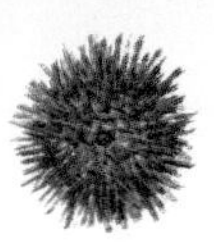

Echinoderms are food of Helmets

Scotch Bonnet 3″

Female Scotch Bonnet lays a tower of eggs

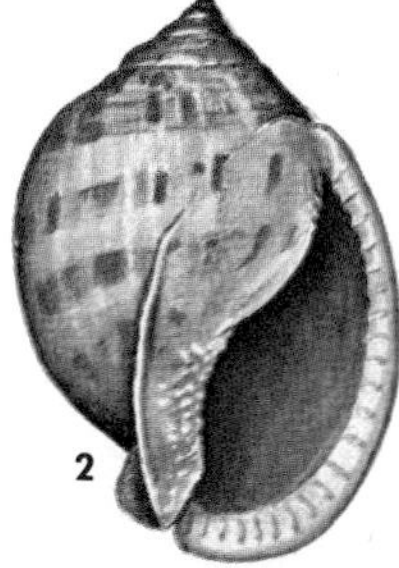

Smooth form of Scotch Bonnet 3″

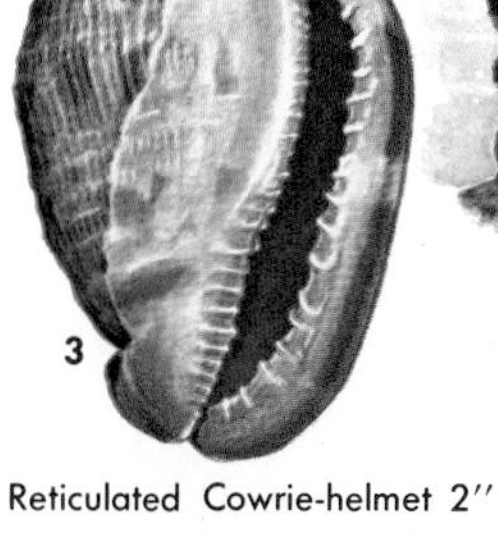

Reticulated Cowrie-helmet 2″

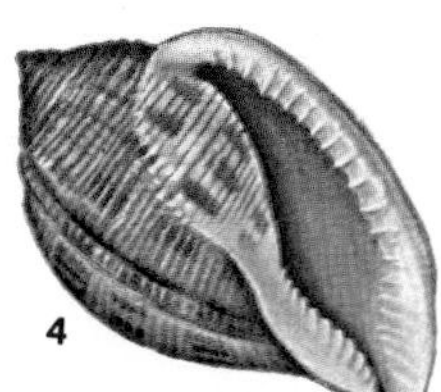

Royal Bonnet 1.5″

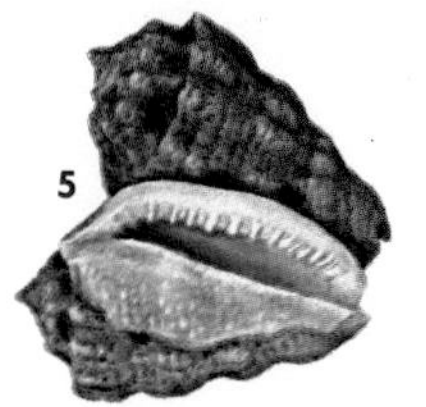

Atlantic Wood-louse 1″

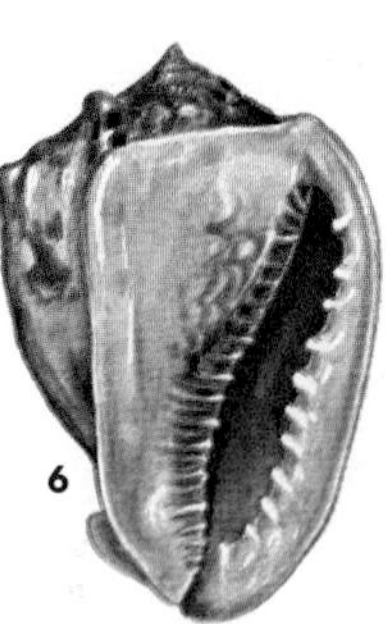

Flame Helmet 4″

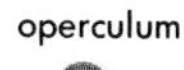

operculum

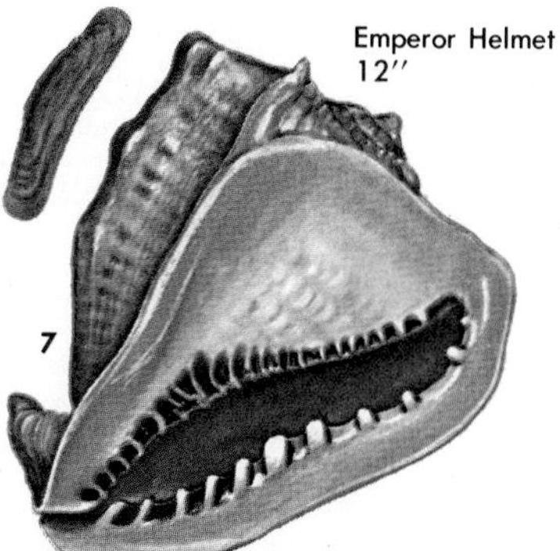

Emperor Helmet 12″

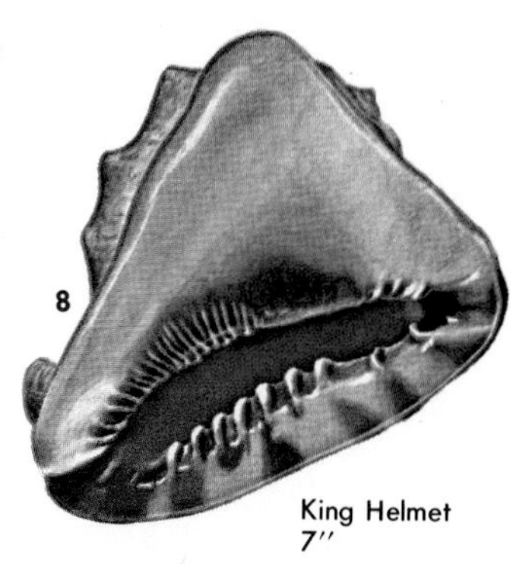

King Helmet 7″

TRITONS

Tritons (family Cymatiidae) are widespread in the oceanic world, some in cold waters, but most in the tropical seas. Nearly all of them grow a very hairy or horny periostracum outer covering, which protects the shell from boring sponges. The aperture is generally armed with shelly "teeth," but there is no posterior canal as is found in the related Frog Shells. Extreme development of the apertural teeth is found in the Distorsios. They need only a small protective operculum, in contrast to the large operculum found in the smooth-mouthed Oregon Triton.

Triton's trumpets of tropical reefs grow to 18 inches and are used as horns in Eastern religious ceremonies and by fishermen in many parts of the world. A small hole is drilled in the side of the spire, and when blown the shell produces a sound resembling a modern cornet. Several other large species of triton's trumpets are known from other seas. They feed on starfish.

1. **KNOBBED TRITON.** *Cymatium muricinum* (Röding). S.E. Fla.–Brazil. Color either gray or dark brown. Aperture bordered by a smooth, cream-colored shield. Interior of aperture usually brown. Siphonal canal long, bent at tip. Moderately common; on intertidal reefs.

2. **ATLANTIC HAIRY TRITON.** *Cymatium pileare martinianum* (Orbigny). S.E. United States–Brazil. Aperture bordered with small, white teeth on a reddish background. Surface irregularly beaded. Periostracum thick, matted, and brown. Varices rounded. Moderately common; on coral reefs.

3. **ANGULAR TRITON.** *Cymatium femorale* (Linné). S.E. Fla.–Brazil. Varices large, wing-shaped, and with white knobs. Surface spirally corded, reddish to orange. Inner lip smooth and white. Periostracum flakes off when dry. Uncommon; in eelgrass, 3–30 ft.

4. **DWARF HAIRY TRITON.** *Cymatium vespaceum* (Lamarck). S.E. United States–Brazil. Shell delicate, with a fairly long, slender siphonal canal. Aperture bordered with small whitish teeth. Periostracum light brown, quite thick. Varices weak. Uncommon; 6–80 ft.

5. **ATLANTIC DISTORSIO.** *Distorsio clathrata* (Lamarck). S.E. United States–West Indies. Exterior cross-hatched with axial and spiral threads. Varices are sharp ridges. The entire shell is slightly distorted, with a hairy periostracum. Uncommon; on coarse sand, 30–400 ft.

6. **FLORIDA DISTORSIO.** *Distorsio mcgintyi* Emerson and Puffer. S.E. United States. Similar to Atlantic Distorsio, but with more distorted whorls and coarser sculpture. Parietal wall has glazed warts. Periostracum hairy. Common; 90–700 ft. This may be a subspecies of *Distorsio constricta* Broderip from E. Pacific.

7. **OREGON TRITON.** *Fusitriton oregonensis* (Redfield). Alaska–S. Calif. Shell light in weight, but strong. Whorls white with numerous, beaded axial riblets. Periostracum heavy, gray-brown, with bristles. Aperture whitish. Common; on sandy and rock bottoms, 20–200 ft.

8. **TRITON'S TRUMPET.** *Charonia variegata* (Lamarck). S.E. Fla.–West Indies. Outer lip has small pairs of white teeth; inner lip has uneven, whitish ridges. Periostracum thin and translucent. Formerly called *C. nobilis*. Moderately common; crevices and hollows in coral reefs.

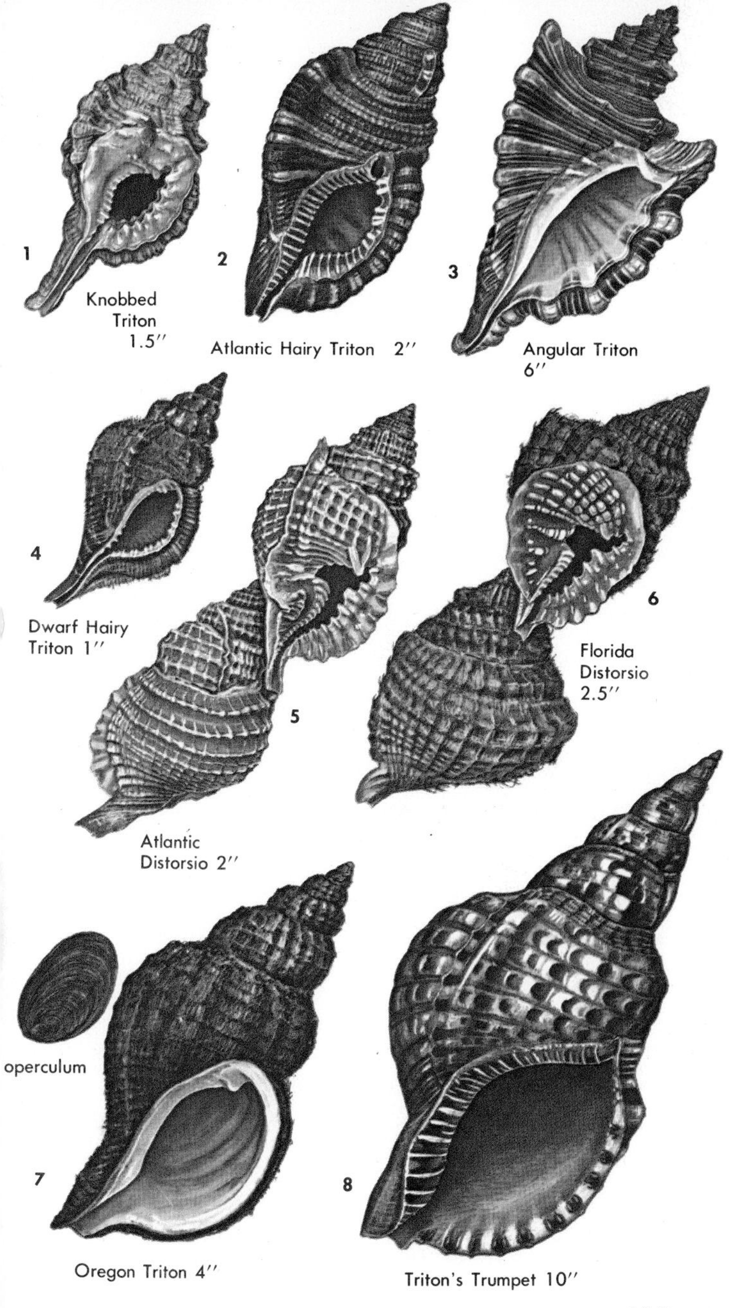
1
Knobbed Triton 1.5″
2
Atlantic Hairy Triton 2″
3
Angular Triton 6″
4
Dwarf Hairy Triton 1″
5
6
Florida Distorsio 2.5″
Atlantic Distorsio 2″
operculum
7
8
Oregon Triton 4″
Triton's Trumpet 10″

FROG, FIG, AND TUN SHELLS

Frog shells (family Bursidae) are carnivorous inhabitants of the rocky, subtidal zone. Several dozen species inhabit tropical water; there are six in southern North America. Eggs are laid in tightly packed spirals of urn-shaped, horny capsules, usually under rocks. The genus *Bursa* has a shelly channel at the upper end of the aperture (the posterior siphonal canal). The outer lip is thick, with small teeth. The operculum is horny.

Fig shells (family Ficidae) are sand-dwellers with a wide, flat foot. The fleshy mantle stretches over the shell to protect it from abrasion. There is no operculum. Figs feed on sea urchins and other echinoderms. They lay eggs in soft capsules arranged in broad, flat bands. In Florida there are one common, shallow-water and two rare deep-sea species.

Tun shells (family Tonnidae) have light but strong shells. The snout of the snail can be expanded to engulf other animals. Acid saliva aids in digesting their prey. The shell of the month-old, swimming veliger consists of horny material different from that of the adult. The operculum in the larva is lost in the adult. There are two species in Florida.

1. **CHESTNUT FROG SHELL.** *Bursa spadicea* (Montfort). S.E. Fla.–Caribbean. Shell flattened laterally (from mouth to back). Surface has rows of fine beads. Posterior siphonal canal long and narrow. Uncommon; rocky bottom, 20–70 ft.

2. **GRANULAR FROG SHELL.** *Bursa granularis* Röding. S.E. Fla.–Caribbean. Shell only slightly flattened laterally. Aperture yellowish with white teeth. Spiral threads poorly beaded; shoulder has knobs. Common; among rocks on coral reefs.

3. **GAUDY FROG SHELL.** *Bursa corrugata* (Perry). S.E. Fla.–Caribbean. Similar to Granular Frog, but orange-brown, with a double row of teeth on outer lip. Apex usually eroded. Uncommon; in coarse sand, 10–40 ft.

4. **CALIFORNIAN FROG SHELL.** *Bursa californica* (Hinds). Monterey, Calif.–Gulf of Calif. Shell heavy, with thick, knobbed varices, between which are 2 low spines. Aperture white. Siphonal canal broad. Common; among rocks, just offshore.

5. **ST. THOMAS FROG SHELL.** *Bursa thomae* (Orbigny). N.C.–Brazil. Characterized by its lavender aperture and 3 rounded knobs between varices. Uncommon; lives in warm, shallow waters over coral reefs.

6. **COMMON FIG SHELL.** *Ficus communis* Röding. N.C.–Mexico. Shell thin but strong. Surface finely criss-crossed. Interior of aperture glossy tan. Common; in shallow, sandy areas. Do not confuse this species with the Pear Whelk (p. 138).

7. **ATLANTIC PARTRIDGE TUN.** *Tonna maculosa* (Dillwyn). S. Fla.–Brazil. Shell thin but strong; whorls with fine spiral grooves. Aperture tan. Nucleus golden brown. Moderately common; lives in warm shallow waters near coral reefs.

8. **GIANT TUN.** *Tonna galea* (Linné). S.E. United States–Brazil. Color light brown; last whorl with 19–21 broad, flattish ribs. Suture forms a deep channel. Uncommon; lives on sandy bottoms in offshore waters that are 3–40 ft. deep.

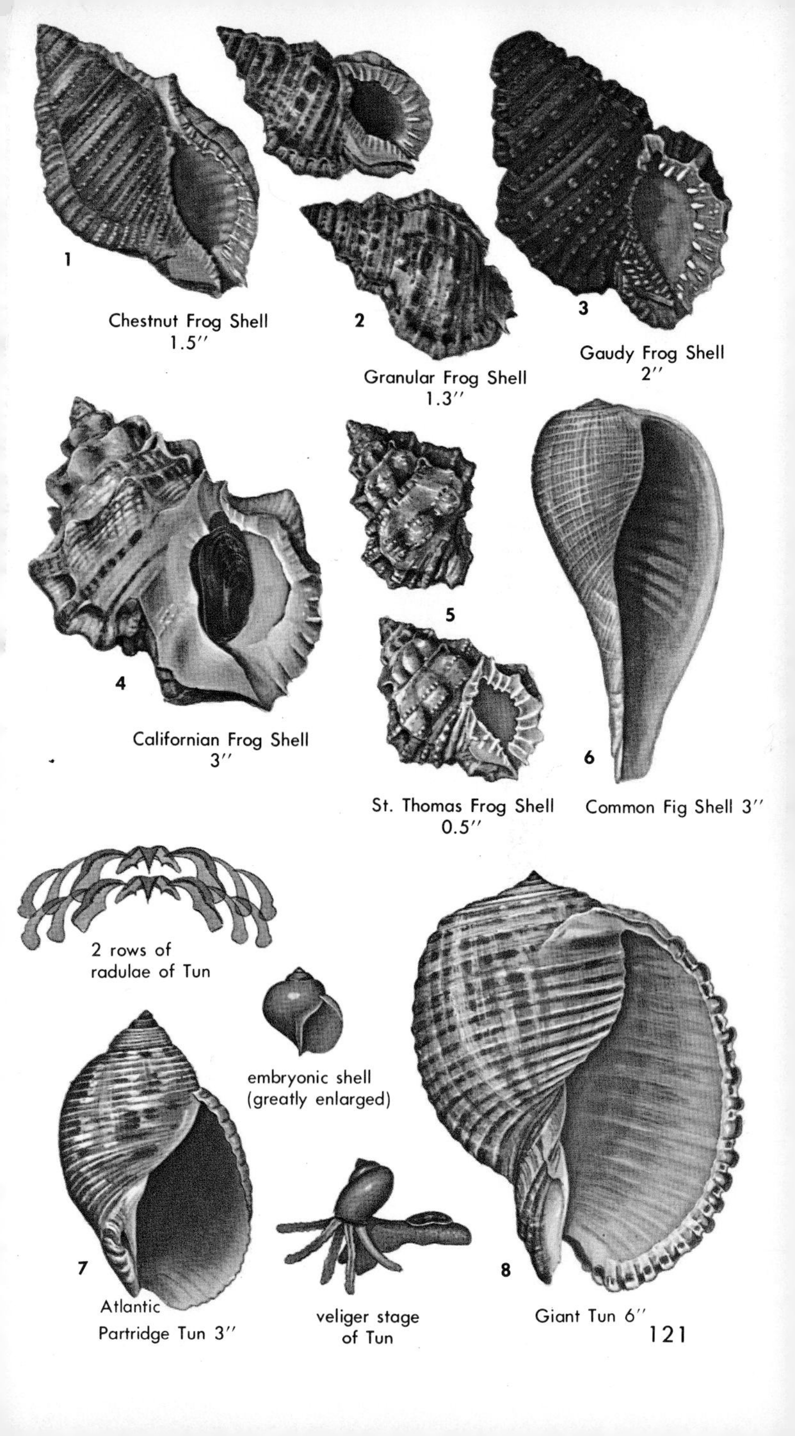
1
Chestnut Frog Shell
1.5″
2
Granular Frog Shell
1.3″
3
Gaudy Frog Shell
2″
4
Californian Frog Shell
3″
5
St. Thomas Frog Shell
0.5″
6
Common Fig Shell 3″
2 rows of
radulae of Tun
embryonic shell
(greatly enlarged)
7
Atlantic
Partridge Tun 3″
veliger stage
of Tun
8
Giant Tun 6″

MUREX SHELLS

Murex or rock shells (family Muricidae) are carnivorous snails consisting of many genera of worldwide tropical and semitropical mollusks characterized by strong, spiny shells. Because of great variability in the form of the shells in this family, the genera are not well defined and, in some cases, are based on such minor characters as the size and number of varices and ribs. Most of the several hundred species live in shallow water. Generally, they prey on live bivalves or freshly killed sea animals. The suction of the foot and the prying action of the strong outer lip of the shell forces open clams and oysters. Their rachiglossate type of radula (p. 25) is well suited for tearing flesh. The operculum is horny, brown, fairly thick, and fits snugly within the aperture. The eggs of Murex, in small, tough capsules about the size of flattened peas, are put in clumps or balls. Adults congregate in shallow water during early summer and lay their eggs under protective rocks and ledges.

A gland attached to the mantle of most species of *Murex* produces a yellowish fluid, which, upon exposure to sunlight, turns into a deep lavender. The Phoenicians and early Romans used this Royal Tyrian Purple to dye their ceremonial robes. The liquid, when exuded from the animal, smells like rotting cabbage. It is believed that this substance is used by the snails to anesthetize bivalves and chitons.

1. **ROSE MUREX.** *Murex rubidus* F. C. Baker. S.C.–Bahamas. Color variable: cream, pink, orange, or reddish. Note 3 small axial ribs between the larger, rounded varices. Common; in shallow, sandy areas.

2. **BEAU'S MUREX.** *Murex beaui* Fischer and Bernardi. S. Fla.–West Indies. The spiny varices usually have prominent, thin, wavy webs, between which are 5 or 6 rows of low, even-sized, tiny knobs. Uncommon.

3. **APPLE MUREX.** *Murex pomum* Gmelin. S.E. United States–Brazil. No long spines. Parietal wall yellow, tan, or orange, with a dark brown spot at the upper end. Outer lip brown-spotted. Common; in shallow water.

4. **GIANT ATLANTIC MUREX.** *Murex fulvescens* Sowerby. N.C.–Fla.–Texas. Shell heavy, with many short, stout spines on the numerous varices. Common offshore; in shallow water to lay eggs.

5. **LACE MUREX.** *Murex dilectus* A. Adams. S. Fla.–West Indies. Aperture small and almost round. Varices very spiny. Color variable, brownish. Apex pink. Formerly *M. florifer arenarius* Clench. Common; near shore.

6. **WESTERN THREE-WINGED MUREX.** *Pteropurpura trialata* (Sowerby). N. Calif.–Baja Calif. The three large, wavy, winglike varices on each whorl are finely laced. Whorls have 4–6 brown bands. Common; 10–60 ft.

7. **CATALINA FORRERIA.** *Forreria catalinensis* Oldroyd. S. Calif. Yellowish to brownish white shell has 7 long, bladelike varices on the last whorl. May have fine, spiral threads. Uncommon; 50–400 ft. A popular shell among collectors.

8. **GIANT FORRERIA.** *Forreria belcheri* (Hinds). Calif. Shell stout, cream-brown, with 10 strong axial blades. Aperture enamel-white. Moderately common; in shallow water; feeds on oysters and mussels.

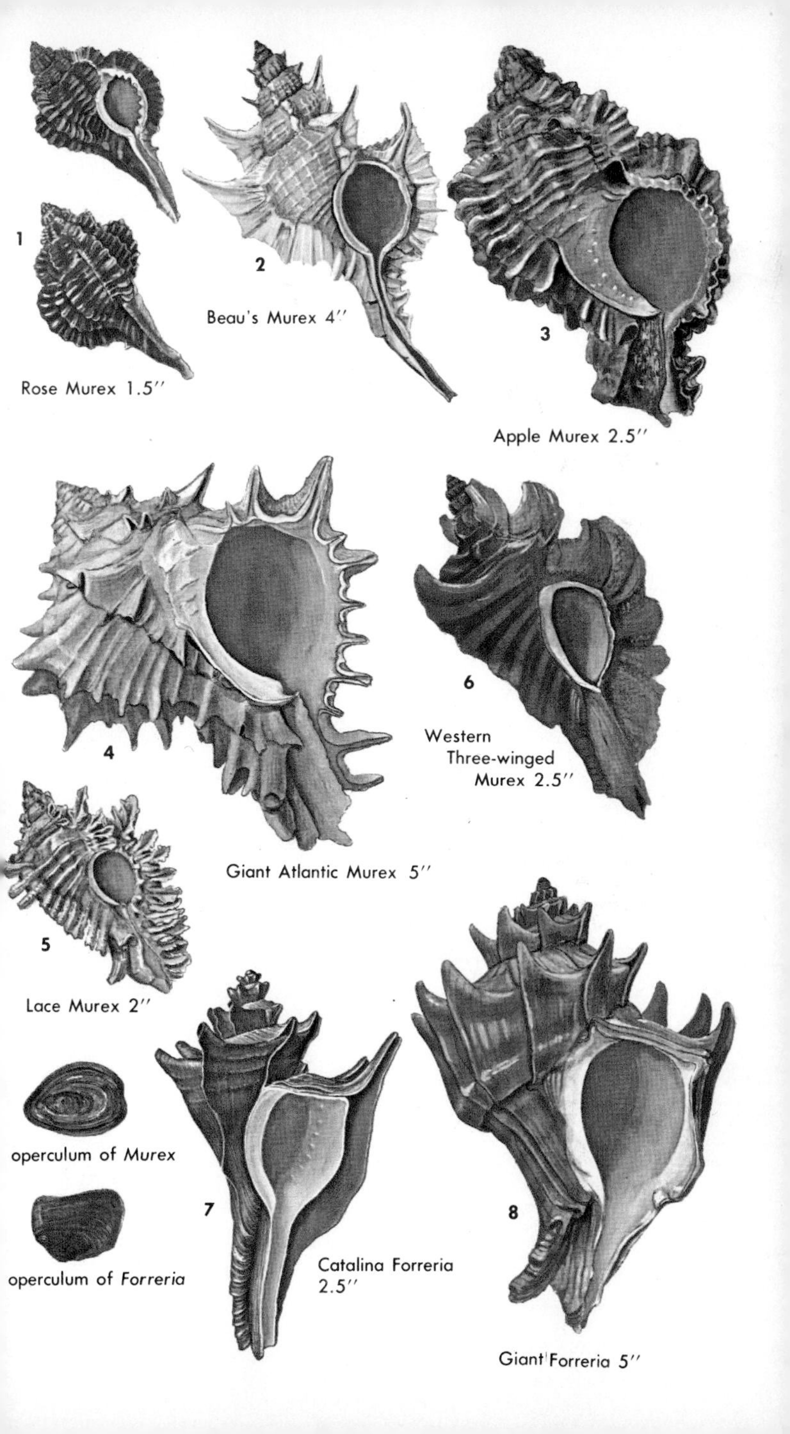
1
Rose Murex 1.5″
2
Beau's Murex 4″
3
Apple Murex 2.5″
4
Giant Atlantic Murex 5″
6
Western
Three-winged
Murex 2.5″
5
Lace Murex 2″
operculum of *Murex*
operculum of *Forreria*
7
Catalina Forreria
2.5″
8
Giant Forreria 5″

MINOR MUREX AND DRUPES

In addition to the large, handsome murex shells, the family Muricidae contains several hundred smaller and less spectacular species belonging to several genera, such as *Muricopsis* and *Acanthina*. They are usually colonial and congregate along rocky shores where they feed on mussels, barnacles, and oysters. Some species hide in crevices and under mats of seaweeds, out of reach of their chief enemies, the sea gulls and crabs. Some groups, like the thorn drupes, bear a strong spine on the lower end of the lip, which is used to pry open barnacles. All produce small amounts of purple dye, have a horny, brown operculum, and lay their eggs in tough urn-shaped capsules in rock crevices.

Many of the normally very spiny or frilled species become quite modified when exposed to a life of rough intertidal waters. The Pitted Murex, for example, is squat and roundish when it grows on oyster banks, but becomes more elongate and with more delicately drawn out spines when in a deep-water habitat.

1. **PITTED MUREX.** *Favartia cellulosa* (Conrad). S.E. United States–West Indies. Shell rough, stout, grayish, minutely pitted, with 5–7 blunt varices, and an upturned siphonal canal. Moderately common; under intertidal stones and in oyster beds.

2. **GEM MUREX.** *Maxwellia gemma* Sowerby. Calif.–Mexico. Last whorl with 6 rounded, smooth varices. Small spiral cords are blue or blackish. Spire with deep, squarish pits. Common; in rocky rubble in shallow water.

3. **FESTIVE MUREX.** *Shaskyus festivus* Hinds. Calif.–Mexico. Has one very large, rounded knob between the thin, lacy varices, which are curled backwards. Common; on rocks near shore.

4. **FOLIATED THORN PURPURA.** *Ceratostoma foliata* (Gmelin). Alaska–Calif. Has 3 spreading, thin varices and numerous spiral cords. Spine on outer lip. Common; on rocks near shore.

5. **NUTTALL'S THORN PURPURA.** *Ceratostoma nuttalli* (Conrad). Calif.–Mexico. Has 3 poorly developed varices between each of which is a swollen axial rib. Long spine on base of outer lip. Common; on rocks.

6. **MAUVE-MOUTH DRILL.** *Muricopsis ostrearum* (Conrad). West Fla. to Keys. Note its mauve to rusty aperture. Do not confuse with the Gulf Oyster Drill (p. 127). Common; in shallow water.

7. **HEXAGONAL MUREX.** *Muricopsis oxytatus* M. Smith. Fla. Keys–West Indies. (Formerly *M. hexagonus* Lamarck.) White to orange-brown, with a high spire. Seven riblike varices are spiny. Common; on shallow-water reefs.

8. **SPOTTED THORN DRUPE.** *Acanthina spirata* (Blainville). Wash.–Calif. Exterior has very fine spiral threads and numerous rows of brown bars. Strong spine on outer lip. Common; on mussel beds near shore.

9. **CHECKERED THORN DRUPE.** *Acanthina paucilirata* (Stearns). Calif.–Mexico. Exterior has large black squares and 4 or 5 smooth, spiral threads. Needlelike spine on lip. Common; on rocks, at high-tide mark.

10. **BLACKBERRY DRUPE.** *Morula nodulosa* (C. B. Adams). S.C.–West Indies. Studded with round, black beads. Aperture purplish black. Outer lip has 4 or 5 lighter teeth. Common; among intertidal rocks.

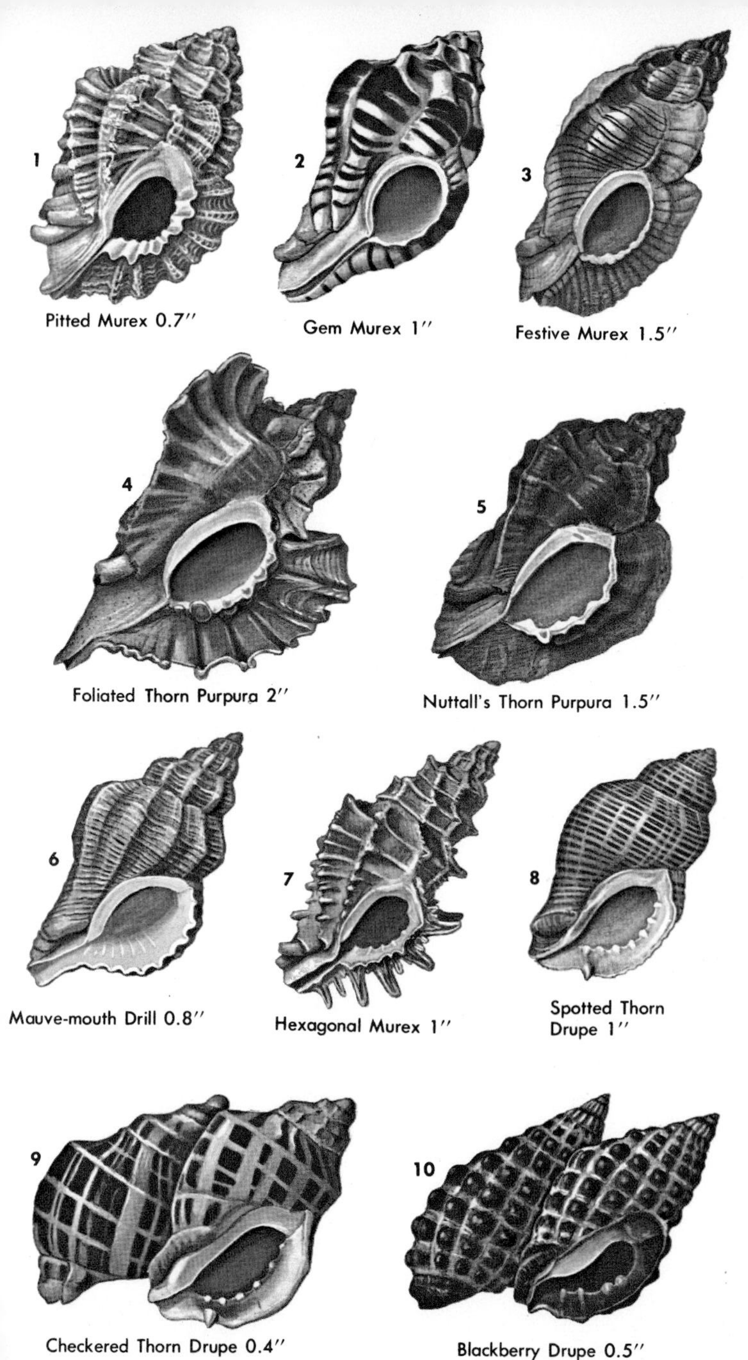

Pitted Murex 0.7″

Gem Murex 1″

Festive Murex 1.5″

Foliated Thorn Purpura 2″

Nuttall's Thorn Purpura 1.5″

Mauve-mouth Drill 0.8″

Hexagonal Murex 1″

Spotted Thorn Drupe 1″

Checkered Thorn Drupe 0.4″

Blackberry Drupe 0.5″

OYSTER DRILLS AND DWARF TRITONS

A large number of small rock-dwelling snails, such as *Urosalpinx*, *Eupleura*, and *Ocenebra* (subfamily Purpurinae), drill holes in the shells of bivalves and especially destroy commercial beds of oysters.

1. **ATLANTIC OYSTER DRILL.** *Urosalpinx cinerea* (Say). Nova Scotia–Fla. Introduced to Calif.–Wash., and northern Europe. Outer lip sometimes thickened on the inside and bearing 2–6 small whitish teeth. Last whorl with 9–12 rounded, axial ribs and many spiral threads. Aperture brownish. Females deposit 2 dozen egg capsules every few weeks during summer months. A well-known oyster pest. Common; from low-tide mark to 25 ft., in rubble of oyster bars.

2. **GULF OYSTER DRILL.** *Urosalpinx perrugata* (Conrad). West Coast of Fla. Last whorl with 6–9 ribs. Aperture rosy brown. Outer lip thick, with about 8 white teeth. Common; near shore and in oyster beds.

3. **TAMPA DRILL.** *Urosalpinx tampaensis* (Conrad). West Coast of Fla. Last whorl with 9–11 sharp, axial ribs crossed by 9–11 strong, spiral threads. Aperture brown. Common; on mud flats and near oyster beds.

4. **CARPENTER'S DWARF TRITON.** *Ocenebra interfossa* (Carpenter). Alaska–Mexico. Spire is half the length of the shell. Last whorl with 8–11 axial ribs crossed by a dozen fine, scaly cords. Common; near shore.

5. **GRACEFUL DWARF TRITON.** *Ocenebra gracillima* (Stearns). Calif.–Mexico. Has rounded axial ribs crossed by a dozen brown-spotted threads. Common; on pilings.

6. **LURID DWARF TRITON.** *Ocenebra lurida* (Middendorff). Alaska–Calif. Smooth, pronounced spiral cords, with 8–10 weak axial ribs. Color variable, gray to brown, rarely banded. Common; on intertidal rocks.

7. **CIRCLED DWARF TRITON.** *Ocenebra circumtexta* (Stearns). Calif.–Mexico. Has about 15 very strong, rough spiral cords crossed by 7–9 low axial ribs on the last whorl. Abundant; on intertidal rocks.

8. **POULSON'S DWARF TRITON.** *Ocenebra poulsoni* (Carpenter). Calif.–Mexico. Shell solid and semiglossy, covered with fine-cut lines. 8 or 9 ribs, with nodules. Aperture whitish. Common; in shallow water.

9. **STIMPSON'S DRILL.** *Eupleura stimpsoni* Dall. Gulf of Mexico. Body whorl has 7 or 8 delicate, thin, pointed varices. Aperture toothed. Color white to gray. Dredged in sandy areas at 50–300 ft.

10. **THICK-LIPPED DRILL.** *Eupleura caudata* (Say). Cape Cod, Mass.–Fla. Shell looks slightly flattened because of broad varices. Has 4–6 riblets between the last 2 varices. Abundant; on oyster beds.

11. **SHARP-RIBBED DRILL.** *Eupleura sulcidentata* Dall. W. Fla. Varices sharp with 2 or 3 riblets between. Color gray to pinkish, rarely with a narrow brown band. Common; in shallow water.

Leathery Capsules (x 5) Containing Eggs of Oyster Drills

Atlantic Oyster Drill

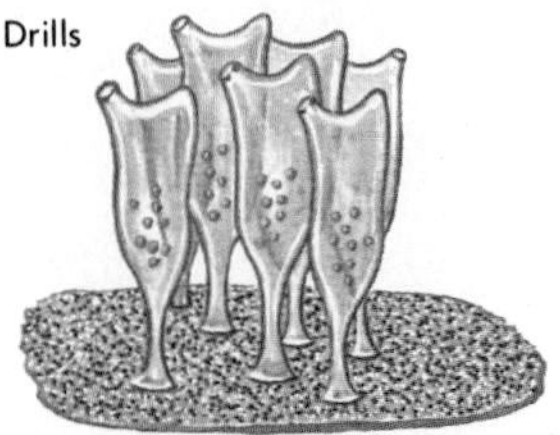

Thick-lipped Drill

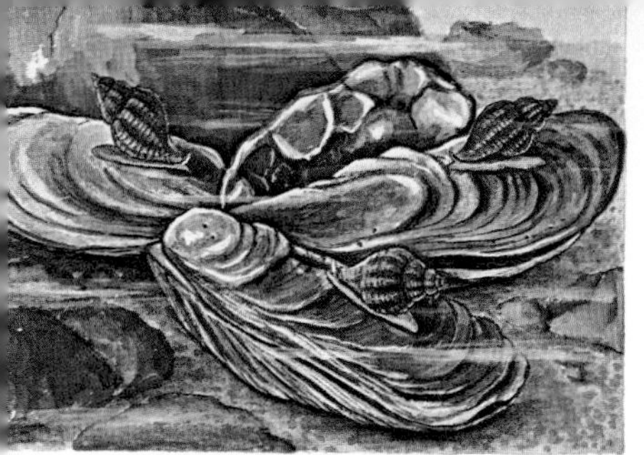

Drills infesting oyster bed

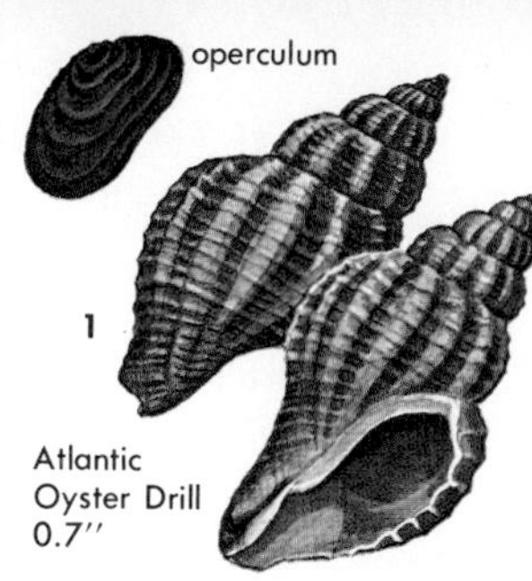

1 Atlantic Oyster Drill 0.7″

2 Gulf Oyster Drill 0.7″

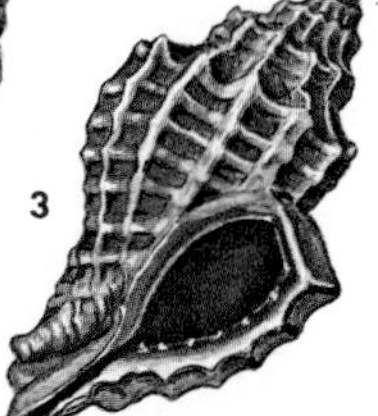

3 Tampa Drill 1″

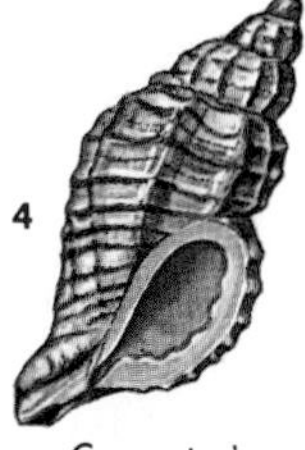

4 Carpenter's Dwarf Triton 0.5″

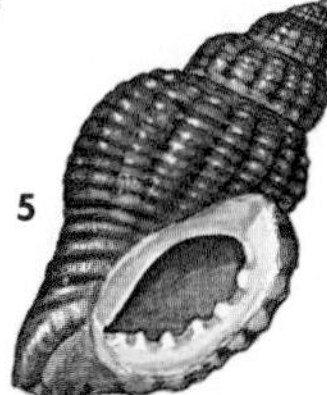

5 Graceful Dwarf Triton 0.4″

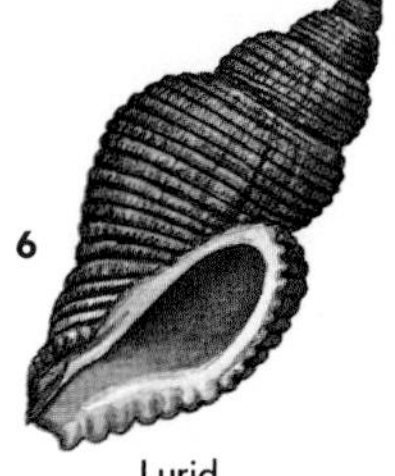

6 Lurid Dwarf Triton 1″

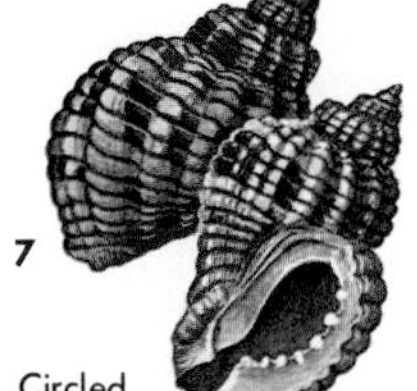

7 Circled Dwarf Triton 0.7″

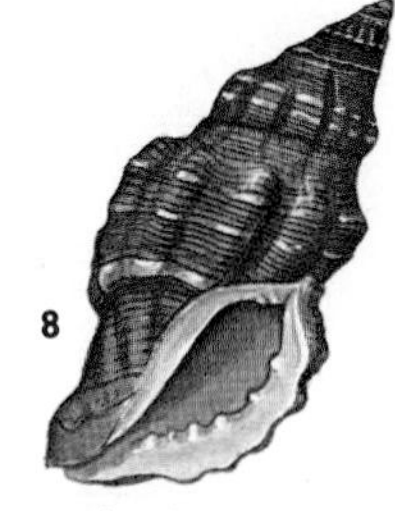

8 Poulson's Dwarf Triton 1.5″

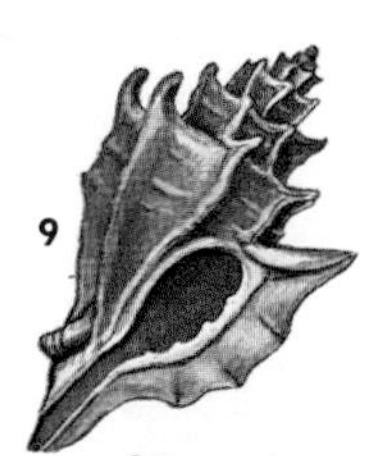

9 Stimpson's Drill 0.5″

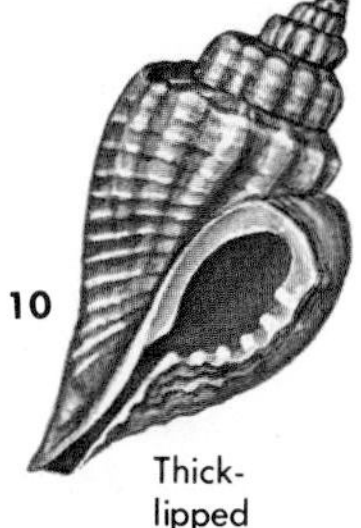

10 Thick-lipped Drill 0.7″

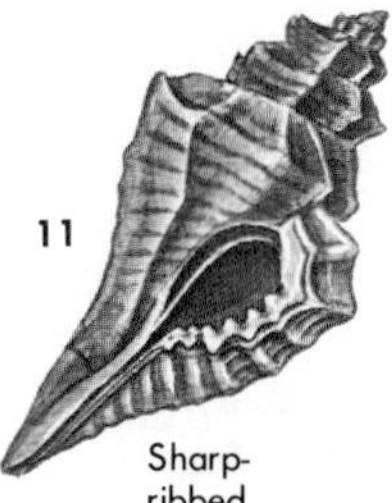

11 Sharp-ribbed Drill 0.5″

DOGWINKLES, TROPHONS, AND CORAL SHELLS

The muricid genera, *Nucella* and *Thais,* contain over a dozen American species, most of which live on rocky intertidal shores. Specimens from rough, shallow waters are generally less spiny than those from deeper, quiet waters. Thais rock shells abound in warm seas. Members of the cold-water genus *Boreotrophon* (about 20 species) are numerous offshore on rubble bottoms.

The coral shells (family Magilidae) contain less than a dozen American species. They live with corals and sea fans and lack a radula.

1. **ATLANTIC DOGWINKLE.** *Nucella lapillus* (Linné). Labrador–N.Y. Color varies from brown to yellow or white; rarely banded. Some have spiral threads, rarely scaled spiral ribs. Common; on intertidal rocks.

2. **FILE DOGWINKLE.** *Nucella lima* (Gmelin). Alaska–N. Calif. Last whorl has 17–20 smooth, round-topped spiral cords, rarely with minute scales. Cords alternate in size. Color whitish to brown. Common; on rocks.

3. **EMARGINATE DOGWINKLE.** *Nucella emarginata* (Deshayes). Bering Sea–Mexico. Spire short. Columella arched and flattened below. Spiral cords are alternately large and small, rarely scaly. Aperture and columella brown. Common; along rocky shores.

4. **CHANNELED DOGWINKLE.** *Nucella canaliculata* (Duclos). Alaska–Calif. Spire high. Suture slightly channeled. Body whorl with 14–16 flat-topped spiral cords. Color white to orange, sometimes banded. Moderately common; on rocks.

5. **FRILLED DOGWINKLE.** *Nucella lamellosa* (Gmelin). Alaska–Calif. Shell variable in sculpture (smooth to very frilled) and color (white to brown). Spire usually high. Rarely spiny. Common on shore rocks.

6. **FLORIDA ROCK SHELL.** *Thais haemastoma* (Lamarck). Two forms: (1) *floridana* (Conrad) (left). N.C.–Brazil. Columella straight and cream-orange. (2) *haysae* Clench (right). Gulf of Mexico. Shell has channeled suture. Abundant near oyster beds.

7. **RUSTIC ROCK SHELL.** *Thais rustica* (Lamarck). S.E. Fla.–West Indies. Shell stout; white columella slightly twisted and has purple stain at lower, inner corner. Aperture always white. Outer lip has brown spots. Common; near shore.

8. **DELTOID ROCK SHELL.** *Thais deltoidea* (Lamarck). Fla.–West Indies. Shell has 2 rows of large blunt spines. Parietal wall violet to rose. Columella has a small ridge at base. Aperture white. Abundant on rocky shores.

9. **CLATHRATE TROPHON.** *Boreotrophon clathratus* (Linné). Arctic Seas–Maine. Chalky white. Shell variable in shape; whorls rounded with many low, leafy ribs. Common; on rubble bottom, 20–200 ft.

10. **MANY-RIBBED TROPHON.** *Boreotrophon multicostatus* (Eschscholtz). Bering Sea–Calif. Suture deep. 8–10 leafy ribs per whorl. Aperture brownish. Common; intertidal in Alaska; 20–200 ft. off Calif.

11. **ORPHEUS TROPHON.** *Boreotrophon orpheus* (Gould). Alaska–Cent. Calif. Pure white; spire high; numerous leafy ribs crossed by spiral cords (3 in upper whorls). Common; 30–500 ft.

12. **ABBREVIATED CORAL SHELL.** *Coralliophila abbreviata* (Lamarck). S.E. Fla.–Brazil. Yellowish white, thick, with numerous spiral threads bearing microscopic scales. Soft parts yellow. Common at base of corals and sea fans.

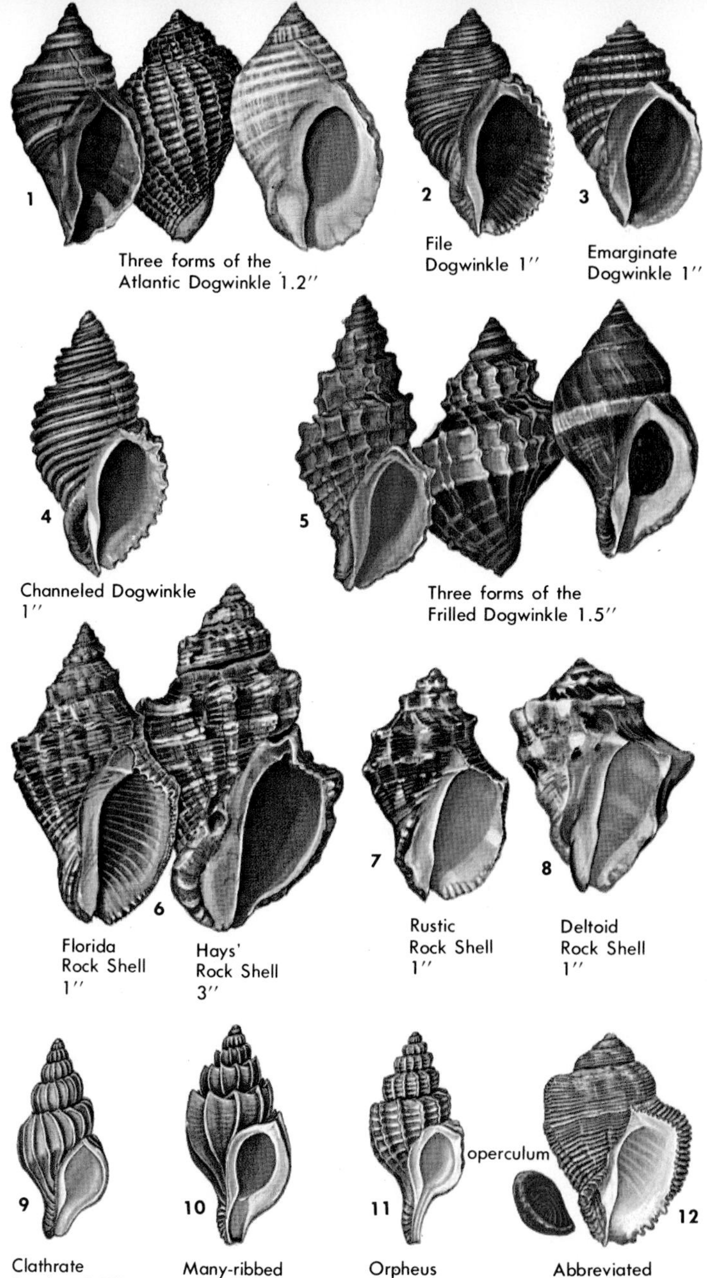

Three forms of the Atlantic Dogwinkle 1.2″

File Dogwinkle 1″

Emarginate Dogwinkle 1″

Channeled Dogwinkle 1″

Three forms of the Frilled Dogwinkle 1.5″

Florida Rock Shell 1″

Hays' Rock Shell 3″

Rustic Rock Shell 1″

Deltoid Rock Shell 1″

Clathrate Trophon 0.7″

Many-ribbed Trophon 1″

Orpheus Trophon 0.6″

Abbreviated Coral Shell 1″

DOVE SHELLS

The approximately 400 dove shells, family Columbellidae, range in size from one tenth of an inch to one inch. They are worldwide in distribution, and locally abundant. About 50 kinds live in North American waters. Although a few are found in deep water, most prefer shallows and are associated with algae and rock habitats. Dove shells feed on soft algae and ingest some animal detritus as they cling to seaweeds. The single, pill-shaped egg capsules, each containing about a dozen eggs, are cemented to seaweeds or to rocks. Dove shells have a slender foot and long, delicate tentacles and siphon. The operculum is horny, brown, and somewhat claw-shaped. The sexes are separate.

Some dove shells exhibit curious polymorphic color patterns, as in the White-Spotted Dove Shell that may be all black or spotted.

1. **COMMON DOVE SHELL.** *Columbella mercatoria* (Linné). S.E. Fla.–West Indies. Shell small, but squat and heavy. Body whorl has about 12 spiral ridges. Color very variable, usually white and brown with broken dark spiral bars. May be spotted with only orange or yellow. Outer lip thick, bearing about a dozen white teeth. Periostracum fairly thick, rough and grayish. Common; 1–30 ft. on seaweed beds and white sand. Used in the making of shell jewelry.

2. **RUSTY DOVE SHELL.** *Columbella rusticoides* Heilprin. West coast of Fla.–Key West. Similar to Common Dove, but more slender, smooth on the center of the body whorl, and with mauve-brown marks between the teeth. Common; 3–20 ft.

3. **WHITE-SPOTTED DOVE SHELL.** *Nitidella ocellata* (Gmelin). Fla. Keys–West Indies. Apex usually eroded. Outer lip with 5 or 6 whitish teeth. Spotting variable. Common; in muddy flats where there are protective rocks.

4. **GLOSSY DOVE SHELL.** *Nitidella nitida* (Lamarck). S.E. Fla.–West Indies. Exterior smooth and shiny. Aperture long, and outer lip has a dozen weak teeth. Very common; on and under rocks at low-tide mark. This species is usually associated with coral reefs in areas where there is an abundance of calcareous algae.

5. **CARINATE DOVE SHELL.** *Nitidella carinata* (Sowerby). Calif.–Mexico. Shell smooth. Color variable: yellowish to brownish. Outer lip with a dozen threads inside. Base with fine, cut lines. Fairly common; in shallow water.

6. **GREEDY DOVE SHELL.** *Anachis avara* (Say). N.J.–Fla. and Texas. Has about a dozen smooth ribs on upper half of each whorl; spiral threads on base of whorl; several weak teeth on inside of inner lip. Abundant; in weed areas.

7. **WELL-RIBBED DOVE SHELL.** *Anachis translirata* (Ravenel). Mass.–N.E. Fla. Has about 20 ribs, which run across each whorl. Spiral, cut lines between ribs. Color brownish gray. Common; from shore to about 100 ft.

8. **PENCILED DOVE SHELL.** *Anachis penicillata* (Carpenter). Calif.–Mexico. Small, slender, with about 15 strong ribs on the last whorl crossed by many spiral threads. Cream with spots of brown. Common; under rocks.

9. **LUNAR DOVE SHELL.** *Mitrella lunata* (Say). Mass.–Texas; West Indies. Very small, glossy, smooth, and marked with zigzag brown or yellow stripes. Outer lip wavy and brown-edged. An abundant species in shallow, weedy bays.

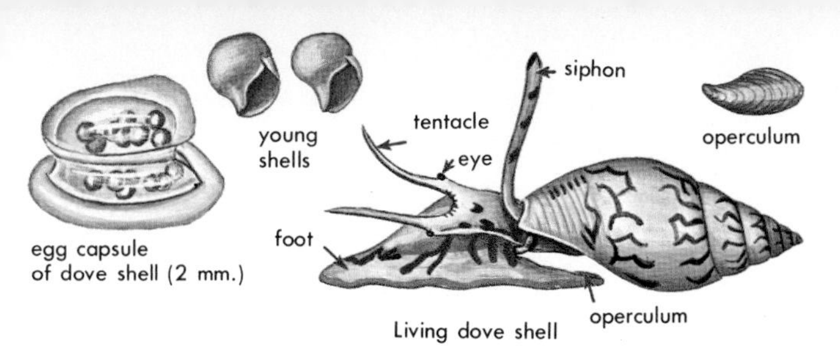

egg capsule of dove shell (2 mm.)

Living dove shell

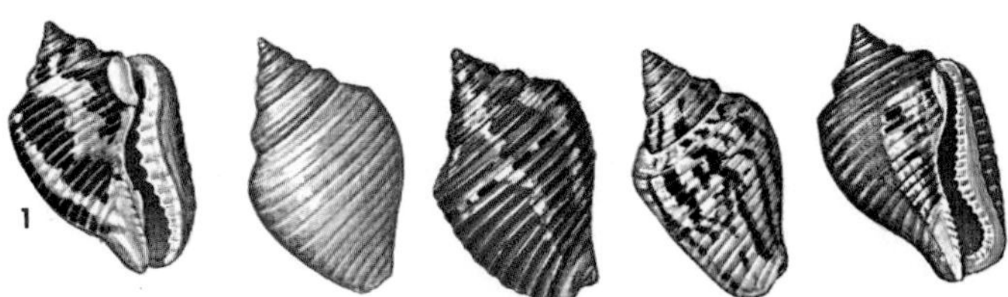

Color variations in the Common Dove Shell 0.3″

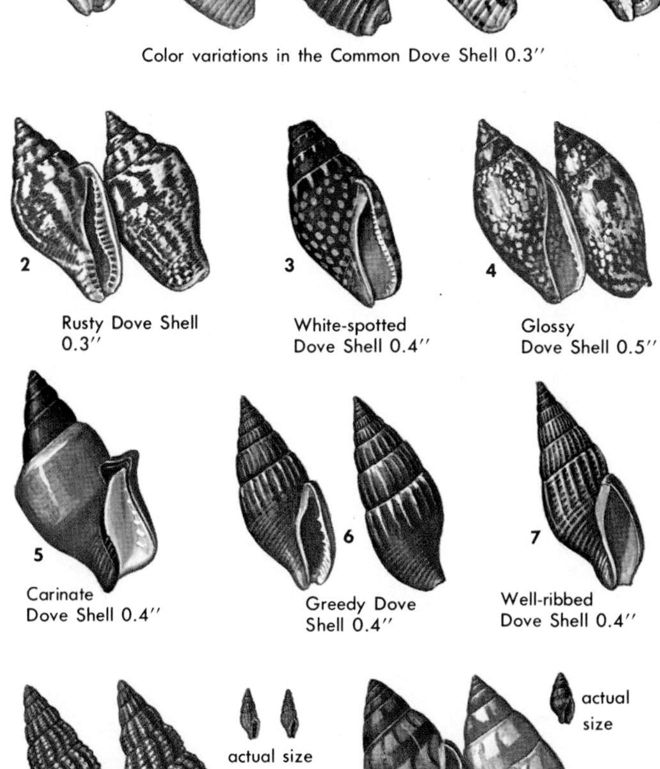

Rusty Dove Shell 0.3″

White-spotted Dove Shell 0.4″

Glossy Dove Shell 0.5″

Carinate Dove Shell 0.4″

Greedy Dove Shell 0.4″

Well-ribbed Dove Shell 0.4″

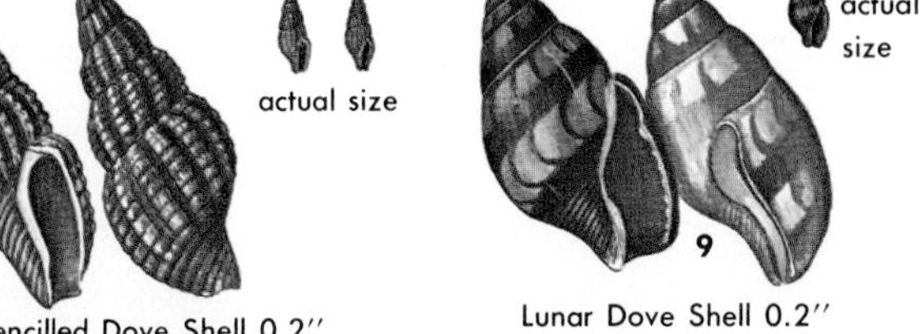

Pencilled Dove Shell 0.2″

Lunar Dove Shell 0.2″

BUCCINUM AND COLUS WHELKS

The whelk family Buccinidae is probably one of the largest and most diversified of all the families of marine snails. Its 70 or more genera are represented by exclusively polar species, as well as by typically tropical groups. There are over 2,000 species in the family, and all are believed to be carnivorous. The radular teeth of the rachiglossate type (see p. 24) are heavily constructed, so as to be effective in tearing away small bits of dead fish. The cold-water whelks are large, very drab in appearance, and usually covered with a heavy periostracum. Most Arctic species have a circumpolar distribution, with some species ranging from northern Canada to Scandinavia and Siberia. Both the genera *Buccinum* and *Colus* have several dozen species in northern waters.

The egg cases of the northern species consist of balls or towers of small, horny capsules, which are formed from a small round pore on the underside of the foot. Some eggs within a capsule are sacrificed as "nurse eggs," with the more advanced young snails feeding on them. The buccinid whelks are very receptive to the odor of dead flesh, and for this reason they may be easily baited to traps. They also feed upon clams, but in turn they serve as a source of food for bottom-feeding fish, such as the cod and haddock. The Northern Whelk is eaten by man in northern Europe.

Some cold-water members of this group exhibit the genetic character of coiling sinistrally, as does Left-handed Buccinum.

1. **COMMON NORTHERN WHELK.** *Buccinum undatum* Linné. Arctic Seas–N.J. Nine–18 slanting, rounded ribs per whorl. Spiral cords weak and numerous. Color dull white to tan, with light brown periostracum. Aperture enamel-white. An abundant species; found just offshore.

2. **GLACIAL BUCCINUM.** *Buccinum glaciale* Linné. Arctic Seas–Wash. and to Gulf of St. Lawrence. Shell fairly thick-walled but lightweight. Outer lip thick and flaring. Two strong wavy cords on last whorl. Color mauve-brown. Common; below low-tide mark.

3. **LEFT–HANDED BUCCINUM.** *Pyrulofusus harpa* (Mörch). Alaska. Shell characteristically sinistral (left-handed). Dextral (right-handed) specimens rare. Smoothish early whorls are indented. Bottom whorls have 6–12 slanting ribs and raised, spiral threads. Fairly common; 50–200 ft.

4. **STIMPSON'S COLUS.** *Colus stimpsoni* (Mörch). Labrador–N.C. Length of aperture is almost half the length of the entire, strong, chalk-white shell. Whorls have numerous fine, spiral lines. Periostracum brown, semiglossy. Common; from shallow to deep water.

5. **HAIRY COLUS.** *Colus pubescens* (Verrill). Gulf of St. Lawrence–N.C. Length of aperture slightly more than entire length of shell. Suture minutely channeled. Whorls rounded. Periostracum brown, sometimes fuzzy. Common; 100–1,000 ft.

6. **PYGMY COLUS.** *Colus pygmaeus* (Gould). Gulf of St. Lawrence–N.C. Shell small, fairly fragile, chalk-white, with numerous spiral, incised lines, and covered with a thin, velvety, olive-gray periostracum. A moderately common species; dredged offshore to 900 ft. This is one of the main foods of the codfish.

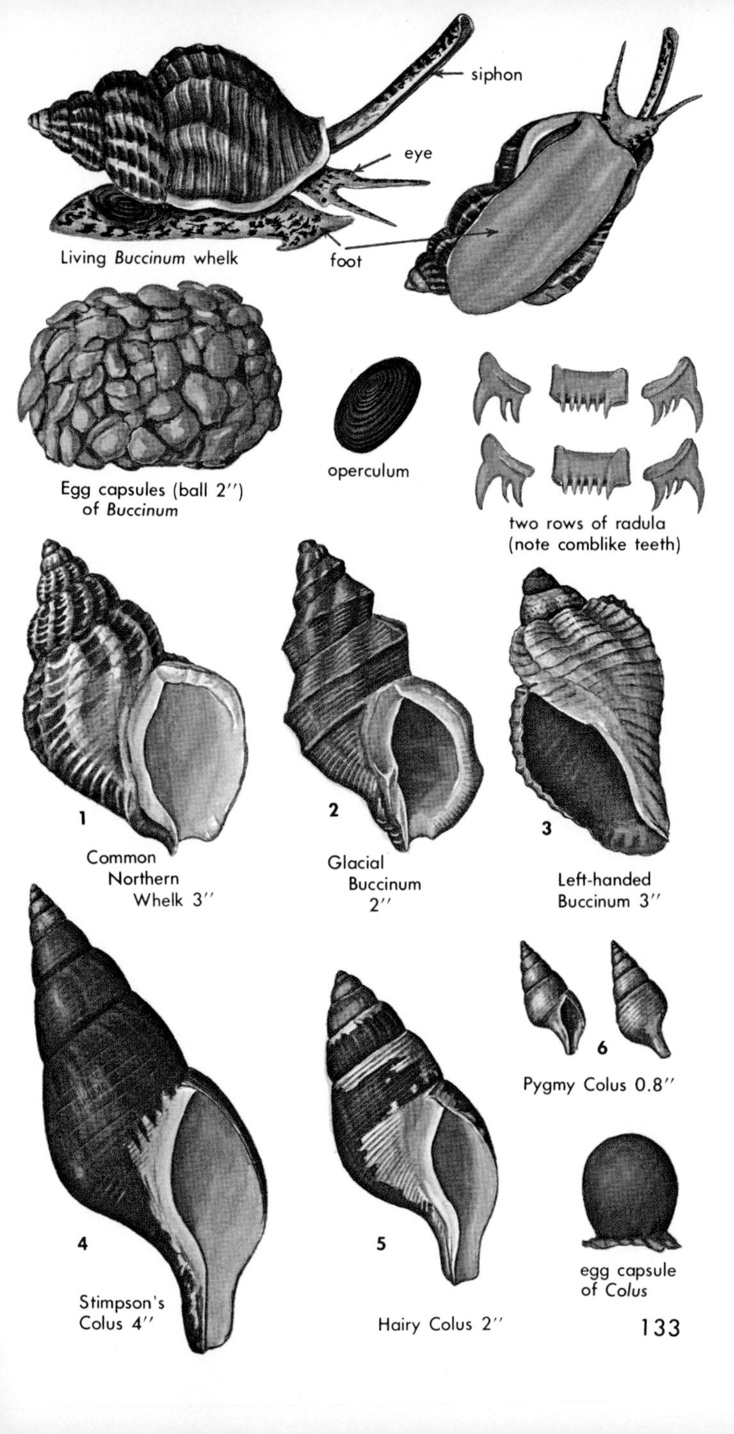
siphon
eye
foot
Living *Buccinum* whelk
Egg capsules (ball 2″) of *Buccinum*
operculum
two rows of radula (note comblike teeth)
1
Common Northern Whelk 3″
2
Glacial Buccinum 2″
3
Left-handed Buccinum 3″
4
Stimpson's Colus 4″
5
Hairy Colus 2″
6
Pygmy Colus 0.8″
egg capsule of *Colus*

SMALL WHELKS AND NEPTUNES

Many unusual genera of many shapes and sizes are found in the family Buccinidae, particularly in warm, tropical waters, where they live under protective rocks during the day and forage for food at night. In the subtidal zones, *Engina, Searlesia,* and *Cantharus* fill the predatory role of the intertidal *Thais* rock shells. Some are serious pests to commercial oyster beds. The tiny but harmless *Engina,* of which there are many species, live under submerged rocks. The Dwarf Tritons, *Colubraria,* are strictly tropical. Their shells superficially resemble those of the large Triton's Trumpets, *Charonia* (p. 118).

On the Californian coast, three cool-water buccinids are quite common. The largest, Kellet's Whelk, is captured in baited traps. The West Coast Livid Macron and the Dire Whelk are common in tide pools. The Neptune Whelks are abundant in cold circumpolar seas, with half a dozen species in Alaskan waters, the largest being the 5- to 7-inch Lyrate Neptune. The New England Neptune, a common carnivorous species, lays a 4 or 5 inch tower of egg capsules.

1. **TINTED CANTHARUS.** *Cantharus tinctus* (Conrad). N.C.–Texas–West Indies. Small canal at upper end of aperture. Lower end of columella white. Color of this common intertidal shell varies. Found in weeds in intertidal muddy area.

2. **FALSE DRILL.** *Cantharus multangulus* (Philippi). N.C.–West Indies. Siphonal canal short and open. Outer lip sharp and wavy. Small single fold at base of columella and 8 or 9 rounded ribs on the last shouldered whorl. Spiral threads weak. Color variable, sometimes mottled with red-brown specks on gray, rarely all yellow or orange. A common shallow-water species found among weeds. Formerly and erroneously placed in genus *Pseudoneptunea.*

3. **WHITE-SPOTTED ENGINA.** *Engina turbinella* (Kiener). Fla. Keys–West Indies. Shell small with about 10 white knobs on the shoulder. Has 4 or 5 white teeth on outer lip. Common; under intertidal rocks.

4. **LIVID MACRON.** *Macron lividus* (A. Adams). Calif.–Mexico. Shell bluish or yellowish white, but covered with a feltlike, brown periostracum. Base of shell has about 6 cut lines. Common; in rocky pools.

5. **DIRE WHELK.** *Searlesia dira* (Reeve). Alaska–Calif. Aperture and columella chocolate-brown. Has low ribs on spire. Spiral threads numerous, fine, and unequal. Outer lip finely scalloped. A common tide-pool species.

6. **KELLET'S WHELK.** *Kelletia kelleti* (Forbes). Calif.–Mexico. Shell white, very heavy, with wavy suture and sharp, crenulated outer lip. Has about 8–10 knobs per whorl. Common; on rock and sand, 6–200 ft.

7. **ARROW DWARF TRITON.** *Colubraria lanceolata* (Menke). N.C.–West Indies. Varix strong and curled back. Former varices show in spire. Ash-gray with orange-brown smudges. Lives under rocks; 6–60 ft.

8. **NEW ENGLAND NEPTUNE.** *Neptunea decemcostata* (Say). Nova Scotia–Mass. Shell grayish white with 7–10 very strong, reddish brown spiral cords. Upper whorls show 2 cords. Common; on rocky bottoms, 8–90 ft.

9. **COMMON NORTHWEST NEPTUNE.** *Neptunea lyrata* (Gmelin). Arctic Ocean–Wash. Whorls have 7–9 poorly developed spiral cords, between which may be smaller spiral threads. Color dirty brown. Common; 3–300 ft.

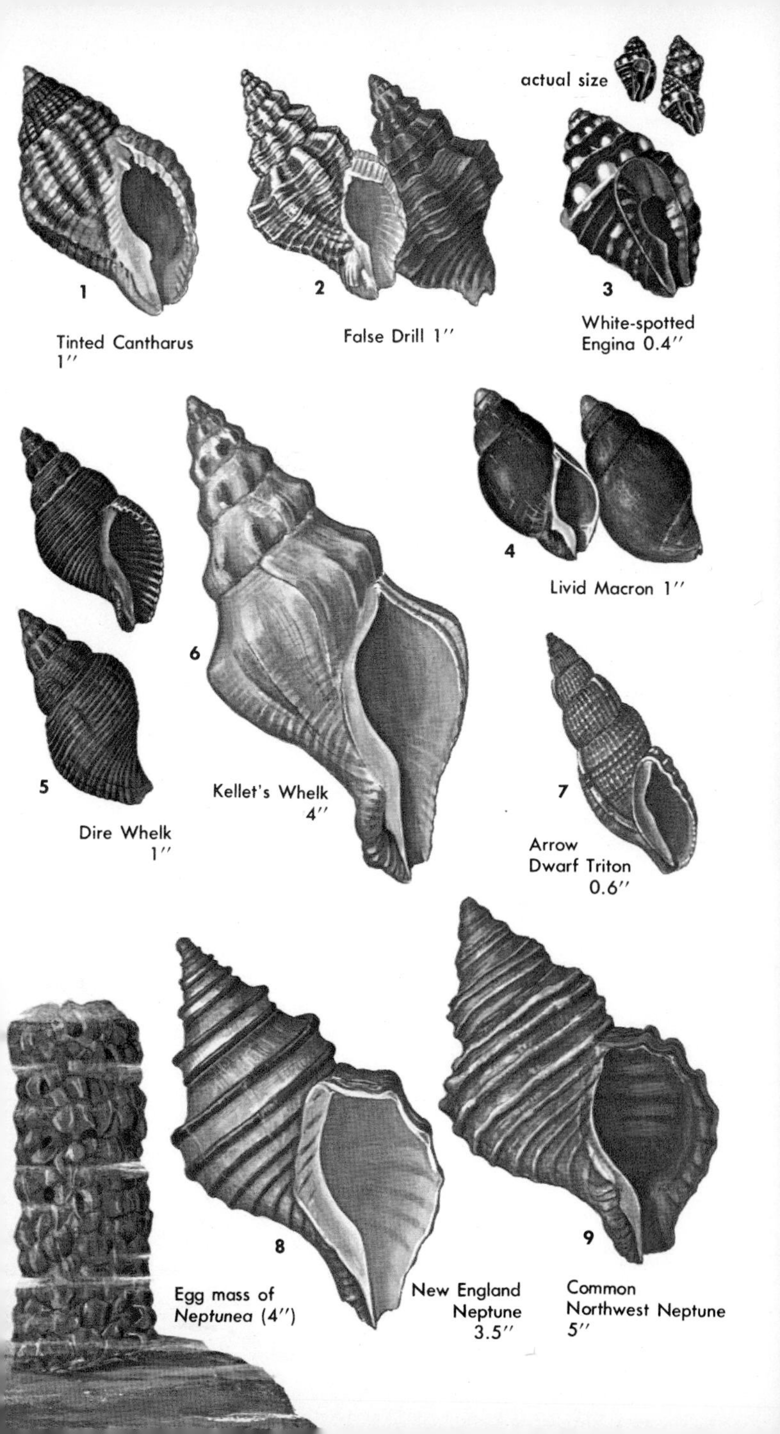

actual size
1
Tinted Cantharus 1″
2
False Drill 1″
3
White-spotted Engina 0.4″
4
Livid Macron 1″
5
Dire Whelk 1″
6
Kellet's Whelk 4″
7
Arrow Dwarf Triton 0.6″
8
Egg mass of Neptunea (4″)
New England Neptune 3.5″
9
Common Northwest Neptune 5″

MELONGENAS

The tropical family Melongenidae contains genera of large, shallow-water snails. In the Americas it is represented by only two genera, the *Melongena* crown conchs and the *Busycon* whelks. The family is unusual in displaying distinct subspeciation and colonial differences among most of its species. They have a broad, claw-shaped operculum that seals the aperture of the shell. The females lay wafer-shaped egg capsules in long strings.

Crown conchs of the genus *Melongena* are limited to three American species, one in Florida, one in the Caribbean, and one on the Pacific side of Central America. The Common Florida Crown Conch abounds in the intertidal areas of Florida and Alabama, usually in the shade of mangrove trees. The snails can tolerate a wide range in water salinity. They are mainly scavengers but will attack live *Fasciolaria* tulip snails and oysters. During the summer, females deposit rows of about a dozen horny capsules on old shells, rocks, and weeds. There are about 200 eggs per capsule, which hatch in about 20 days. Females are usually larger than the males.

1. **FLORIDA CROWN CONCH.** *Melongena corona* (Gmelin). Fla.–Alabama; Yucatan. Usually has 1 or more rows of open spines on shoulder; color variable, rarely albino. Colonies usually have uniform characteristics, some due to inheritance, others because of environment and food. Many experts prefer to recognize only one species in Florida. Others accept such subspecies as the high-spired, northern form *altispira* Pilsbry and Vanatta (fig. 1d), and the dwarf Lower-Fla.-Key form, *bicolor* (Say) (fig. 1b). Spineless forms (fig. 1c) are the result of dietary deficiency. Rarely, shells are found with 2 rows of spines (fig. 1a).

FULGUR WHELKS

Fulgur whelks of the genus *Busycon* have been endemic to the southeastern United States for over 60 million years. Today there are about 6 well-known species living in eastern American waters. Some coil clockwise as they grow and are called dextral, or right-handed; others (Perverse Whelk, opposite page) are coiled sinistrally and are called left-handed. Fulgur whelks, or conchs, feed almost exclusively on hard-shelled clams. The snail envelops the clam in its massive, muscular foot, and by exerting a strong, steady pull forces the clam's valves ajar. At this moment, the snail wedges the edge of its shell between the clam's gaping valves. Then the snail plunges its long proboscis deep into the soft parts of the clam. The snail feeds on about one large clam per month.

2. **PERVERSE WHELK.** *Busycon perversum* (Linné). Off West Fla.–Yucatan. Shell very heavy, sinistral (left-handed), with large shoulder spines. Swollen ridge runs around middle. Uncommon; on sand bottom, 20–100 ft., usually far offshore.

3. **LIGHTNING WHELK.** *Busycon contrarium* (Conrad). N.J.–Fla. Shell is sinistral. Under 7 in. it has axial brown streaks; larger shells are usually white. Small shoulder knobs. Common in Fla. shallows; rare off N.J. in deep water.

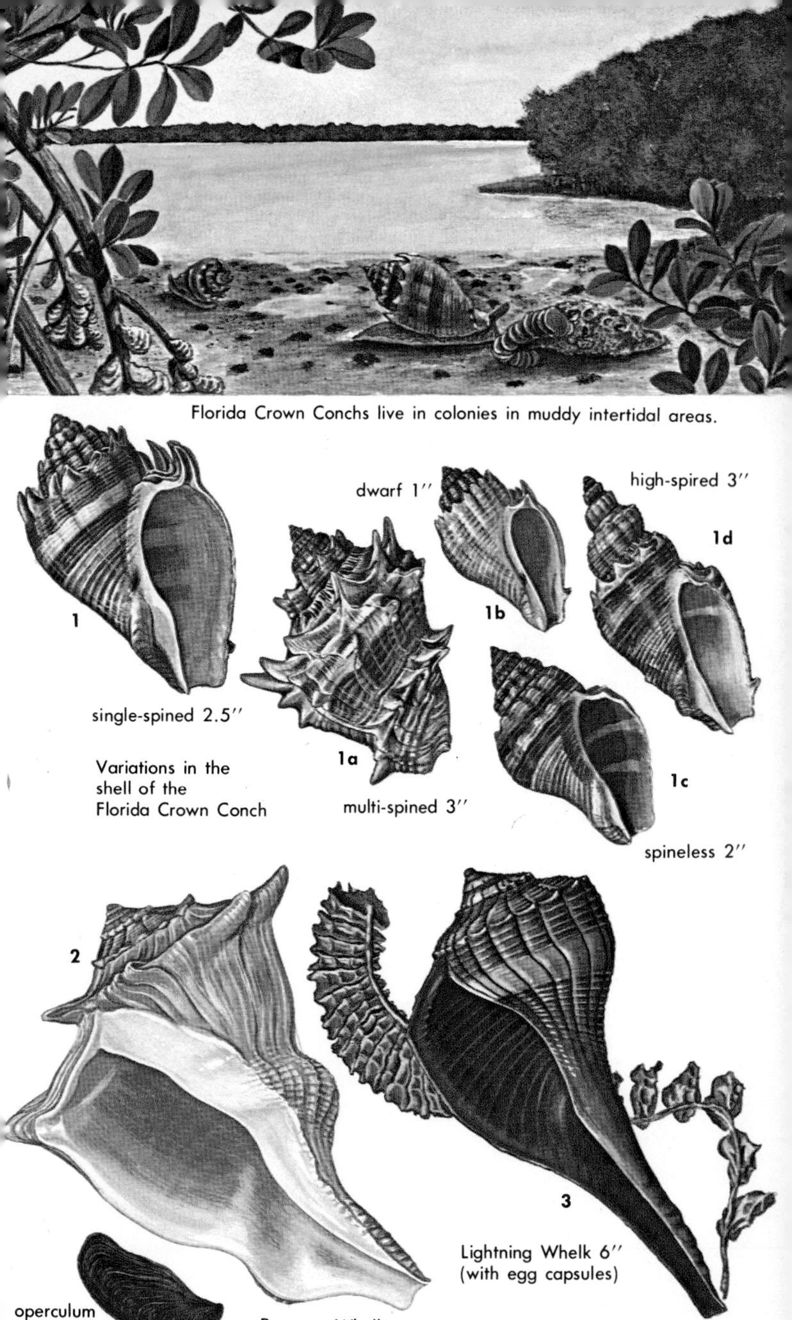

Florida Crown Conchs live in colonies in muddy intertidal areas.

FULGUR WHELKS—continued

Speciation in the fulgur whelks still remains a matter of conjecture and contention, mainly because the biology and geological history of this group have yet to be thoroughly studied. Some authorities consider Kiener's Whelk *(B. eliceans)* as a right-handed Atlantic subspecies of the Gulf of Mexico Perverse Whelk (see previous page); we consider it a southern subspecies of the New England Knobbed Whelk. A smooth-shouldered race of the Lightning Whelk from Sarasota, Fla., is named *aspinosum* Hollister. Only intense natural-history research will solve these problems.

Female fulgur whelks lay long strings of egg capsules. Each rubbery capsule is formed in a round pore located near the front end of the sole of the foot. Twenty to 100 minute eggs are sealed in each egg capsule. The early end of the egg strand is cemented to a stone or dead shell. The size of the capsules depends upon the age and size of the mother snail.

1. **KNOBBED WHELK.** *Busycon carica* (Gmelin). Cape Cod, Mass.–Georgia. Low knobs on the shoulders. Aperture cream to brick-red. Young shells are axially streaked with brownish purple. Common; from shore to 30 ft.

2. **KIENER'S WHELK.** *Busycon carica* subspecies *eliceans* (Montfort). N.C.–N.E. Fla. Spines heavy; shell heavy, with a semiglossy surface and with a swelling around the lower middle of the body whorl. Intergrades exist in N.C. and Virginia. Common; from shore to 30 ft.

3. **TURNIP WHELK.** *Busycon coarctatum* (Sowerby). Gulf of Mexico off Yucatan and south of Texas. Shell turnip-shaped; whorls well rounded and bearing a row of small, brown, sharp spines. Axial streaks predominant. Aperture golden yellow. Uncommon; 12–60 ft.

4. **CHANNELED WHELK.** *Busycon canaliculatum* (Linné). Cape Cod, Mass.–N.E. Fla.; Calif. Suture is broadly and deeply channeled. Aperture large and brown within. Shoulders keeled. Periostracum thick and fuzzy. Common; in sand, 3–50 ft.

5. **PEAR WHELK.** *Busycon spiratum* (Lamarck). N.C.–Texas. Shoulders usually smooth and rounded. Suture channel V-shaped. Periostracum fuzzy and tan. Siphonal canal quite long. From N.C. to both sides of Fla. the subspecies *pyruloides* (Say) has smoothly rounded whorls.

A form with a keel on the shoulder (as illustrated on the opposite page) is common from Alabama to Texas and south to Yucatan, Mexico. It is the subspecies *plagosum* (Conrad). High-spired, narrower shells from Texas have the names *texanum* and *galvestonense* Hollister.

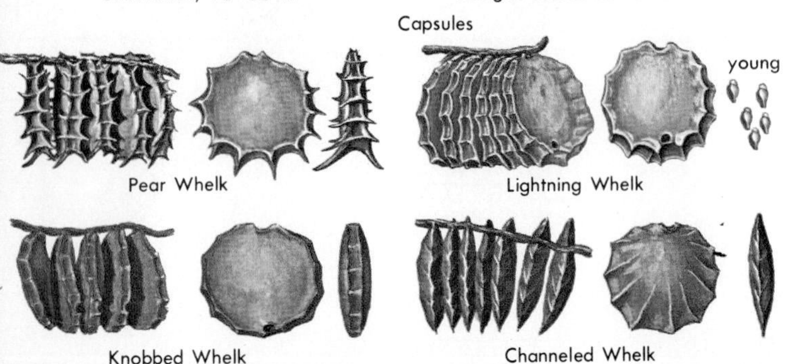

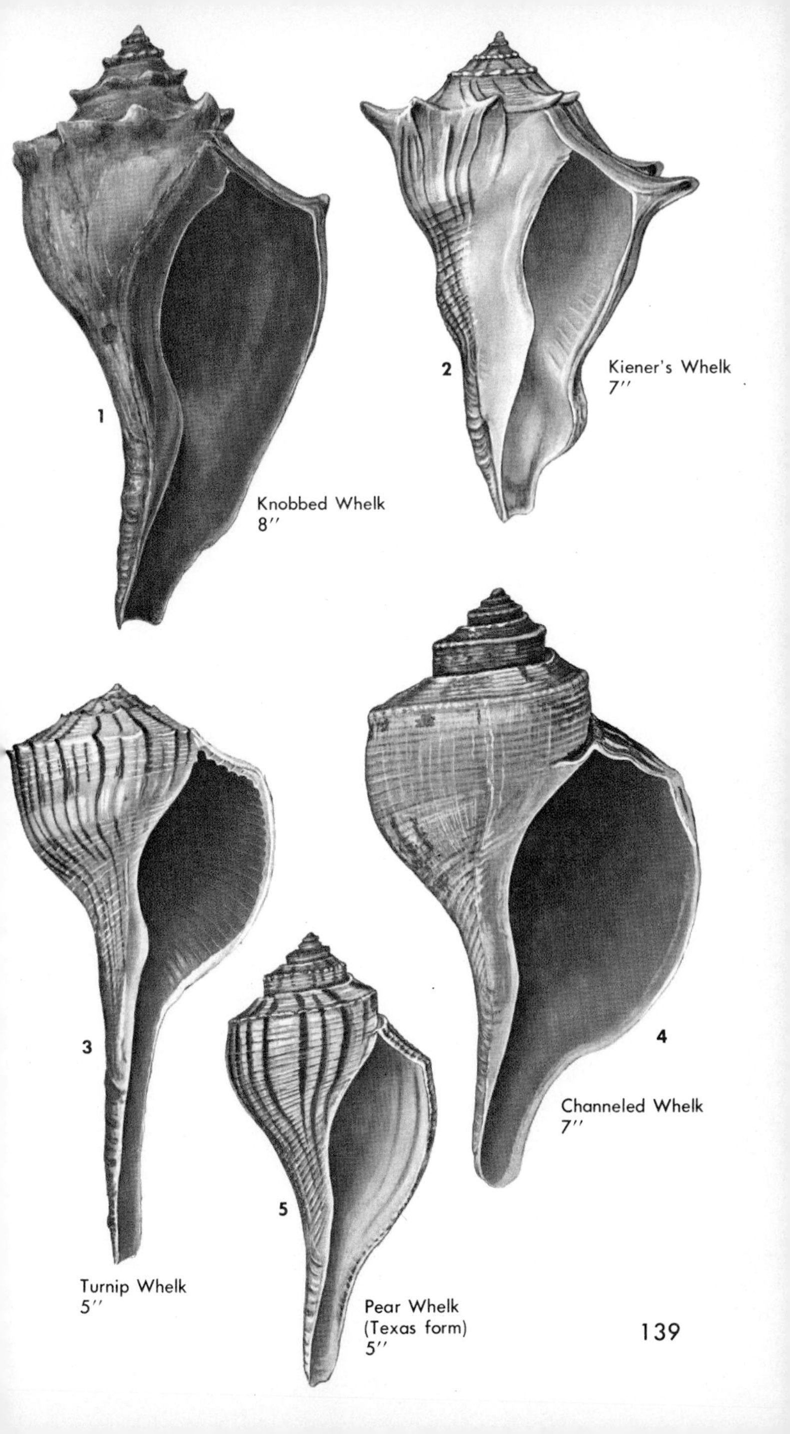
1
Knobbed Whelk
8″
2
Kiener's Whelk
7″
3
Turnip Whelk
5″
4
Channeled Whelk
7″
5
Pear Whelk
(Texas form)
5″

NASSA MUD SNAILS

The large family of nassa snails, Nassariidae, is very widely distributed, with most of the hundred or more species occurring in warm, intertidal waters, although a few live in cold seas or at depths of over 3,000 feet. About seven Pacific Coast and five Atlantic Coast species are known, all of which are very active scavengers. They are gregarious and occur in great numbers in some places. In the genus *Nassarius,* the thin foot is flaring in front, and at the posterior end it bears two cirri, or fleshy prongs. The small horny operculum usually has small spines. The siphonal canal is long and draws water in over a sensitive smell organ, the osphradium. Intertidal forms can live out of water for four or five days. Unlike many marine snails, nassas are attracted toward light.

The sexes are separate, with the shells of the males usually being smaller. Egg capsules, containing about 50 eggs, are laid in rows on algae, shells, or stones, or, rarely, on the underside of moon-snail sand collars. After hatching, the free-swimming veligers take about two weeks to develop into snails.

When several species of nassas occur in the same general area, the differences between them can be readily determined by observing their rate of crawling, their preferences for certain types of mud or sand substrates, the coloration of the foot and siphon, the preferred distance from the low-tide limits, and other seemingly unimportant but significant behavioral characters.

1. **COMMON EASTERN NASSA.** *Nassarius vibex* (Say). Mass.–West Indies. Parietal shield thickly glazed with yellowish ivory. A dozen axial ribs are coarsely beaded. Spiral threads weak. Color gray, sometimes with narrow, broken, brown bands. Common; on muddy sand in shallow water.

2. **VARIABLE NASSA.** *Nassarius albus* (Say). Carolinas–Caribbean. Parietal shield white and narrow. Color whitish, sometimes with brown bands. Whorls shouldered, with 8–12 ribs per whorl. Abundant; on clear sand, 6–100 ft. Formerly called *ambiguus* (Pulteney).

3. **SHARP-KNOBBED NASSA.** *Nassarius acutus* (Say). West coast Fla.–Texas. Spire pointed. Surface shiny, with numerous small, pointed beads. Color gray, but sometimes with fine, brown spiral lines. Adults vary from 0.2–0.4 in. Moderately common; in sand in shallow water to 20 ft.

4. **NEW ENGLAND NASSA.** *Nassarius trivittatus* (Say). Nova Scotia–Georgia. Parietal shield thinly glazed. Suture channeled. Outer lip sharp but strong. Whorls neatly beaded, the last one having about 11 rows. Common; on clean sand in offshore waters, 6–200 ft.

5. **EASTERN MUD NASSA.** *Ilyanassa obsoleta* (Say). Canada–northern Fla. W. Canada–Calif. (introduced). Spire usually badly eroded. Rounded whorls with weak, flattened beads or spiral ridges. Columella has a single, strong ridge near the base. Parietal wall thickly glazed with purple-brown. Posterior end of foot without cirri. Operculum horny, blackish, oval, and without sawteeth. A rare white-banded form occurs in some areas in New Jersey. Abundant; on muddy, intertidal flats in quiet bays. This species is omnivorous and is unique among the American nassas in having a digestive crystalline rod.

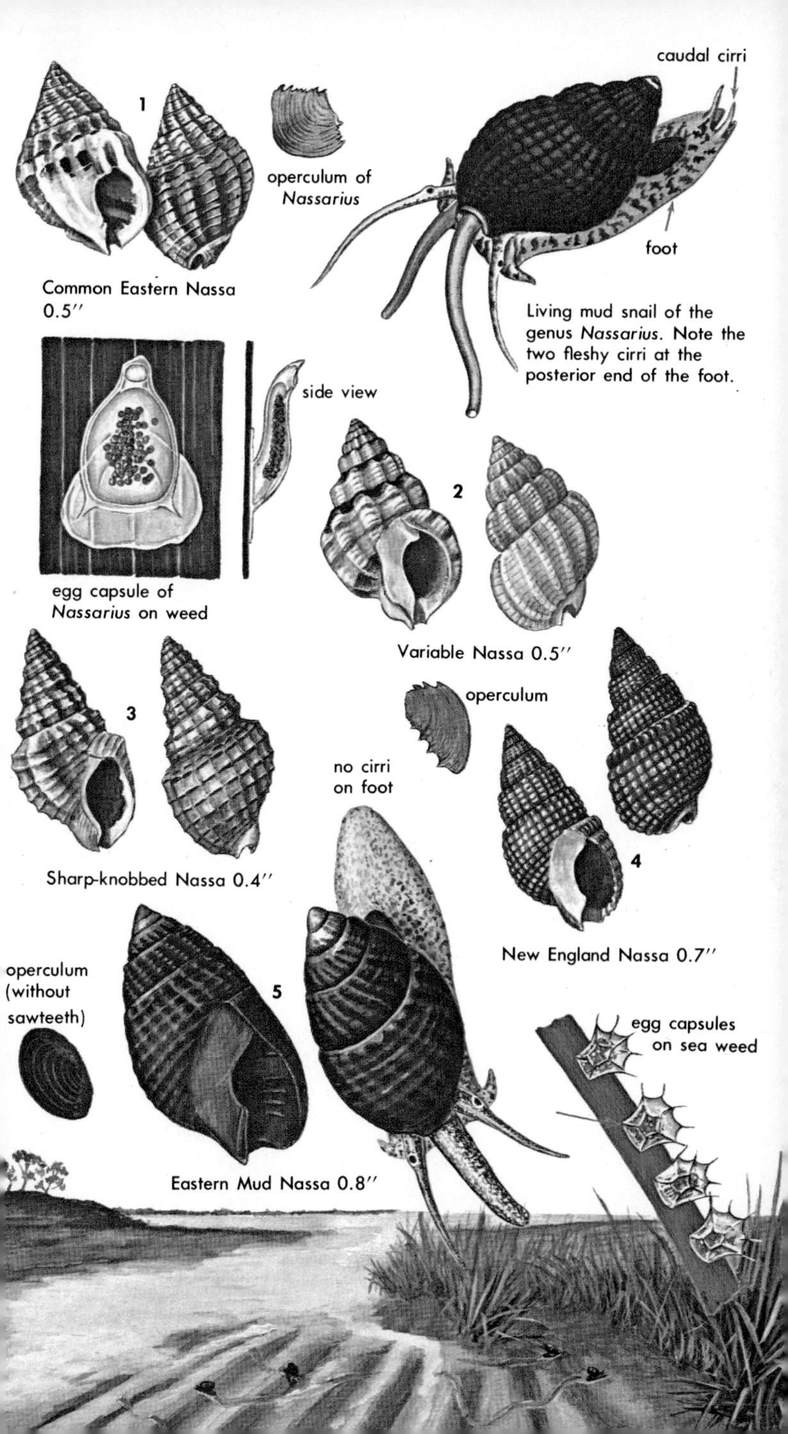
1
Common Eastern Nassa
0.5″
operculum of
Nassarius
caudal cirri
foot
Living mud snail of the
genus *Nassarius*. Note the
two fleshy cirri at the
posterior end of the foot.
side view
egg capsule of
Nassarius on weed
2
Variable Nassa 0.5″
3
Sharp-knobbed Nassa 0.4″
operculum
no cirri
on foot
4
New England Nassa 0.7″
operculum
(without
sawteeth)
5
Eastern Mud Nassa 0.8″
egg capsules
on sea weed

NASSA MUD SNAILS—continued

Nassa mud snails are among the most active and responsive of the scavengers of the sea. Their sense of "smell" is, in actuality, a keen ability to taste the chemical decomposition of dead flesh. Flavors are spread through the water, drawn in through the snail's siphon, and passed over the osphradium, an elongated, gill-like taste organ located on the roof of the mantle cavity. Within a few seconds the snail is activated to head directly for the source of the dead-meat flavor. Nassa mud snails contribute to the cleanliness of sunny mud flats by eating decaying crabs, fish, and other invertebrates exposed at low tide. Some species of nassa feed on egg masses of polychaete worms by sucking them in through their probosces.

The identification of nassa is sometimes difficult because of the great variation in the shells. Poor food supply, changing salinity of the water, and overcrowding may dwarf or otherwise modify the shells. In most species the thickly glazed parietal shield, to the left of the aperture, is not developed until the snail has fully matured. Minor shell characters, such as those of sculpture or coloration, may vary geographically, with members of colonies in the north, such as the Western Lean Nassa, having more and stronger axial ribs than do those of colonies in the south. For instance, in southern California these colonies are classified as the form or subspecies *cooperi* (Forbes).

PACIFIC NASSA SNAILS

1. **SMOOTH WESTERN NASSA.** *Nassarius insculptus* (Carpenter). Calif.–Mexico. Parietal shield whitish, glazed, and narrow. Body whorl has fine, spiral threads. Early whorls ribbed. Periostracum greenish yellow. Spire eroded. Common; offshore, 20–200 ft.

2. **WESTERN FAT NASSA.** *Nassarius perpinguis* (Hinds). Puget Sound–Mexico. Outer lip thin but strong. Surface finely crisscrossed or minutely beaded; 2 or 3 spiral bands of orange-brown. Parietal shield not well developed. Common; from shore to 300 ft.

3. **WESTERN MUD NASSA.** *Nassarius tegula* (Reeve). N. Calif.–Mexico. Parietal shield heavily glazed. Body whorl rather smooth around the middle. Has row of knobs below suture. Color dirty brown, sometimes with 3 weak bands. Common; on intertidal mud flats.

4. **WESTERN LEAN NASSA.** *Nassarius mendicus* (Gould). Alaska–Mexico. Spire high. Last whorl has about a dozen beaded ribs; rarely, banded with dark brown. In California, the form *cooperi* (Forbes) has 7–9 beaded ribs. Outer lip not thickened. Common; in shallow water, in sandy areas.

5. **GIANT WESTERN NASSA.** *Nassarius fossatus* (Gould). Vancouver Is.–Mexico. The largest and one of the most common intertidal Nassas on Pacific Coast. Outer lip jagged along edge and constricted at top. Color orange-brown to brownish white.

6. **CALIFORNIAN NASSA.** *Nassarius rhinetes* Berry. Ore.–Baja Calif. Parietal shield reduced. Outer lip thin and sharp. Numerous beads arranged in 20–30 slanting axial ribs. Uncommon; offshore, 20–200 ft. Was called *N. californicus Conrad,* which, however, is a fossil species.

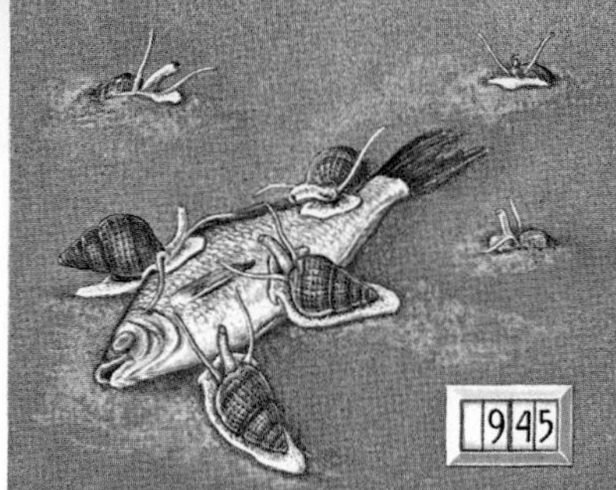

Nassa mud snails constantly keep their siphons above the sand and inhale water. Any odor of dead flesh attracts them to the source within a few minutes.

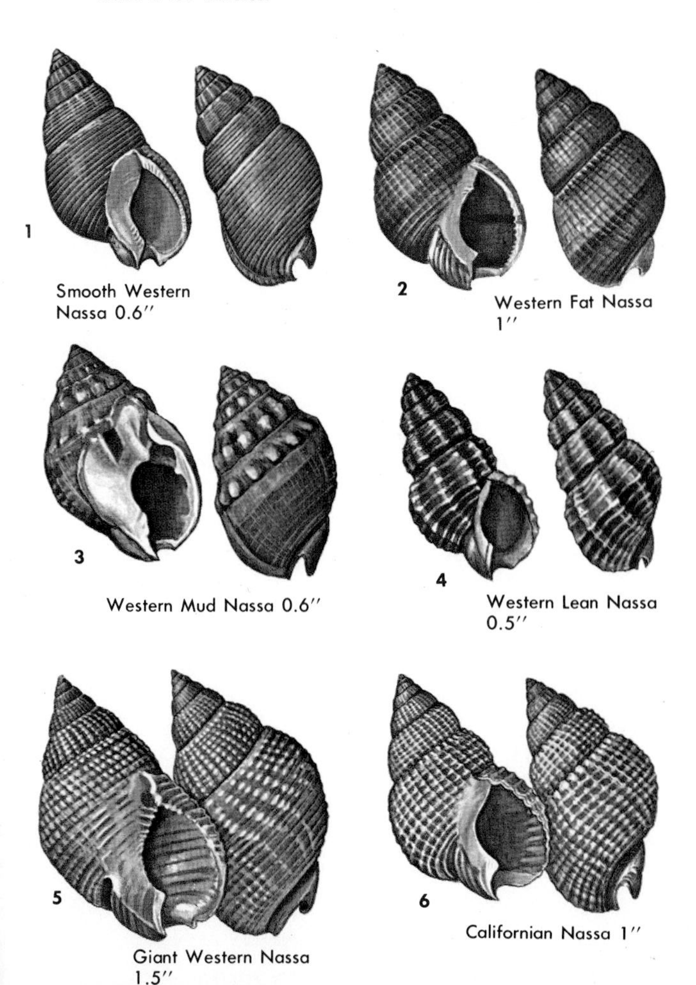

Smooth Western Nassa 0.6″

Western Fat Nassa 1″

Western Mud Nassa 0.6″

Western Lean Nassa 0.5″

Giant Western Nassa 1.5″

Californian Nassa 1″

HORSE AND TULIP CONCHS

The family Fasciolariidae of shallow tropical shores contains a wide variety of groups, such as the *Fasciolaria* tulips, the giant *Pleuroploca* Horse Conchs, and the *Fusinus* spindle shells. All are aggressive carnivores, usually feeding on bivalves and other snails. The siphonal canal is well developed. A large, brown, horny operculum completely seals the shell when the animal is withdrawn.

The Florida Horse Conch, one of the largest living gastropods, can grow to a length of 2 feet, but because of overcollecting, giants are now rarely found. The orange flesh, although edible, is reported to have a peppery taste. The true and banded tulips are common species in very shallow water in the vicinity of muddy, weedy shores of Florida.

The horny, pliable egg capsules of tulip conchs are formed from a pore in the front end of the sole of the female's foot. Several dozen eggs are placed within each capsule as it is formed and is attached to a rock or old shell. Most of the eggs are unfertilized or are arrested in development. These serve as "nurse" cells or food eggs and are eaten by the normal, surviving young snails within each capsule. The young emerge from the capsule in about one month. The sides of the capsules of the Florida Horse Conch have five or six circular rims, whereas those of the Banded Tulip are smooth. Capsules are laid in clumps, sometimes 6 inches in diameter.

1. **FLORIDA HORSE CONCH.** *Pleuroploca gigantea* (Kiener). N.C.–Mexico. Shell grayish white to salmon, covered with a brown, flaky periostracum. Whorls usually knobbed, rarely smooth. Young shells orange. Albino shells rare. A knob-less form occurs in West Florida. Common; 3–20 ft., on sand and weeds.

2. **TRUE TULIP.** *Fasciolaria tulipa* (Linné). N.C.–West Indies. Color pattern variable, usually with greens and browns, but rarely reddish. Two or three crinkled threads follow just below the suture. Moderately common; in shallow water of quiet bays with sand and weed bottoms.

3. **BANDED TULIP.** *Fasciolaria hunteria* (Perry). N.C.–Texas. Whorls smooth. Color dark blue-green or light mauve-tan, with narrow, brown spiral lines. The thin spiral lines do not extend onto the siphonal canal in the typical form. Anterior tip sometimes tinted with orange. Common; in sandy mud and weeds; 3–40 ft. of water.

4. **BRANHAM'S TULIP.** *Fasciolaria hunteria* subspecies *branhamae* Rehder and Abbott. Replaces the typical species in deeper water from Texas to Mexico. It is a larger animal. Siphonal canal is longer and with spiral color lines.

Egg Capsules

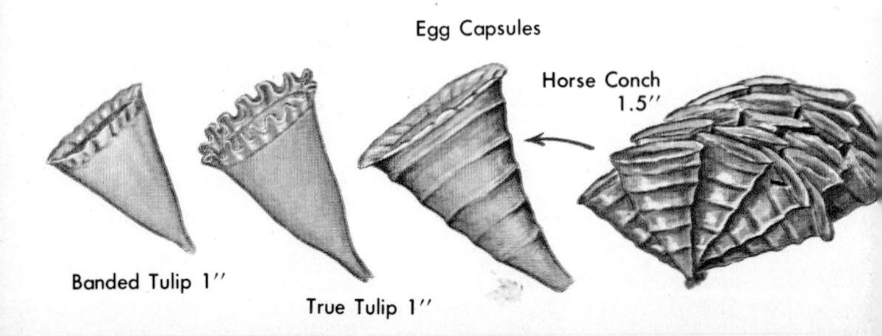

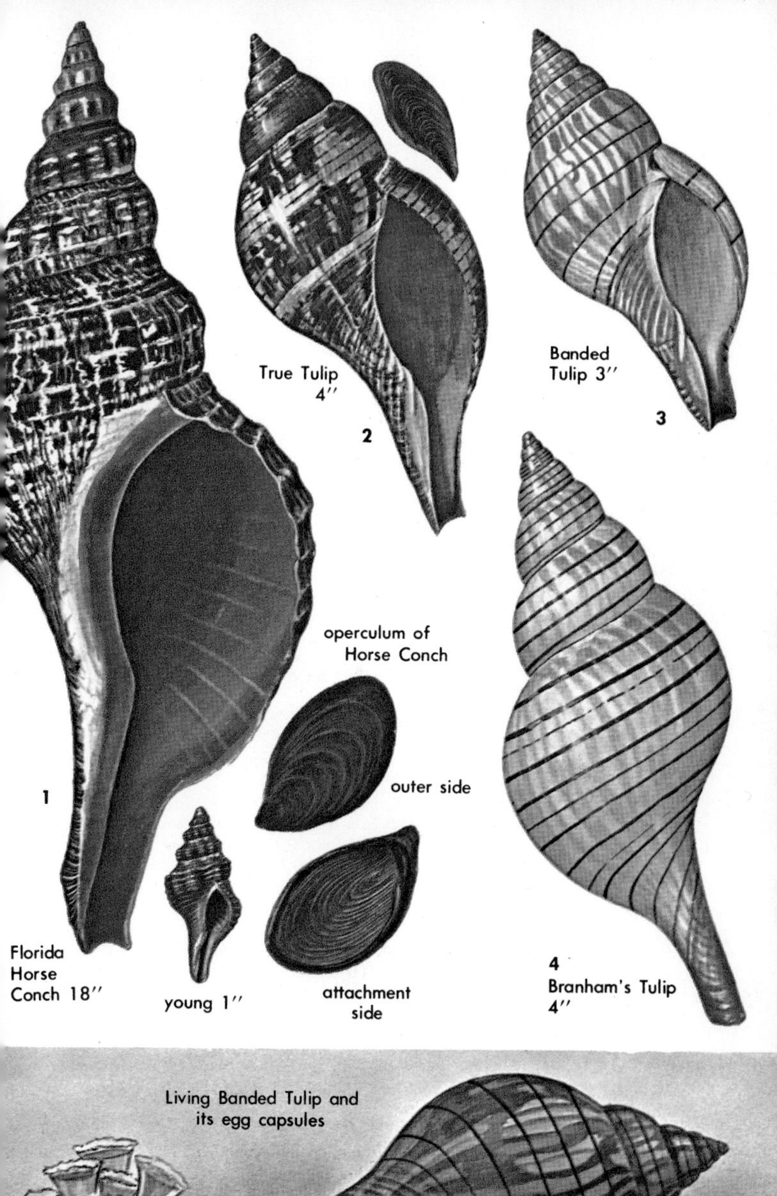
True Tulip
4″
2
Banded
Tulip 3″
3
operculum of
Horse Conch
outer side
1
4
Branham's Tulip
4″
Florida
Horse
Conch 18″
young 1″
attachment
side

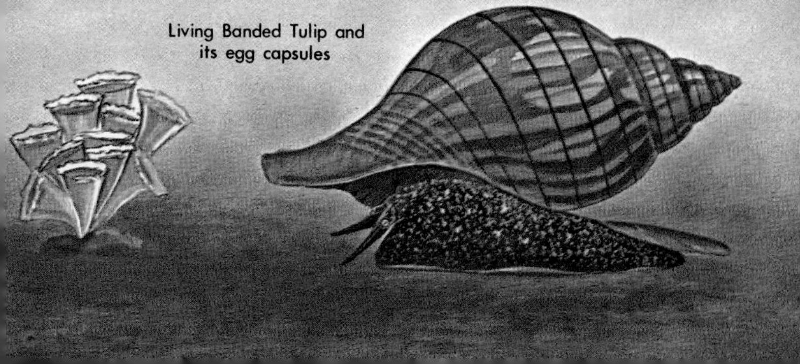
Living Banded Tulip and
its egg capsules

SPINDLES AND LATIRUS

The family Fasciolariidae also includes the spindles *(Fusinus)* and similar shells. All are tropical, carnivorous snails with a rather heavy shell and generally with a long siphonal canal. The dark-brown, horny operculum can completely seal the aperture of the shell. The spindles lack distinct folds on the arching columella. Most species are sparse, usually occurring in pairs on sand. They feed on marine worms. About 60 living spindles and a great number of fossils are known.

Latirus and the closely related genus *Leucozonia* are shallow-water rock-dwelling snails. The heavy shells have three or four folds on the columella base and strong nodes on the shoulder of the whorls.

1. **KOBELT'S SPINDLE.** *Fusinus kobelti* (Dall). S. Calif. Shell heavy, white, with several orange-brown spiral cords and 8–10 axial ribs. Siphonal canal moderately long. Periostracum light brown, moderately thick. Fairly common; 20–200 ft.

Harford's Spindle. *Fusinus harfordi* (Stearns) (not illus.). W. Canada–Calif. A similar orange-brown, 2 in., with 11 or 12 wide, rounded axial ribs. Aperture white. Periostracum shiny and reddish brown. Rare; 20–200 ft.

2. **ORNAMENTED SPINDLE.** *Fusinus eucosmius* (Dall). Gulf of Mexico. Shell strong and white with 8 large, rounded axial ribs crossed by strong, sharp spiral threads. Apex sometimes slightly askew. Periostracum heavy, yellowish. Not uncommon; 60–800 ft. on sand.

3. **SANTA BARBARA SPINDLE.** *Fusinus barbarensis* (Trask). Ore.–Calif. Shell large, grayish white; periostracum tan to light brown. Early whorls have about 10 low axial ribs, crossed by cords. The ribs are weak or absent in the body whorl of adults, which are very rare in collections. Uncommon; 300–1,600 ft.

4. **CHESTNUT LATIRUS.** *Leucozonia nassa* (Gmelin). Fla.–Texas; Caribbean. Shell heavy, usually somewhat encrusted with lime. About 9 large nodules are very pronounced on the shoulder of each whorl, which also has many faint, spiral threads. Color is rich chestnut-brown, with a lighter narrow, spiral band. Aperture has spiral threads. Columella has 3 or 4 weak folds. Animal is red. Common; on rocks from low tide to 20 ft.

5. **WHITE-SPOTTED LATIRUS.** *Leucozonia ocellata* (Gmelin). Fla.–Caribbean. Shell small, heavy, dark brown, with 8 or 9 large, whitish nodules at the periphery of each whorl, and 3 or 4 rows of small white squares at the base of the last whorl. Intertidal; common under rocks.

6. **PAINTED SPINDLE.** *Aptyxis luteopicta* Dall. S. Calif.–Mexico. Shell small, with numerous axial ribs and fine spiral cords. Outer lip thin. Color dark brown, usually with an indistinct, cream-colored band on the periphery of each whorl. Common; low-tide mark to 120 ft.

7. **MCGINTY'S LATIRUS.** *Latirus mcgintyi* Pilsbry. S.E. Fla. Shell strong, elongate, heavy. Cream-colored with a yellow-brown aperture. Body whorl has 8 low, rounded axial ribs, crossed at the periphery by 4 strong spiral cords. Uncommon; 6–60 ft., near reefs.

8. **BROWN-LINED LATIRUS.** *Latirus infundibulum* (Gmelin). S. Fla.–West Indies. Shell tan, with numerous small, orange-brown, raised spiral cords and 7 or 8 strong axial ribs. Columella has 3 weak folds. Aperture has white cords. Animal red with white speckles. Rare in Fla. Under rocks from 6–60 ft.

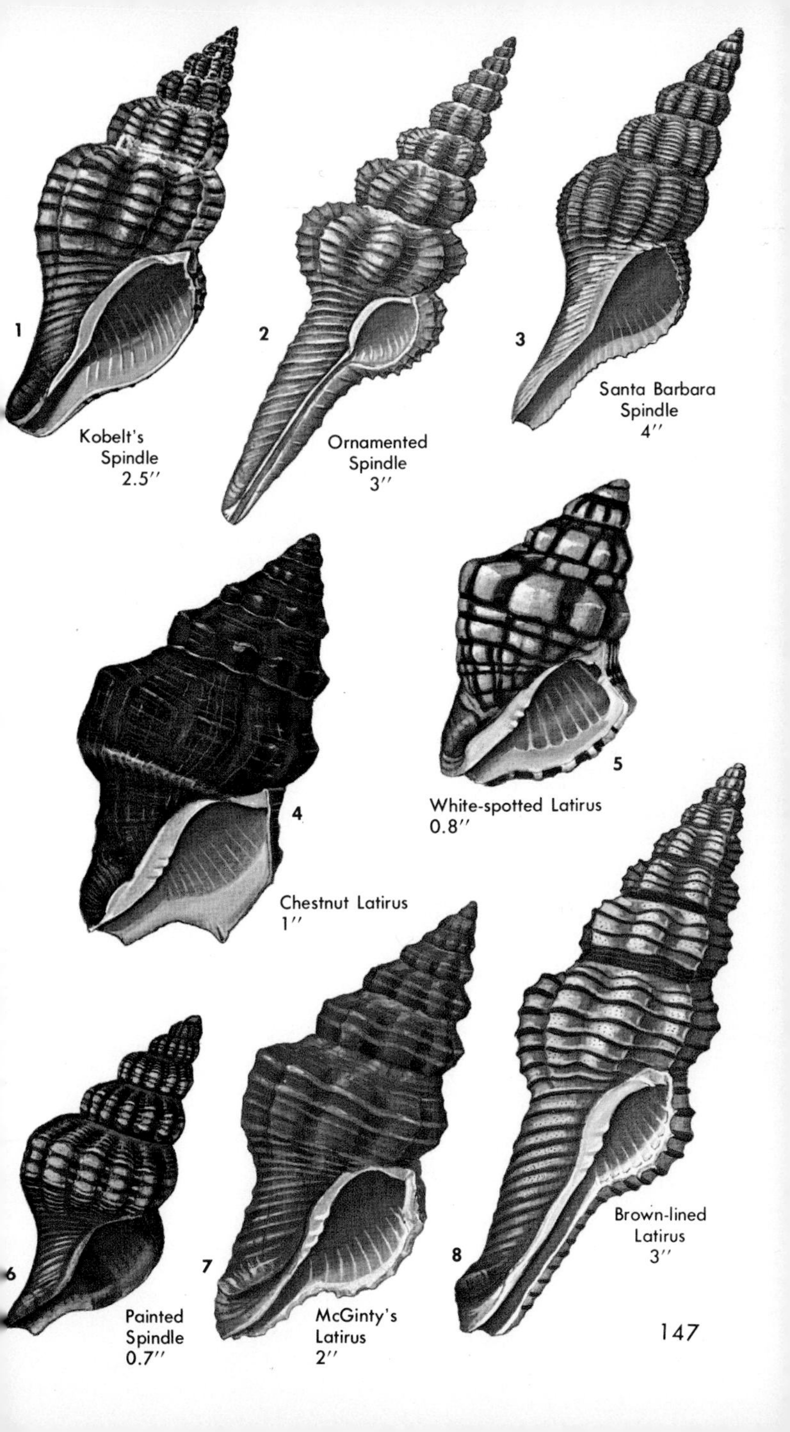
1
Kobelt's
Spindle
2.5″
2
Ornamented
Spindle
3″
3
Santa Barbara
Spindle
4″
4
Chestnut Latirus
1″
5
White-spotted Latirus
0.8″
6
Painted
Spindle
0.7″
7
McGinty's
Latirus
2″
8
Brown-lined
Latirus
3″

OLIVE SHELLS

The many hundreds of olive shells, family Olividae, are characterized by highly glossy shells with numerous fine wrinkles and folds on the columella. The foot and mantle partially envelop the shell, giving it its shiny finish. All members of the family are carnivorous and sand-burrowers. Larger specimens may be caught with baited hook and line. Only two species of *Oliva* occur in the United States. The genus lacks an operculum. The closely related genus *Olivella* has a shell of less than half an inch long and with a long, horny operculum.

1. **VARIABLE OLIVELLA.** *Olivella mutica* (Say). S.E. United States–Caribbean. Shell is small, stubby, variably colored, with strong, glossy callus on columellar wall which has no striae. Common; in shallows on sandy mud.

2. **SAN PEDRO OLIVELLA.** *Olivella pedroana* Conrad. Ore.–Baja Calif. Shell fairly stout, with a heavy columellar callus. Base of shell and callus may be white. Moderately common; 5–100 ft.

3. **PURPLE DWARF OLIVE.** *Olivella biplicata* Sowerby. W. Canada–Baja Calif. Shell quite large and heavy. Color variable, usually bluish gray. Columellar wall has heavy callus. Common intertidally; to 20 ft.

4. **RICE OLIVELLA.** *Olivella floralia* Duclos. N.C.–Caribbean. Shell slender, white, shining, often brown-flecked. Upper whorls white to blue or purplish. Columellar wall has fine plaits. Common; 3–30 ft.

5. **TINY FLORIDA OLIVELLA.** *Olivella pusilla* Marrat. Fla. Similar to Variable Olivella but smaller, more slender. Variably colored. Several plaits on columellar wall. Callus often brownish. Common; in shallows where mud and sand are mixed together.

6. **WEST INDIAN OLIVELLA.** *Olivella nivea* (Gmelin). S.E. United States–Caribbean. Shell usually cream-colored, with brownish markings. Apex sharply pointed. Base of shell white. Common; intertidal zone to 150 ft.

7. **LETTERED OLIVE.** *Oliva sayana* Ravenel. S.E. United States. Shell large, shiny, cream to grayish tan with brown tentlike markings. Sides of whorls only slightly concave. Common; intertidal zone to 20 ft.

8. **NETTED OLIVE.** *Oliva reticularis* Lamarck. S. Car.—Caribbean. Similar to Lettered Olive but smaller and lighter. Sides of whorls more convex. Note rare albino. Common; to 20 ft.

VASE SHELLS

9. **CARIBBEAN VASE.** *Vasum muricatum* (Born). S.E. Fla.–Caribbean (family Vasidae). Shell heavy, covered with a thick, gray-brown periostracum. Columella has 4 or 5 strong spiral cords. Aperture white, rarely tinged with pink or purplish. Operculum narrow, thick, and brown. Foot cream, speckled with gray. Mantle coral-pink near the edges. Feeds on clams and worms. Fairly common; 2–30 ft.

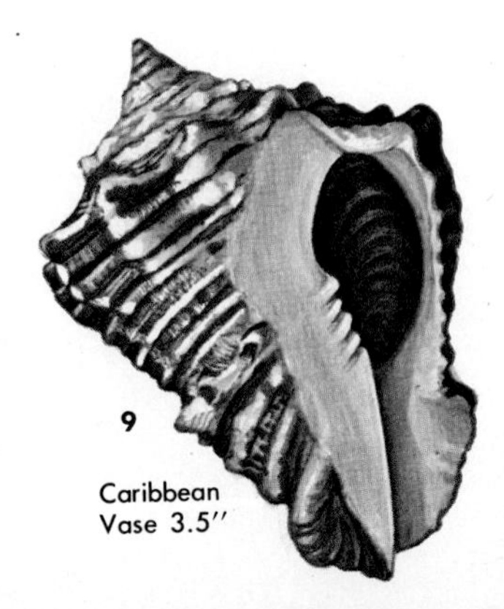

Caribbean Vase 3.5″

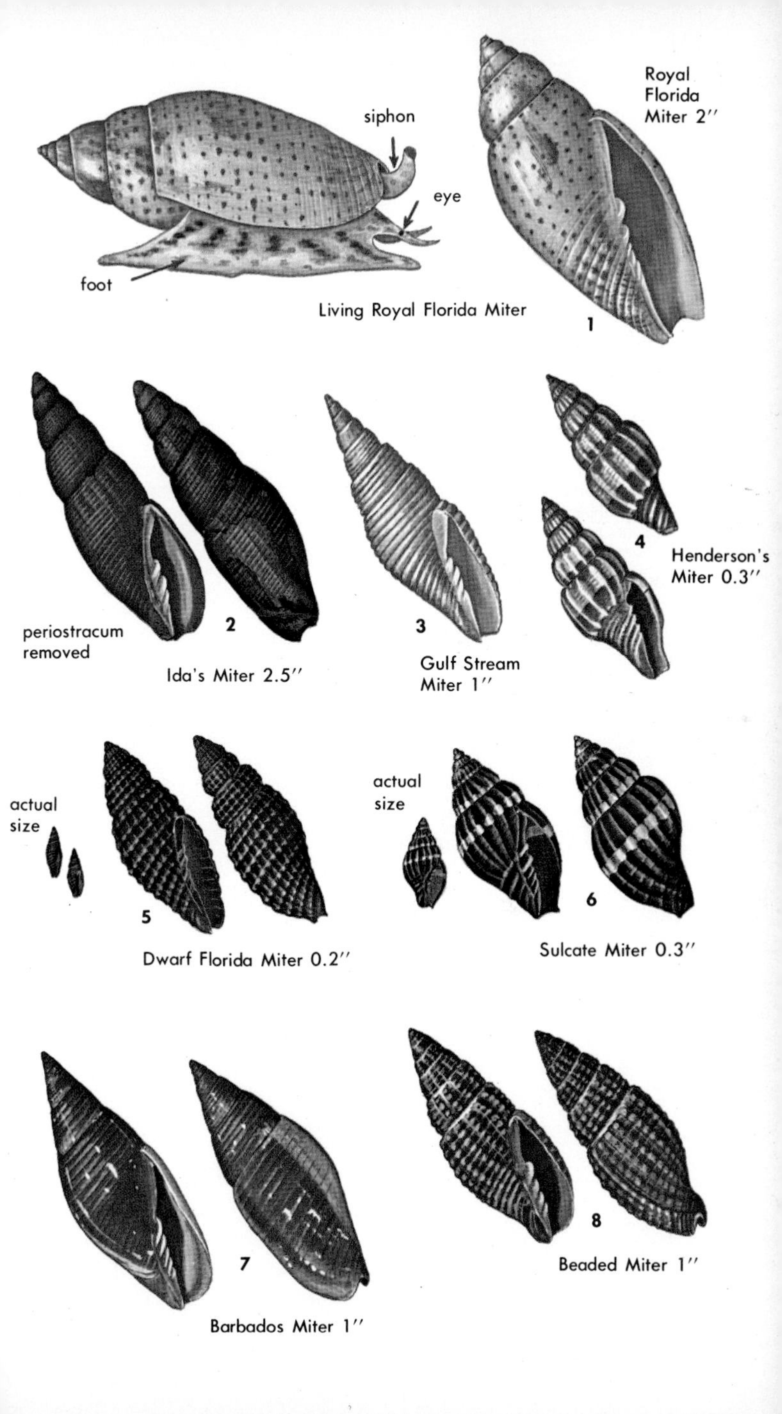
siphon
eye
foot
Living Royal Florida Miter
Royal
Florida
Miter 2″
1
periostracum
removed
2
Ida's Miter 2.5″
3
Gulf Stream
Miter 1″
4
Henderson's
Miter 0.3″
actual
size
5
Dwarf Florida Miter 0.2″
actual
size
6
Sulcate Miter 0.3″
7
Barbados Miter 1″
8
Beaded Miter 1″

VOLUTES

The 200-or-so species of volutes, famous for their exquisite beauty, are considered by many to be the aristocrats of the shells. Most are tropical. Several genera are found in American waters. In most volutes, the nuclear whorl consists of a small pimple-like structure, usually smooth. All volutes are carnivorous, feeding on small marine invertebrate animals.

Scaphella, the major American genus, is characterized by its elongate, spotted shell and lack of an operculum.

In the cold-water, northern *Arctomelon,* there are two prominent columellar folds.

1. **JUNONIA.** *Scaphella junonia* (Shaw). N.C.–Fla. (both coasts) and Texas. Shell rather large, solid, smooth and cream-colored, with numerous spiral rows of roundish spots, purple to dark brown in color, reaching maximum size on the body whorl. Columella has 4 folds. Animal colored much like the shell. Uncommon; at 6–180 ft.

The subspecies *S. j. butleri* Clench (Butler's Volute), with a larger, whiter shell and with smaller spots, has been dredged from deep water (60–200 ft.) off Yucatan. Another yellow-stained form, *S. j. johnstoneae* Clench, has been found off Alabama.

2. **DOHRN'S VOLUTE.** *Scaphella dohrni* (Sowerby). S. Fla. Similar to Junonia, but lighter-shelled and more slender, with 9 or 10 rows of square brown spots. Surface has very fine, incised spiral lines; somewhat coarser toward base. Rare; 400–860 ft.

3. **KIENER'S VOLUTE.** *Scaphella kieneri* Clench. Gulf of Mexico. Shell large, thin, pale creamy-tan, with about 6 rows of square or rectangular, reddish-brown spots. Surface microscopically striate. Aperture white, rarely violet-tinged. Common; at 100–300 ft.

4. **DUBIOUS VOLUTE.** *Scaphella dubia* (Broderip). S. Fla.–Gulf of Mexico. Similar to Dohrn's Volute, but more slender and with only 6 or 7 rows of square brown spots. Shoulder smooth or ribbed. Exterior cream to pinkish. Columella with or without 2 folds. Rare; 500–600 ft.

Gould's Volute, *Scaphella gouldiana* (Dall) (not illus.), N.C.–Fla.; Bahamas, is similar; fatter, short ribs on shoulder, with broad spiral bands of brown (instead of spots). May be subspecies only. Uncommon; 400–3,000 ft.

5. **TORRE'S VOLUTE.** *Bathyaurinia torrei* (Pilsbry). Off E. Fla.–Cuba. Shell long, highly polished; grayish cream with a row of red-brown spots just below the suture and on the base of the whorl. Shoulder has short axial ribs. No columellar folds. Rare; 60–1,400 ft.

6. **STEARNS' VOLUTE.** *Arctomelon stearnsi* (Dall). Alaska. Shell large, strong, semiglossy, bluish white, clouded. Columella has 2 large and 1 small fold. Aperture brownish, glossy. Nucleus bulbous, chalky white. Uncommon; in about 600 ft. of water.

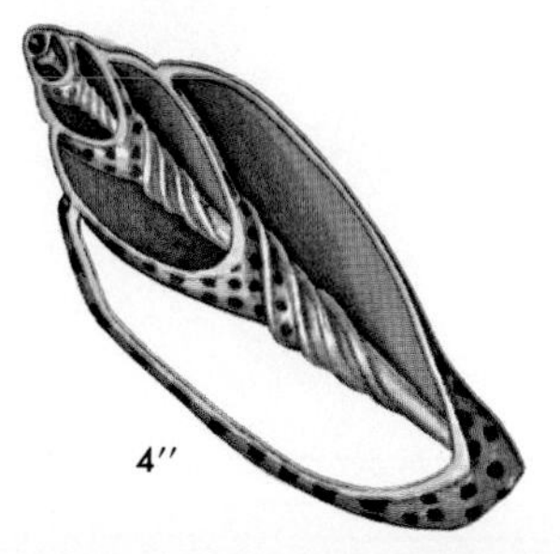

SECTION of Junonia Volute, cut parallel to the axis, showing the columella with its four folds and the relative size and shape of the body whorl through five revolutions of growth.

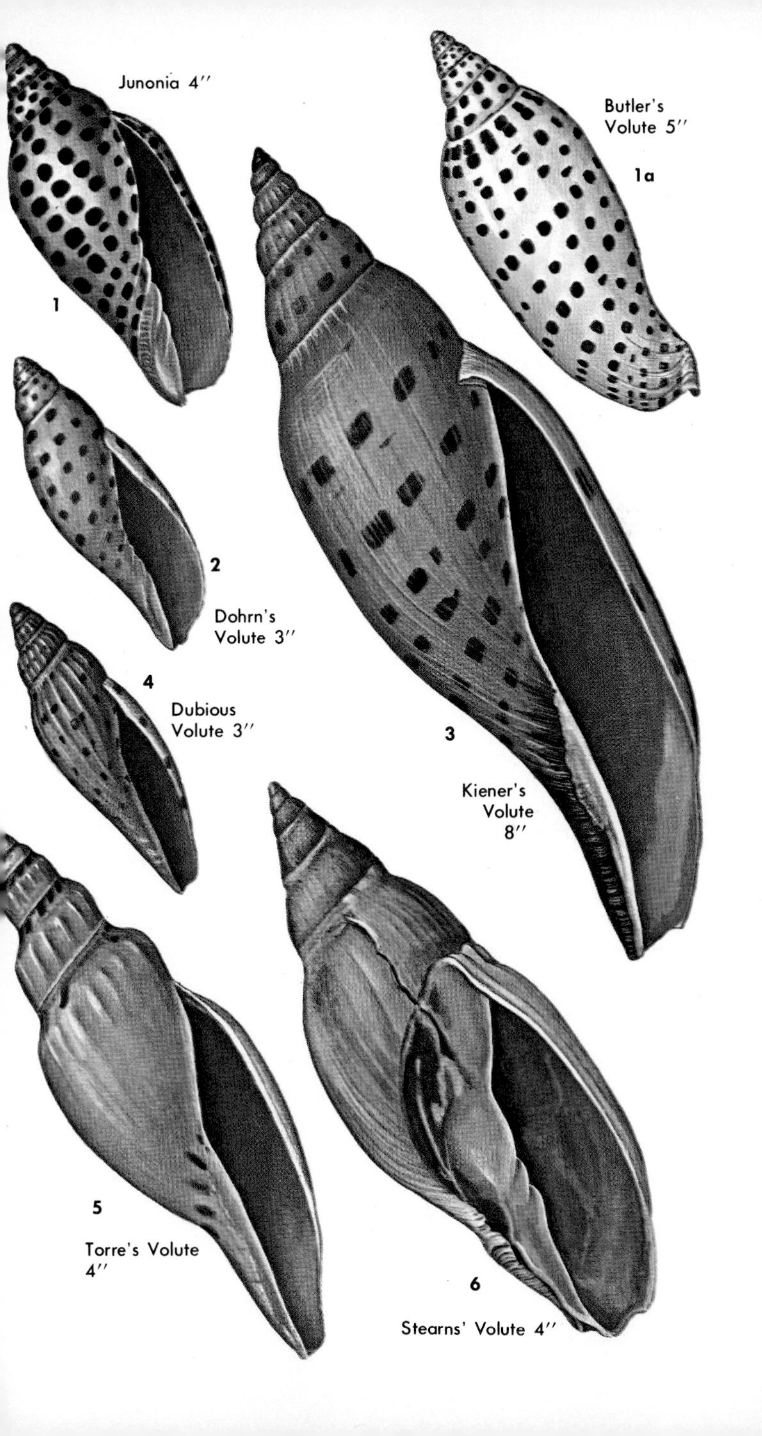
Junonia 4″
1
Butler's
Volute 5″
1a
2
Dohrn's
Volute 3″
3
4
Dubious
Volute 3″
Kiener's
Volute
8″
5
Torre's Volute
4″
6
Stearns' Volute 4″

NUTMEG SHELLS

Nutmegs, family Cancellariidae, have no operculum and employ a mixture of mucus and sand to seal the aperture of the shell. The shells exhibit a crisscross sculpturing on the early whorls. The large aperture ends in a short anterior canal. In the genus *Cancellaria* the spire is short to moderately long, the columellar folds are very strong and well developed, and the outer lip is lined within. In *Trigonostoma* the shell is low-spired and loosely coiled, with a triangular aperture. In *Narona* the axial sculpture predominates and the three columellar teeth vary from weak to moderately strong. *Admete* is characterized by an obliquely cut columella and the three barely visible columellar folds.

Most of the 200 species in this rather small, carnivorous family live in fairly deep water. Many have no radulae; others have weak teeth. There are relatively few living species in North America, but the family blossomed into numerous species in the tropical waters along the Pacific coast of Central America.

1. **COMMON NUTMEG.** *Cancellaria reticulata* (Linné). N.C.–Texas. Shell strong, cream to gray, banded or marked with orange-brown. Numerous, somewhat beaded spiral cords crisscross with the weak axial ribs. The columella has 2 or 3 folds, the upper one very strong and furrowed. Inner lip strongly lined. Albino (fig. 1b) specimens occur on the west coast of Fla. Common; in sand, 6–30 ft. A subspecies, *C. r. adelae* Pilsbry (fig. 1c), from lower Fla. Keys, differs in having a smooth body whorl. Aperture faintly tinged with pink. Uncommon; in shallow water.

2. **PHILIPPI'S NUTMEG.** *Trigonostoma tenerum* (Philippi). S. Fla.–Caribbean. Shell fairly thin, with 4 strongly shouldered, bluntly beaded whorls. Umbilicus deep and funnel-shaped. Uncommon; 12–80 ft. Not uncommonly found in sandy mud along the southwest coast of Florida.

3. **RUGOSE NUTMEG.** *Trigonostoma rugosum* (Lamarck). S. Fla. (rare)—Caribbean (common). Shell large, dirty white to gray, with about 14 strong axial ribs crossed by numerous spiral cords. Columella has 4 or 5 strong folds. In sand, 6–60 ft. Common in isolated areas in the West Indies.

4. **AGASSIZ'S NUTMEG.** *Trigonostoma agassizi* (Dall). Gulf of Mexico–Caribbean. Shell small, rather high-spired, with strong axial ribs crossed by fine spiral threads. Uncommon; offshore, 100–300 ft.

5. **SMITH'S NUTMEG.** *Trigonostoma smithi* (Dall). Off Carolinas. Similar to Agassiz's Nutmeg but a bit darker in color, longer, higher spired, and more loosely coiled; with more spiral threads. Uncommon; 66–300 ft.

6. **COMMON NORTHERN ADMETE.** *Admete couthouyi* (Jay). Arctic Seas–Mass., and to S. Calif. Outer lip thin. Shell stout, coarsely reticulate, dull white. Periostracum thick, gray-brown. Common; 20–900 ft.

7. **COOPER'S NUTMEG.** *Narona cooperi* (Gabb). Central–Baja Calif. Shell large, fairly heavy. Spire turreted. Color brownish cream, with many narrow, brown spiral bands. Aperture light orange. Uncommon; 20–200 ft.

8. **CRAWFORD'S NUTMEG.** *Crawfordia crawfordiana* Dall. S. Calif. Shell fairly large, heavy, white to pale brown, with numerous fine, raised spiral threads. Axial ribs rounded. Uncommon; 100–1,200 ft.

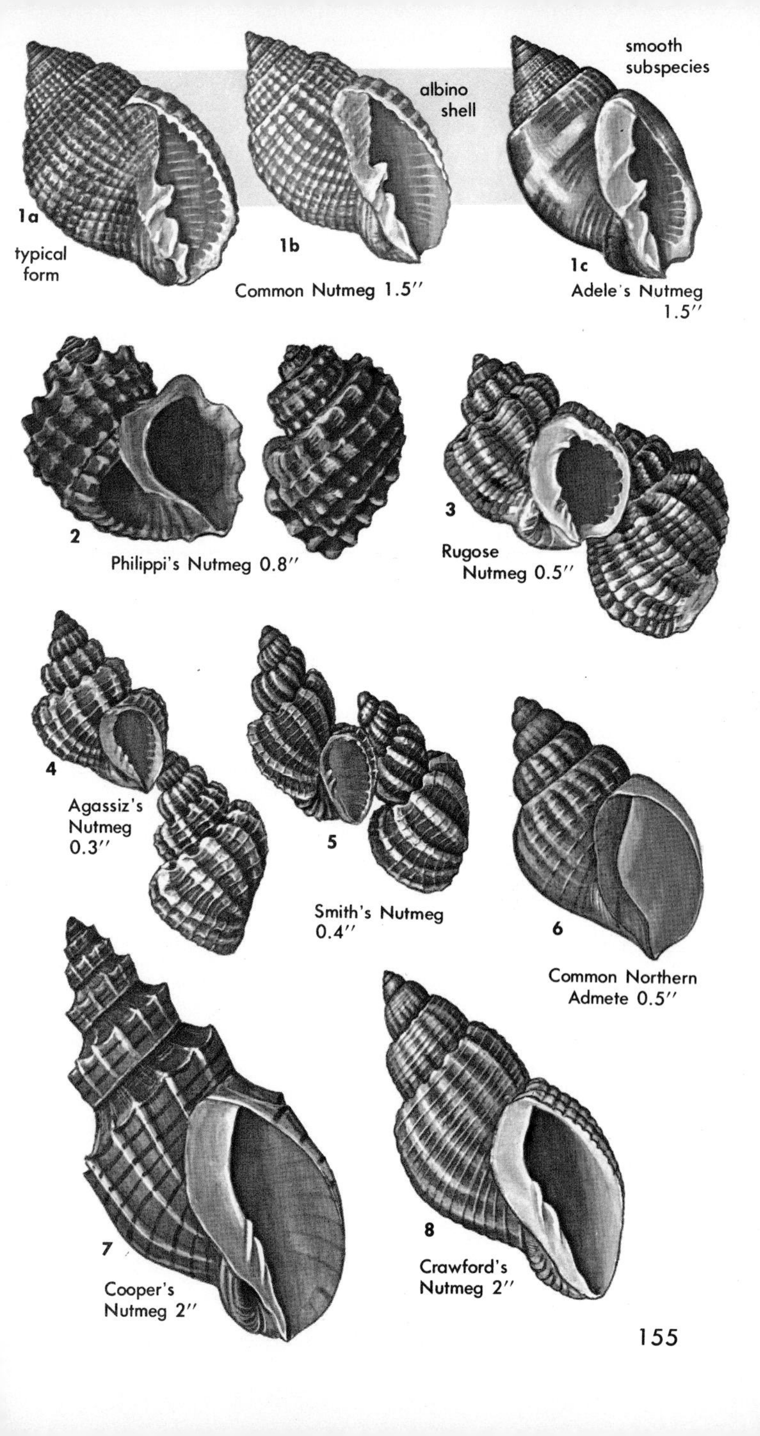
smooth
subspecies
albino
shell
1a
typical
form
1b
Common Nutmeg 1.5″
1c
Adele's Nutmeg
1.5″
2
Philippi's Nutmeg 0.8″
3
Rugose
Nutmeg 0.5″
4
Agassiz's
Nutmeg
0.3″
5
Smith's Nutmeg
0.4″
6
Common Northern
Admete 0.5″
7
Cooper's
Nutmeg 2″
8
Crawford's
Nutmeg 2″

MARGINELLAS

Most marginellas, or margin shells (family Marginellidae), are small but very colorful. The beauty of these shells has made them very popular with collectors. They have no operculum. In the genus *Marginella* the shell is pear-shaped, with a high spire and a thick outer lip. The rather thin shell of *Hyalina* is shaped somewhat like a grain of rice. In *Persicula* the spire is low, sometimes almost sunken, and there are 6 to 8 columellar teeth. The mantle covers the shell while the snail is crawling. Most species are tropical, but the Jersey Marginella ranges into cool waters along eastern United States.

1. **CARMINE MARGINELLA.** *Marginella haematita* Kiener. S. Car.—Caribbean. Shell glossy, bright, deep rose or pink. Outer lip thickened, with about 15 small, rounded teeth on inner edge and 4 strong columellar teeth opposite. Uncommon; in water 100–550 ft. deep.

2. **TAN MARGINELLA.** *Marginella eburneola* Conrad. N.C.–Caribbean. Similar to the Carmine Marginella but narrower and usually yellowish tan. Outer lip thickened, with 7–9 teeth on the inner edge. Spire elongated. Uncommon; 8–1,000 ft. This species was formerly called *M. denticulata* Conrad.

3. **GOLD-LINED MARGINELLA.** *Marginella aureocincta* Stearns. N.C.–Caribbean. White to cream; body whorl has 2 narrow, orange-tan spiral bands. Only 4 small teeth on outer lip. Very common; 2–550 ft.

4. **SNOWFLAKE MARGINELLA.** *Persicula lavalleeana* (Orbigny). S. Fla.–Caribbean. Shell small, pure white. Outer lip thin, curled inward. Columella has 3 oblique folds. Common; 3–250 ft. Formerly called *P. minuta* (Pfeiffer). The similar *P. jewetti* (Carpenter), S. Calif., has a lower spire.

5. **PRINCESS MARGINELLA.** *Persicula catenata* (Montagu). S.E. Fla.–Caribbean. Tan to cream; 7 or 8 spiral rows of long white spots outlined with reddish brown, and 3 subdued brownish spiral bands. Uncommon; 20–500 ft. The similar *P. pulcherrima* (Gaskoin), Fla. Keys–West Indies (not illus.), 0.5 in., has spiral bands of white dots and vertical brown lines.

6. **FLUCTUATING MARGINELLA.** *Persicula fluctuata* (C. B. Adams). Lower Fla. Keys–Caribbean. Longitudinal wavy brown lines form rows of "arrows." Columella has 3 folds. In the Princess Marginella the "arrows" point in opposite direction. Rare; in shallow water.

7. **ORANGE-BANDED MARGINELLA.** *Hyalina avena* (Kiener). N.C.–Caribbean. Shell small, long and narrow. Shell whitish to deep cream, usually with 4 orange-tan spiral bands. Spire short. Aperture wide at base. Columella has 3 or 4 oblique teeth. Animal whitish, with a few bright-orange blotches. Common; intertidal to 30 ft.

8. **CALIFORNIA MARGINELLA.** *Hyalina taeniolata* (Mörch). Central Calif.–Mexico. Shell small, slender, gray to whitish, with 3 or 4 narrow, orange spiral bands. Lower columella white, with 4 folds. Outer lip smooth, incurved. Common; under stones below low tide.

9. **TEARDROP MARGINELLA.** *Bullata ovuliformis* (Orbigny). S.E. United States–Caribbean. Shell very small, glossy, blue-white. Aperture narrow, curved, extending beyond apex. Columella has 3 or 4 oblique folds. Outer lip thickened, with numerous minute teeth within. Common; from low-tide mark to 250 ft. Usually found in sand.

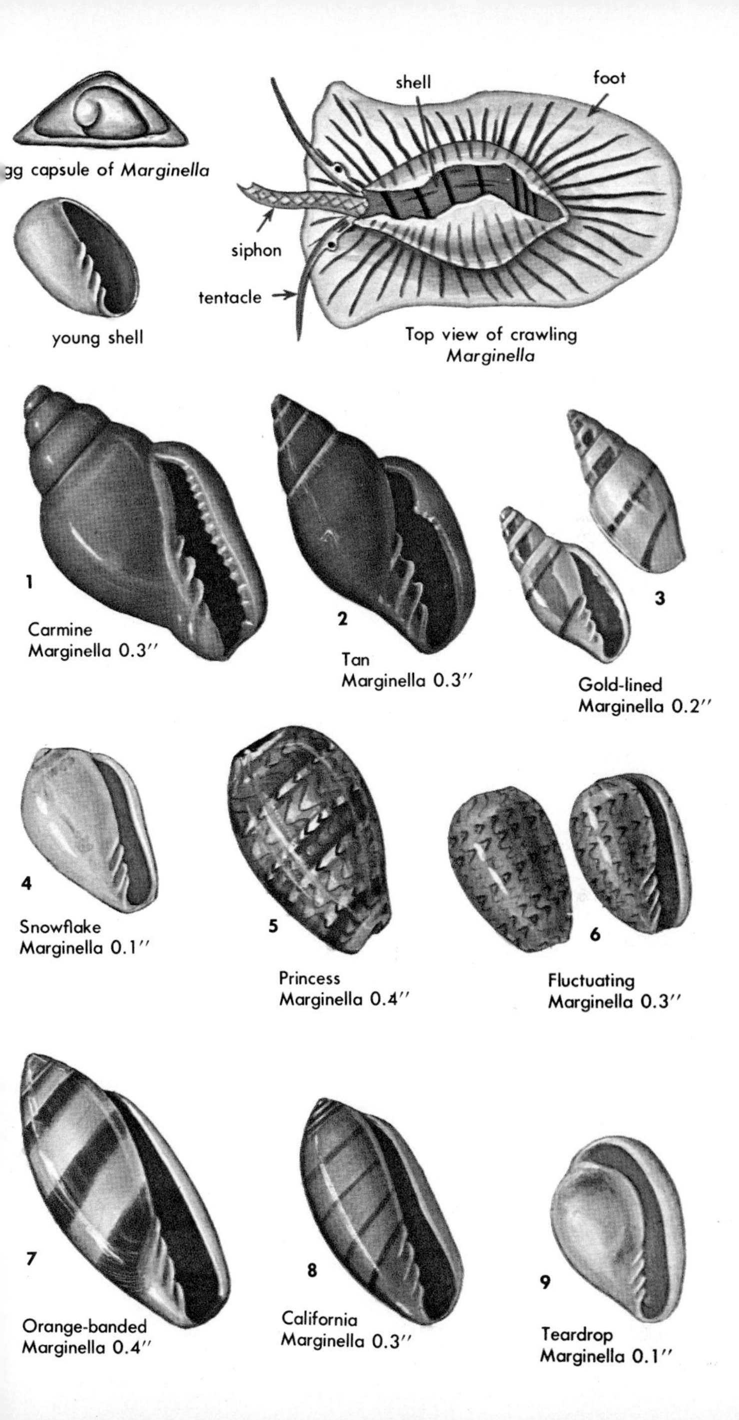
gg capsule of *Marginella*
young shell
shell
foot
siphon
tentacle
Top view of crawling *Marginella*
1
Carmine Marginella 0.3″
2
Tan Marginella 0.3″
3
Gold-lined Marginella 0.2″
4
Snowflake Marginella 0.1″
5
Princess Marginella 0.4″
6
Fluctuating Marginella 0.3″
7
Orange-banded Marginella 0.4″
8
California Marginella 0.3″
9
Teardrop Marginella 0.1″

MARGINELLAS—continued

Marginellid species total over 300, including over 30 in the Caribbean, many of which live in Florida waters. Many large species live near West Africa and Australia. Marginellas are very active and glide over the bottom at a remarkable speed for so small a snail. The foot is large and rather square in front. Animals are often spotted or striped in vivid colors. The genus *Prunum* (below) has a shell with a stubby spire, partly or entirely covered with enamel—an extension of the thickened lip. The genus *Hyalina* contains species with long slender shells. They live under the shelter of dead coral blocks and can burrow rapidly through the sand. Some abundant species are used in shell jewelry.

1. **COMMON ATLANTIC MARGINELLA.** *Prunum apicinum* (Menke). N.C.–West Indies; Gulf states. Shell small, extremely glossy, usually golden to brownish orange, with 2 or 3 subdued brown spots on the outer lip. A gray variety (fig. 1a) is found in the Florida Keys. Animal whitish, with gray-black speckles. Very common in shallow, grassy areas. Rarely, sinistral (fig. 1b).

2. **JERSEY MARGINELLA.** *Prunum roscidum* (Redfield). N.J.–S.C. Shell small, similar to Common Atlantic Marginella but somewhat narrower. Light cream-white, with 3 obscure spiral bands. Outer lip straight and usually marked with brown spots. Columella has 4 folds. Uncommon; intertidally to 30 ft.; on sandy bottom.

3. **QUEEN MARGINELLA.** *Prunum amabile* (Redfield). N.C.–Fla. Shell small, white to tan, with 3 faint orange bands and 2 well-marked, red-brown spots on the outer lip. The shoulder of the last whorl may be heavily suffused with orange. A well-developed callus may partially cover the apex. Uncommon; on sand at 150–750 ft.

4. **PALLID MARGINELLA.** *Hyalina tenuilabra* Tomlin. Fla.–West Indies. Shell small, thin, rather cylindrical, frequently with an almost flat apex. Color milky white. Aperture rather wide. Outer lip thin and sharp. Lower part of columella arched, with 3 or 4 well-developed folds. Not uncommon; 6–100 ft. of water.

5. **VELIE'S MARGINELLA.** *Hyalina veliei* (Pilsbry). S.C.–west coast of Fla. Shell rather thin, but strong; usually pale yellowish, but sometimes grayish white, somewhat translucent. Outer lip thickened, white, strongly pushed-in at the middle. Columella has 3 or 4 folds. Spire quite high. Common; in shallows, sometimes inside dead Pen shells.

6. **ORANGE MARGINELLA.** *Prunum carneum* (Storer). S.E. Fla.–West Indies. Exterior glossy, bright orange, with 2 narrow, whitish bands on the middle of the body whorl and just below the suture. Apex partially covered with an enamel callus. Outer lip thickened, smooth, whitish. Uncommon; on coral reefs from below low tide to 36 ft.

7. **ROYAL MARGINELLA.** *Prunum labiatum* (Kiener). Gulf of Mexico. Shell large; similar to Orange Marginella, but 1.5 in., stouter, and with 3 dark spiral bands. Rare; off Mexico and off S. Texas. This is the largest Gulf of Mexico species, and is brought in by shrimp fishing boats.

8. **WHITE-SPOTTED MARGINELLA.** *Prunum guttatum* (Dillwyn). S.E. Fla.–Caribbean. Exterior cream-white, with 3 faint, pinkish bands on the body whorl. Irregularly spotted with small white flecks. Outer lip smooth, white, with 5 brown spots. Columella has 4 teeth. Not uncommon; found in shallow water, 3–30 ft. deep. usually in soft coral sand.

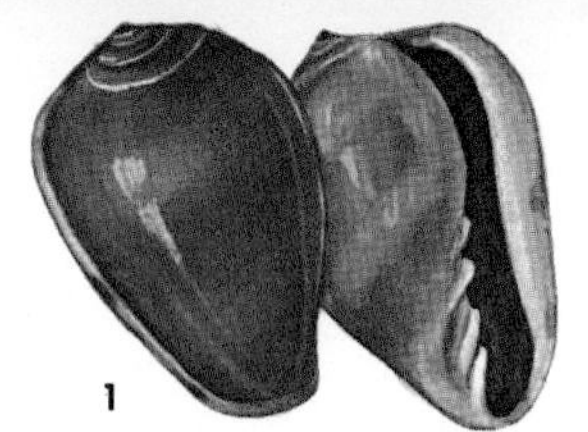
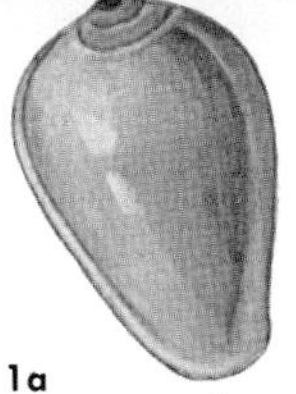
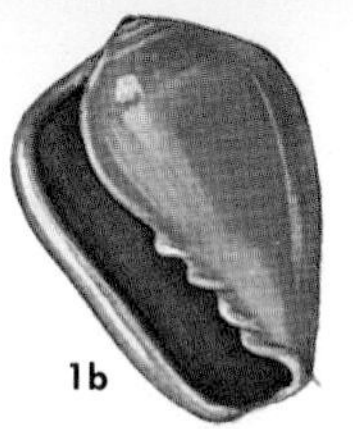

1 typical form

1a gray form

1b sinistral form

Common Atlantic Marginella 0.3″

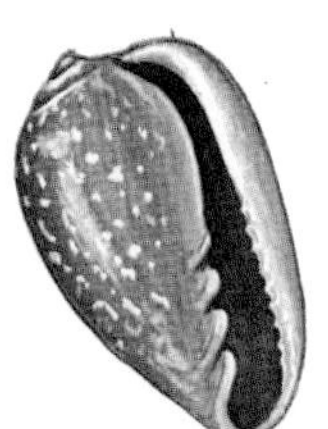
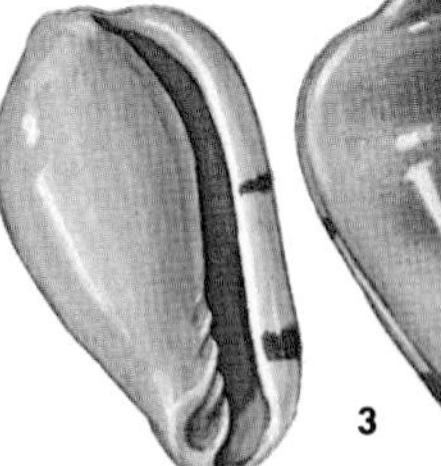
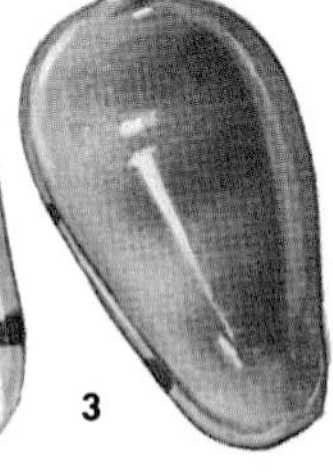

2 Jersey Marginella 0.3″

3 Queen Marginella 0.6″

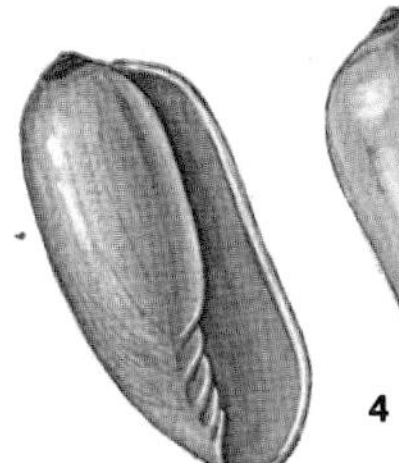
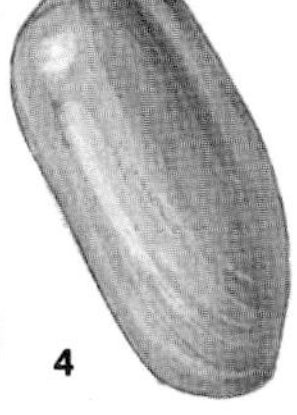
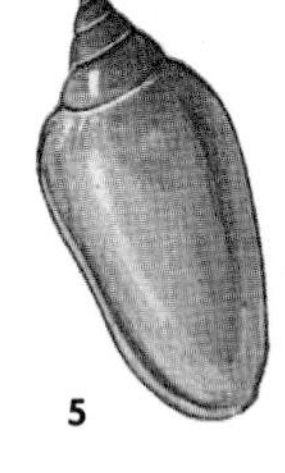

4 Pallid Marginella 0.6″

5 Velie's Marginella 0.4″

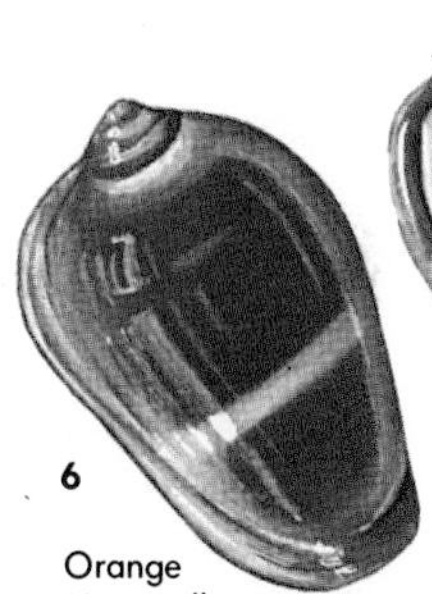
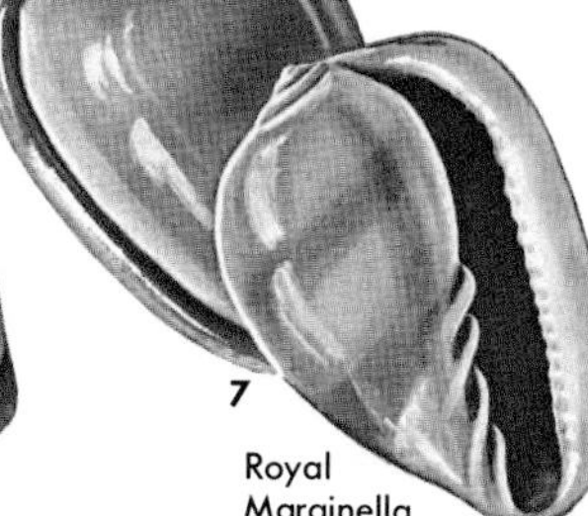

6 Orange Marginella 0.7″

7 Royal Marginella 1″

8 White-spotted Marginella 0.7″

CONE SHELLS

The family Conidae contains about 500 living species, the majority from the warm waters of the Indian and Pacific oceans. However, there are about a dozen species from southeastern United States, but only one from California.

Readily recognized by their general shape, these colorful shells are favorites with collectors and include some of the most highly valued shells in the world. Cone shells are generally solid, with a long aperture and typical conic outline. A horny periostracum, and sometimes marine growths, cover the natural, vivid colors of the shell. The horny operculum is elongated and usually only one fifth the length of the aperture. The foot and siphon are usually brightly colored.

The sexes are separate, and the females lay eggs in the warmer months of the year. The minute eggs, a few to several hundred, are placed in flattened, leathery capsules which are attached in neat rows to the underside of rocks. Hatching occurs in about two weeks. The veliger remains in a free-swimming, pelagic stage for a week or more.

Cones are of considerable interest because of the poison apparatus that all species possess. Only a few in the Indo-Pacific area are known to have inflicted fatal stings on humans, yet it is advisable to handle all large living American specimens with care. The tiny "harpoon," or radular tooth, along with a few drops of a neurotoxic venom from the coiled poison duct, is injected into the victim by the proboscis. Cones normally feed on marine worms and other mollusks, although a few Pacific species capture and eat live fish. The sting may be used as a defensive mechanism, particularly against the mollusk-eating octopuses. Cones feed actively at night, particularly at low tide.

The venom is manufactured in a long, coiled tube lying posterior to the pharynx. A long, muscular bulb at the end of the tube functions to eject the venom. Radular teeth are formed in two rows within the radular sac. As the teeth become formed and when each has been sufficiently impregnated with calcium to give rigidity, each tooth is moved forward into a chamber where it awaits use. Once a tooth is ejected and spent, a new one is moved into the mouth.

1. **ALPHABET CONE.** *Conus spurius* Gmelin. Fla.–Caribbean. Shell cream-white, with a variety of patterns that are usually of spiral rows of irregular, squarish, orange-yellow spots. Spire elevated at the center, giving a concave contour to the top. Sides of whorls smooth; base has cut-in lines. Aperture white; operculum horny, elongated; periostracum thin and translucent tan. Feeds on worms. Rather common on sand from 1–50 ft.

The typical form, *spurius,* with large, dark spots, is commonest in the Caribbean. The Fla. form, *atlanticus* Clench, has smaller, lighter spots. A rare color abnormality, *aureofasciatus* Rehder and Abbott, has spiral bands of yellow or orange-brown. A smaller form with white background and numerous purplish spots is common in Yucatan and Dry Tortugas.

This species and its forms are popular among collectors.

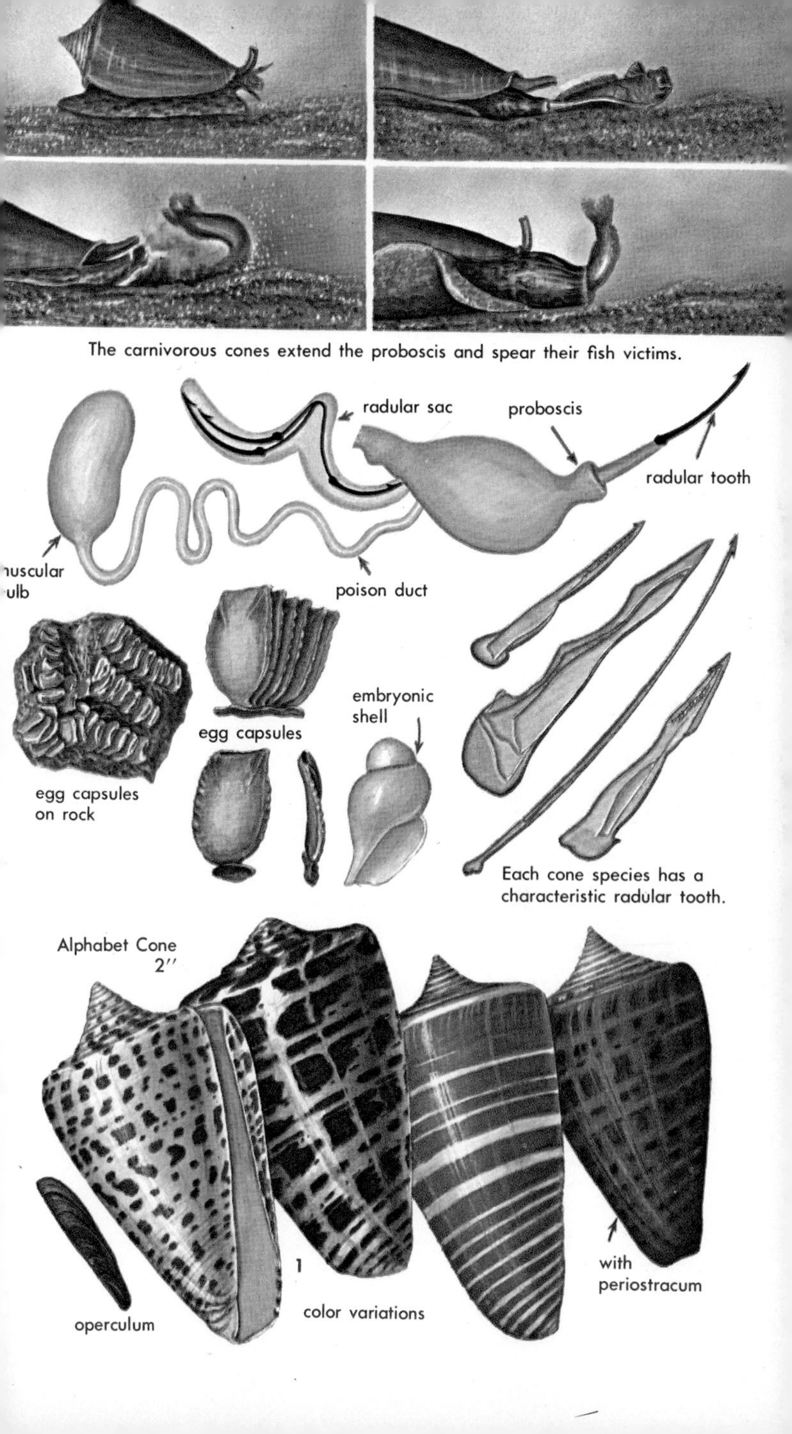

The carnivorous cones extend the proboscis and spear their fish victims.

CONES—continued

The warm waters of Florida and the Caribbean support several dozen species of colorful cones. So do the warm shore waters on the Pacific from Mexico south through Panama to Ecuador, which has many superb species. Some kinds show great variability in color and to a lesser extent in sculpturing. The Florida Cone has several forms considered by some experts to be subspecies, such as the darkly hued Burry's Cone. Although Atlantic cones are armed with harpoon teeth and a venom duct, none is considered harmful to man.

1. **FLORIDA CONE.** *Conus floridanus* Gabb. N.C.–both coasts of Fla. Sides flat and smooth. Tops of the whorls in spire are rather concave. Color variable, often white with wide patches of orange or yellow and a white band on body whorl. Common; on sand; low tide to 40 ft.

C. f. floridensis Sowerby (Floridensis Cone—fig. 1a) is a form with darker coloration, rows of brown dots predominating. Moderately common; on sand. Same range.

C. f. burryae Clench (Burry's Cone—fig. 1b), lower Fla. Keys, has whorls with brown lines instead of dots. Base dark brown. Uncommon; 40–90 ft.

2. **THE SENNOTTS' CONE.** *Conus sennottorum* Rehder and Abbott. Gulf of Mexico. Turnip-shaped, smooth, whitish, with a few rows of small yellowish-brown dots. Uncommon; 30–90 ft.

3. **SOZON'S CONE.** *Conus sozoni* Bartsch. S.C.–Fla. Top of whorls concave. Body whorl has 2 whitish bands. Uncommon; 30–200 ft. Alias *C. delessertii* Récluz.

4. **CROWN CONE.** *Conus regius* Gmelin. S. Fla.–Brazil. Spire has knobs. Color variable, usually mottled, rarely all yellow-brown. Common; on reefs.

5. **MAZE'S CONE.** *Conus mazei* Deshayes. Gulf of Mexico–Caribbean. Shell slender, high-spired. Whorls flat-sided and adorned with rows of yellow squares. May be smooth or with spiral lines (form *mcgintyi* Pilsbry), depending on environment. Rare; 60–400 ft.

6. **JASPER CONE.** *Conus jaspideus* Gmelin. S. Fla.–Caribbean. Shell spindle-shaped; spire rather high. Numerous spiral lines cut around the sides. Color variable. Nuclear whorls may be white, tan, or purple. This is a very common species; found on sandy bottoms in warm shallows.

7. **STEARNS' CONE.** *Conus jaspideus stearnsi* Conrad. N.C.–west coast of Fla. Shell slender, with flat sides; brownish gray to black with white spots, rarely mottled or streaked. Abundant; in weedy, muddy shallows. Some recognize Stearns' Cone as an inshore subspecies of the more southerly Jasper Cone.

8. **GLORY-OF-THE-ATLANTIC CONE.** *Conus granulatus* Linné. Fla. Keys–West Indies. Rounded whorls in spire have spiral scratches. Spiral grooves on sides. Aperture rosy pink. Rare; on seaward reefs. This is a greatly sought species that is occasionally collected on the reefs off the lower and middle Florida Keys. It is more common in the Caribbean.

9. **MOUSE CONE.** *Conus mus* Hwass. S.E. Fla.–West Indies. Shoulder of whorls have weak, white knobs, with brown patches between. Interior of aperture has 2 wide tan bands. Common; found on flats or shore reefs in the intertidal zone.

10. **CALIFORNIAN CONE.** *Conus californicus* Hinds. Central Calif.–Mexico. Shell smooth, light brown, with slightly rounded shoulders. Aperture tinted with brown. Moderately common; among subtidal rocks.

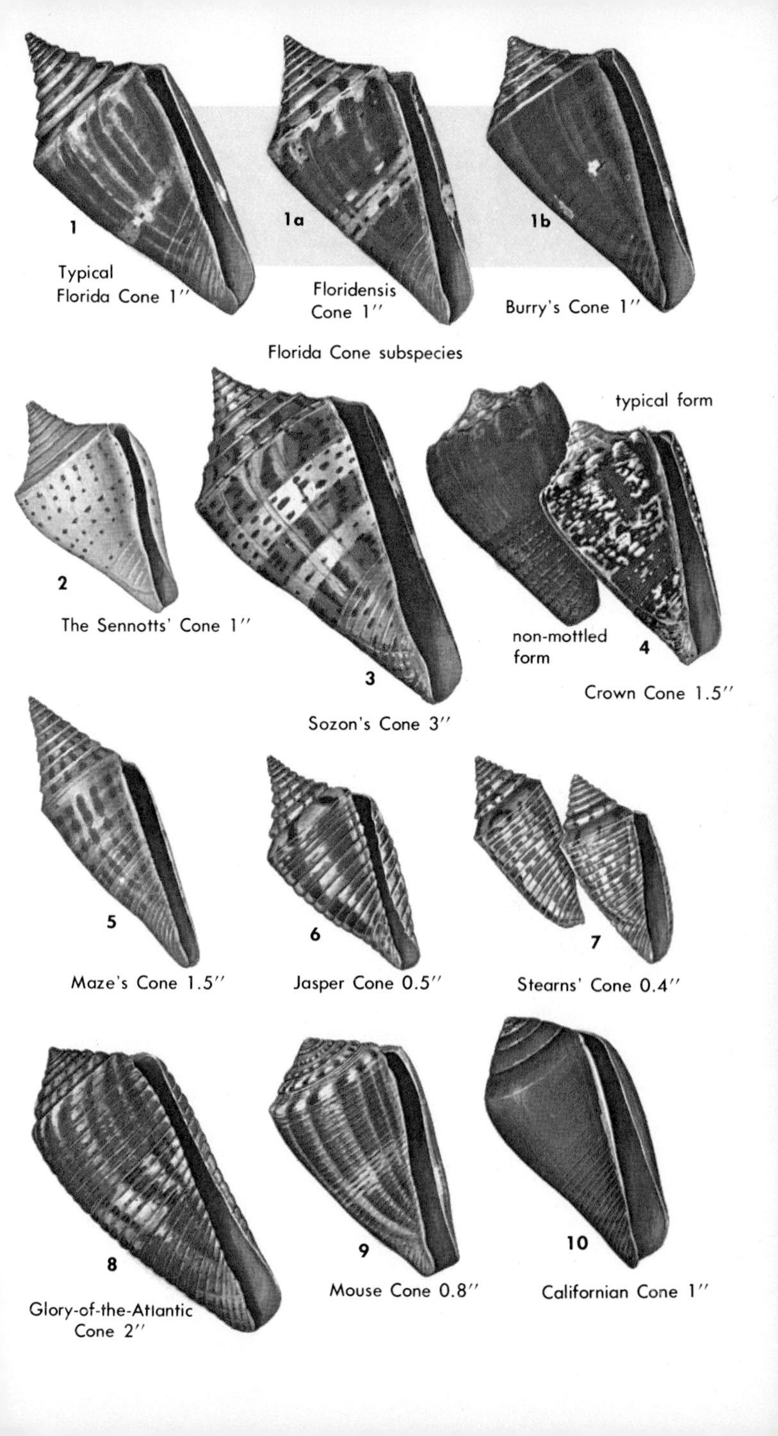
1
1a
1b
Typical
Florida Cone 1″
Floridensis
Cone 1″
Burry's Cone 1″
Florida Cone subspecies
typical form
2
The Sennotts' Cone 1″
non-mottled
form
4
3
Crown Cone 1.5″
Sozon's Cone 3″
5
6
7
Maze's Cone 1.5″
Jasper Cone 0.5″
Stearns' Cone 0.4″
10
9
8
Mouse Cone 0.8″
Californian Cone 1″
Glory-of-the-Atlantic
Cone 2″

AUGER SHELLS

The long, slender auger shells of the family Terebridae are tropical, sand-dwelling carnivores. There are about 300 Recent species, but only about a dozen in North American waters. The columella usually bears a weak spiral ridge. The operculum is clawlike in contrast to the round, spiral operculum of the similarly shaped shells of the family Turritellidae. The augers are closely related to the cones in having a poison gland and a harpoon-like radular tooth. Presumably they feed on marine worms. The eggs and reproductive habits of this family are unknown. Most augers live in shallow waters in sandy mud or coral sand. Intertidal species feed on young clams. Starfish *(Astropecten)* swallow live augers.

1. **FLAME AUGER.** *Terebra taurinum* Lightfoot. S.E. Fla.–Gulf of Mexico; Caribbean. Shell long, heavy, slender; cream-colored with 3 rows of rectangular, orange-brown spots on the body whorl. Upper whorls have faint axial ribbing. Upper half of each whorl slightly swollen. A single incised spiral line separates the halves. On muddy sand, 12–20 ft. This is a choice collector's item but is becoming more difficult to find because of overcollecting.

2. **ATLANTIC AUGER.** *Terebra dislocata* Say. Va.–Caribbean. Shell slender. Each whorl has about 20–25 axial ribs, divided somewhat below the shoulder by a deeply cut spiral line. Small, cut spiral lines show between ribs on lower portions of each whorl. Columella has 2 spiral ridges. Color pale gray with darker bands of purplish gray and reddish brown. Intertidal; common on sand beaches and sand flats. In the northern part of its range the shell is dirty gray.

3. **GRAY ATLANTIC AUGER.** *Terebra cinerea* Born. S.E. Fla.–Brazil. Shell slender; whorls flat-sided, with 45–50 small, slightly raised riblets extending partway down the whorls. Color cream, grayish tan, or bluish brown, commonly with dark-brown markings below the suture. Shell surface has numerous rows of microscopically fine, incised spiral lines. Moderately common on sandy or rocky bottoms at 3–30 ft.

4. **FLORIDA AUGER.** *Terebra floridana* Dall. S.C.–S. Fla. Shell very long, slender, and rather thin. Upper part of each whorl has 2 rows of about 20 somewhat slanting, smooth axial ribs; the upper row is noticeably larger than the lower and is separated from it by a deeply impressed line. The lower part of the body whorl has 3 or 4 raised spiral threads. Columella has a single fold. Rare; 20–200 ft.

5. **SHINY ATLANTIC AUGER.** *Terebra hastata* Gmelin. S.E. Fla.–Caribbean. Shell smooth, glossy, golden tan, with a whitish band below the suture. Nucleus white. Columella white and smooth. Uncommon; 6–60 ft.

6. **SALLÉ'S AUGER.** *Terebra salleana* Deshayes. N. Fla.–Texas. Smaller than Gray Atlantic Auger, and brown or whitish. Surface has microscopic pinpoints. About 30 ribs per whorl. Nucleus tan or purple. Common; intertidal; on sand.

7. **FINE-RIBBED AUGER.** *Terebra protexta* Conrad. S.E. United States. Shell small, long, and slender, with about 15 whorls. Color grayish or gray-brown. Whorls slightly concave, usually with about 22 fine axial ribs crossed by incised spiral lines. Nucleus brown. A form called *T. p. lutescens* E. A. Smith has about 30 axial riblets. *T. p. limatula* Dall, another form, has the ribs and spiral threads about equal-sized. All forms are common; 12–90 ft.

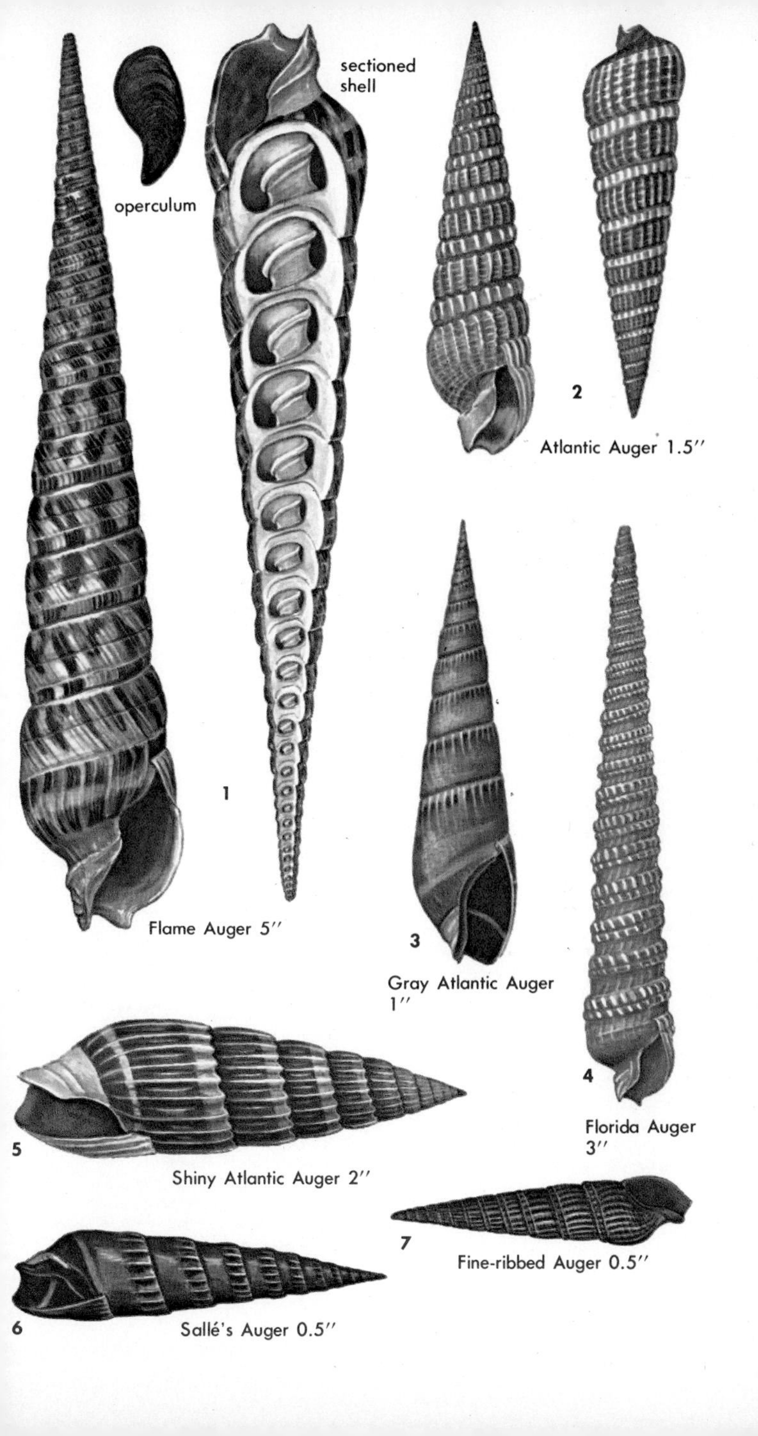
sectioned shell
operculum
1
Flame Auger 5″
2
Atlantic Auger 1.5″
3
Gray Atlantic Auger 1″
4
Florida Auger 3″
5
Shiny Atlantic Auger 2″
6
Sallé's Auger 0.5″
7
Fine-ribbed Auger 0.5″

TURRIDS

Turrids (family Turridae) are a highly evolved group, well established over a hundred million years ago, with hundreds of living and fossil species. At present they are classified according to general shell characters, but when more is known about the biology and life histories, a more satisfactory classification can be expected.

Turrids are characterized by an anal notch, or sinus, the turrid notch, in the posterior margin of the outer lip of the aperture. They are usually spindle-shaped, with a long spire and a straight anterior canal. In some species the radula is like that of the cones, with harpoon-like teeth and a functioning neurotoxic poison gland. Other turrids have radulae of several rows of teeth similar to those found in the volutes. The horny operculum, if present, generally has an apical nucleus; in many species, however, the operculum is degenerate or absent. These characteristics have little bearing on the generic classification of turrids now in current use. Because there are so many species in American waters, identification is difficult except in the case of larger species. Some of turrid genera lack an operculum.

1. **DOLEFUL TURRID.** *Pseudomelatoma moesta* (Carpenter). S. Calif.–Mexico. Shell spindle-shaped, with a beaded band just below the suture, with 9 or 10 slightly curved axial ribs, and with numerous faint, incised spiral lines. In some specimens the axial ribs are almost lacking. Canal short. Moderately common; found under stones in the intertidal zone.

2. **DELICATE GIANT TURRID.** *Polystira tellea* (Dall). Off S.E. Fla. Shell grayish white, spindle-shaped, with numerous fine, slightly raised spiral cords. Sutures indistinct. Periostracum light brown. Not uncommon; in water 20–200 ft. deep. Do not confuse with deep-water *Fusinus* spindle shells (p. 147), which lack the turrid notch in the outer lip.

3. **WHITE GIANT TURRID.** *Polystira albida* (Perry). S. Fla.–Gulf of Mexico–Caribbean. Shell pure white and spindle-shaped; sculptured with numerous sharply keeled spiral cords and fine spiral threads. Canal long and open. Periostracum light brown. Operculum clawlike, brown. Not uncommon; 30–300 ft.; often taken by shrimp boats.

4. **PERVERSE TURRID.** *Antiplanes voyi* (Gabb). Alaska–S. Calif. Shell smooth, whitish, left-handed, with a greenish brown, thin periostracum. Operculum leaf-shaped. Protoconch smooth, globular, and with only 2 whorls. Growth lines exceedingly fine. Uncommon; 30–300 ft.

5. **FILOSE TURRID.** *Mitromorpha filosa* (Carpenter). Central Calif.–Mexico. Shell small, light orange-brown to black, with numerous distinct spiral cords which may be slightly beaded. Strongly resembles a *Mitra* (p. 151) but lacks the columellar ridges and has the typical turrid notch of this family on the outer lip. Uncommon; 20–100 ft.

6. **CARPENTER'S TURRID.** *Megasurcula carpenterianus* (Gabb). Central Calif.–Baja Calif. Shell spindle-shaped; spire conical, golden tan, with numerous narrow, reddish brown bands. Turrid notch weak. Shoulder of whorls smooth or slightly angular. Surface has numerous fine growth lines. Uncommon; 30–300 ft., on mud bottom. Enters lobster pots. Several varieties have been named. The shouldered form with small nodes is *M. c. tryoniana* (Gabb).

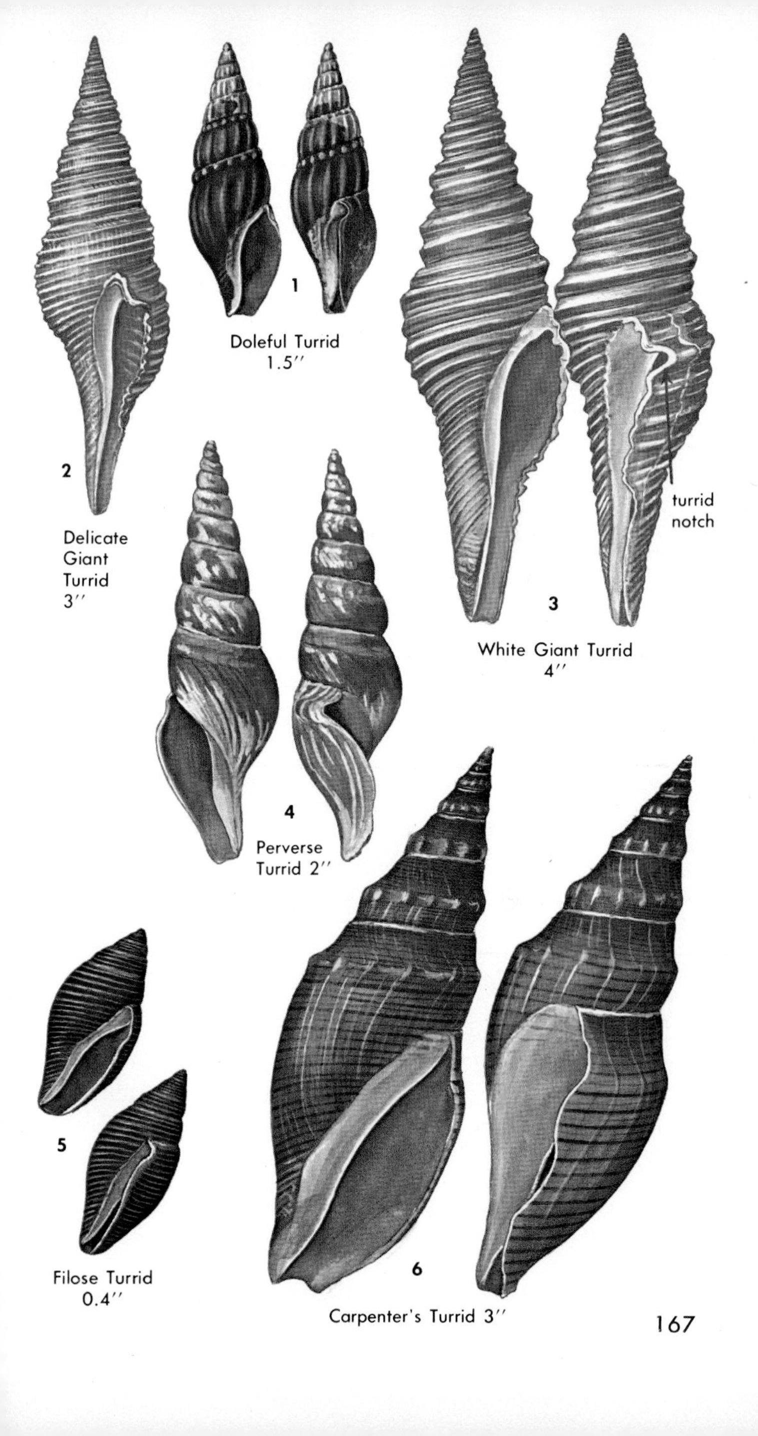
1
Doleful Turrid
1.5″
2
Delicate
Giant
Turrid
3″
turrid
notch
3
White Giant Turrid
4″
4
Perverse
Turrid 2″
5
Filose Turrid
0.4″
6
Carpenter's Turrid 3″

1. **COMMON STAR TURRID.** *Cochlespira radiata* Dall. S. Fla.–Gulf of Mexico–Caribbean. Shell is small, delicate, translucent. Shoulder sharply keeled, with numerous small, somewhat triangular spines. Common; 180–1,000 ft. Alias *Ancistrosyrinx*.
  Elegant Star Turrid. *C. elegans* Dall (not illus.). Off Key West. Shell is similar, 1–2 in., with many more blunt spines. Rare; 1,500 ft.

2. **VOLUTE TURRID.** *Daphnella lymneiformis* (Kiener). N.C.–Caribbean. Shell is small, with about 8 whorls. Nuclear whorls smooth; following whorls have strong axial ribs; final whorls have only fine spiral threads. Growth lines very fine. Aperture cream-colored, with yellowish brown, wavy markings. Uncommon; 12–150 ft.

3. **SANIBEL DRILLIA.** *Crassispira sanibelensis* Bartsch and Rehder. Lower west coast of Fla. Spire elongate, anterior canal short. Has about 9 long, fairly wide axial ribs crossed by numerous raised spiral threads. Uncommon; 6–20 ft., in weedy areas.

4. **OYSTER DRILLIA.** *Crassispira ostrearum* (Stearns). N.C.–Fla. Shell strong, long, with about 20 weakly beaded axial ribs crossed by spiral threads. A smooth spiral cord follows the suture. Common below low tide to 30 ft.

5. **WHITE-KNOBBED DRILLIA.** *Monilispira leucocyma* (Dall). S. Fla.–Texas–West Indies. Shell light to dark gray-brown, with smooth, broad spiral cord below suture and 2 white-beaded cords on each whorl. Common; on grass below low tide to 30 ft.

6. **WHITE-BANDED DRILLIA.** *Monilispira albinodata* (Reeve). S.E. Fla.–West Indies. Shell small, dark brown-black, with white, beaded bands of small nodules. Body whorl has 2 or 3 white spiral bands. Common; under rocks, intertidal, to 12 ft.

7. **BARTLETT'S MANGELIA.** *Mangelia bartletti* (Dall). S. Fla.–Caribbean. Body whorl is ⅔ of shell length; has many rounded axial ribs crossed by fine spiral threads. White to tan, brown at suture. Common; in sand, 20–300 ft.

8. **PLICOSE MANGELIA.** *Mangelia plicosa* (C. B. Adams). Mass. Caribbean. Shell dark reddish brown, with 11 or 12 strong axial ribs crossed by strong spiral cords, giving a crisscrossed pattern. Common 6–30 ft.

9. **STELLATE MANGELIA.** *Stellatoma stellata* (Stearns). Fla. west coast and Keys. Shell high-spired, turreted, with about 11 axial ribs, with fine spiral striae between. Color yellowish to brown. Common; in sand and weeds, 1–18 ft.

10. **THEA DRILLIA.** *Cerodrillia thea* (Dall). West coast of Fla. Shell strong, rather glossy, brown, with short, slanting, cream-colored ribs. Outer lip has a prominent, deep, U-shaped sinus. Uncommon; below low-tide mark in shallow bays.
  Perry's Drillia, *Cerodrillia perryae* Bartsch and Rehder (not illus.), from the West Coast of Florida is similar, ½ inch, with a golden-brown band around the periphery and with 8 or 9 axial ribs per whorl. Common; in shallows.

11. **HARPLIKE LORA.** *Lora harpularia* (Couthouy). Labrador–R.I. Shell stout, elongate, with 7 or 8 buff or flesh-colored whorls. Angled above periphery, with about 16 rounded axial ribs crossed by fine spiral threads. Sinus small. Common; on rocky bottoms, 60–90 ft.

12. **TURRICULATE LORA.** *Lora scalaris* (Möller). Greenland–Mass. Shell thin but broader and larger than the Harp-like Lora, with many sharp axial ribs crossed by faint spiral threads. Common; 30–1,200 ft., usually on rocky bottoms.

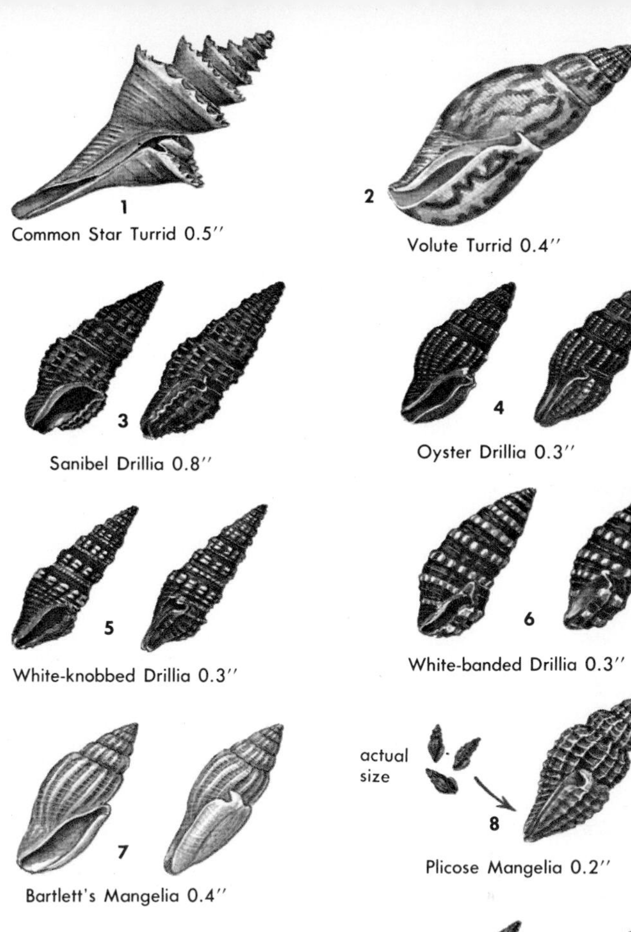

1 Common Star Turrid 0.5″

2 Volute Turrid 0.4″

3 Sanibel Drillia 0.8″

4 Oyster Drillia 0.3″

5 White-knobbed Drillia 0.3″

6 White-banded Drillia 0.3″

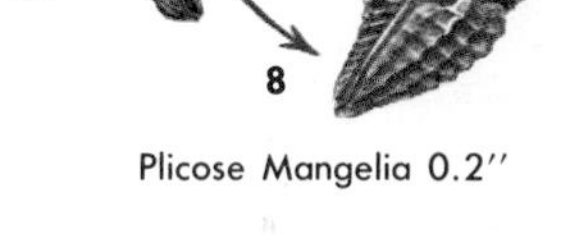

actual size

8 Plicose Mangelia 0.2″

7 Bartlett's Mangelia 0.4″

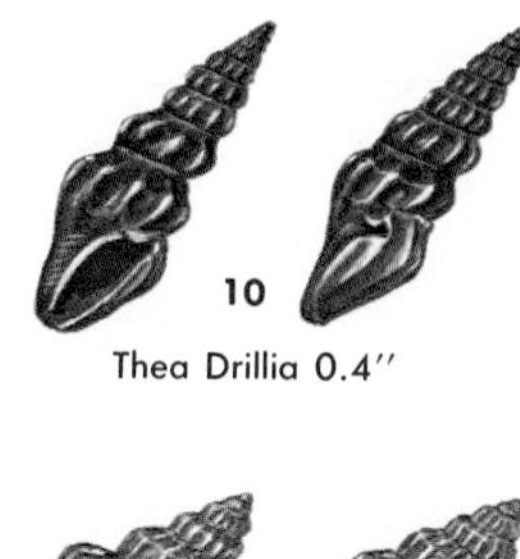

9 Stellate Mangelia 0.4″

10 Thea Drillia 0.4″

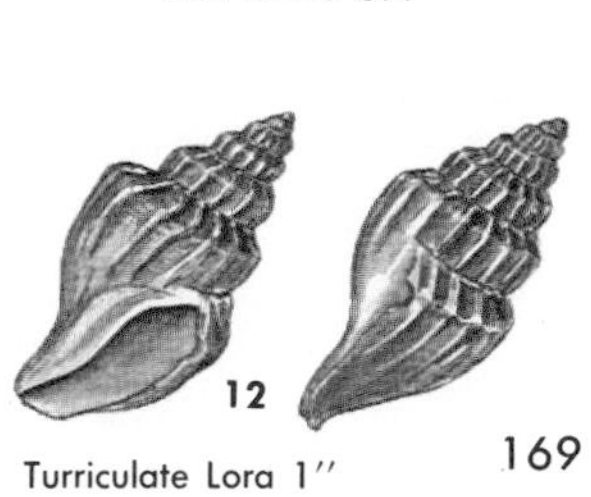

11 Harplike Lora 0.5″

12 Turriculate Lora 1″

BUBBLE SHELLS

The subclass Opisthobranchia contains a wide assemblage of marine snails. A number of related families and genera are included in the broad term "bubble shells." All have rather large bodies that more or less envelop the shell, which in many species is fragile and even rudimentary. Bubble shells are generally carnivorous, swallowing their prey alive, after which it is crushed by strong, limy plates lining the gizzard. Some genera, notably *Haminoea,* occasionally vary their diet by feeding on algae. The bubble shells are hermaphroditic, laying jellylike ribbons in which are embedded many thousands of small eggs. The families Bullidae, Hydatinidae, and Scaphandridae all belong to the same order.

1. **CALIFORNIAN BUBBLE.** *Bulla gouldiana* Pilsbry. S. Calif. Shell large, fragile, smooth and rounded, grayish brown with dark, angular streaks bordered with white or cream. Periostracum dark brown with fine wrinkles. Spire depressed. Commonly found at night; on mud flats at low tide.

2. **WEST INDIAN BUBBLE.** *Bulla occidentalis* A. Adams. N.C.–S.E. Fla. and Caribbean. Shell smooth, varying from fragile to quite strong, and from cylindrical to fairly swollen. Color whitish with irregular brownish markings. Common; on grassy mud flats.

3. **WATSON'S CANOE BUBBLE.** *Scaphander watsoni* Dall. N.C.–Fla. Keys. Shell large, thin, but strong. Upper third narrow, with the edge of the outer lip extending beyond the apex. Lower shell swollen, with a broad white aperture. Surface marked with brown spiral lines. Uncommon; 100–2,000 ft.

4. **GIANT CANOE BUBBLE.** *Scaphander punctostriatus* (Mighels). Arctic Sea–Caribbean. Shell quite large, thin, moderately strong, white. Apex constricted posteriorly; shell widely rounded at base. Surface smooth with rows of microscopic dots. Periostracum straw-colored. Moderately common; 60–6,000 ft.

5. **GOULD'S PAPER BUBBLE.** *Haminoea vesicula* (Gould). Alaska–Mexico. Shell small, very fragile, pale greenish yellow, with a tiny apical perforation. Periostracum rusty brown or yellowish orange. Common; intertidal; in bays.

6. **EASTERN PAPER BUBBLE.** *Haminoea solitaria* (Say). Mass.–N.C. Shell thin, globose, spirally grooved, whitish to amber color. Outer lip thickened. Aperture arises to right of perforation. Common; intertidally to 30 ft.

7. **SOWERBY'S PAPER BUBBLE.** *Haminoea virescens* (Sowerby). Puget Sound–Mexico. Shell small, greenish, fragile, with large, wide aperture. No apical perforation. Outer lip high, winged posteriorly. Uncommon; on rocks.

8. **MINIATURE MELO.** *Micromelo undata* (Bruguière). Fla. Keys–Caribbean. Shell ovate, moderately thin, and fragile. White to cream, with 3 red spiral and many branching, wavy axial lines. Uncommon; at low tide under green, feathery algae.

9. **BROWN-LINED PAPER BUBBLE.** *Hydatina vesicaria* (Lightfoot). S. Fla.–Caribbean. Shell large, globose, moderately thin, and fragile. Color white to tan, with numerous close-set, somewhat wavy, brown spiral lines. Animal very large and colorful, rust-red with a narrow azure border. Foot large and broad. Moderately common locally; burrowing in sand in shallow water.

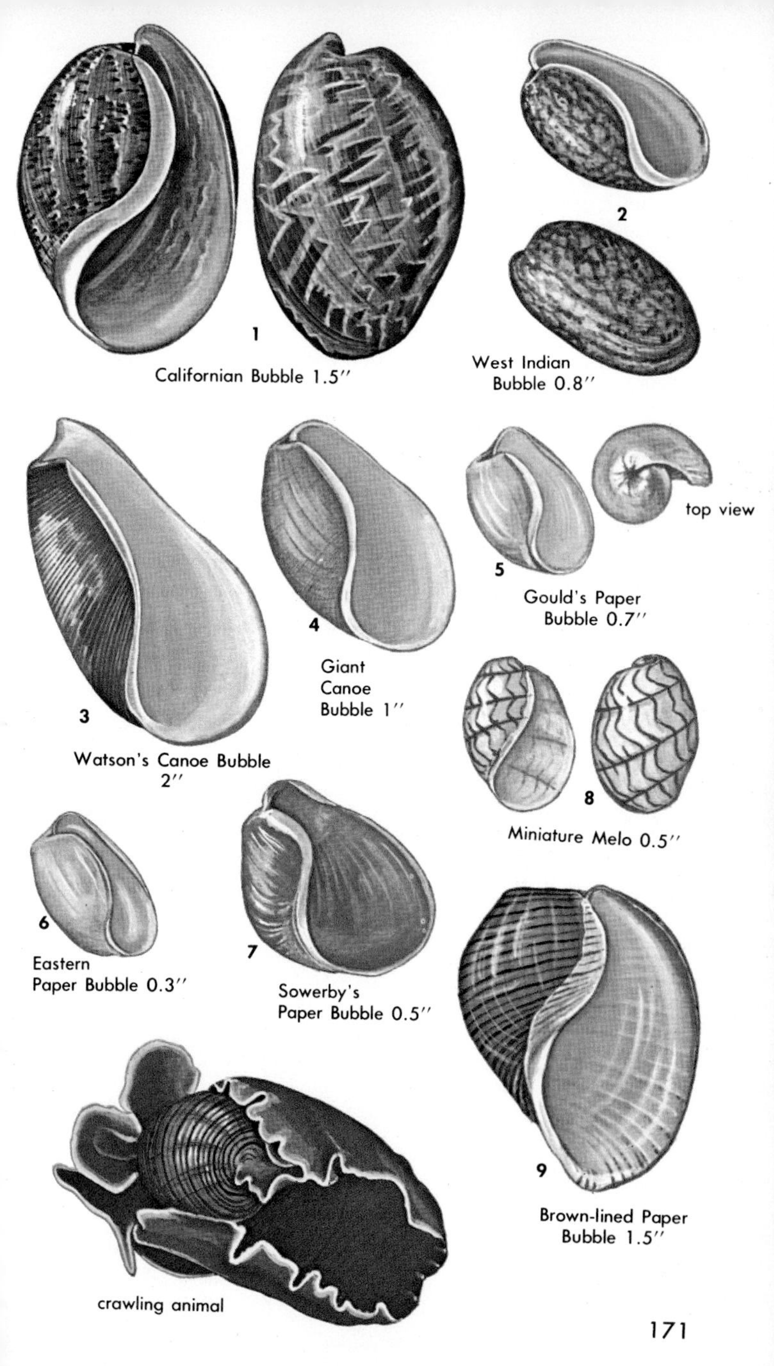
1
Californian Bubble 1.5″
2
West Indian
Bubble 0.8″
3
Watson's Canoe Bubble
2″
4
Giant
Canoe
Bubble 1″
5
top view
Gould's Paper
Bubble 0.7″
8
Miniature Melo 0.5″
6
Eastern
Paper Bubble 0.3″
7
Sowerby's
Paper Bubble 0.5″
9
Brown-lined Paper
Bubble 1.5″
crawling animal

SEA HARES

The tectibranch gastropods have evolved into several different stocks, some characterized by a reduction in, or even a loss of, the shell. There is a plumelike gill on the right side of the body, which is usually protected by a flap of skin. All species are hermaphroditic, and some are capable of self-fertilization.

One genus of sea slugs, *Berthelinia,* has a modified shell of two valves, resembling a pelecypod. These small green gastropods have been recently discovered in the Caribbean and the Gulf of California. In all cases they live on a green alga, *Caulerpa*. Tentacles, a mouth with a radula, and a single muscle scar separate these gastropods from the pelecypods.

The *Navanax* (family Aglajidae) from California lack a shell in the adult stage. They live among beds of eelgrass and feed on Paper Bubbles. These slugs breed most of the year. Each may lay several hundred thousand eggs inside light yellow, stringy coils.

Sea hares of the family Aplysidae are found in intertidal grass beds. Most live for only one year. The flat, horny, semicalcareous shell is internal and protected by two flaplike extensions of the foot, which help the large sea slug to swim slowly through the water. The color of *Aplysia* is affected by the kinds of algae upon which they feed. They lay an extraordinary number of eggs, estimated at over 85 million. Animals give off harmless, deep-purple fluid if disturbed.

1. **GREEN BERTHELINIA.** *Berthelinia chloris* (Dall). S. Calif.(?)–Baja Calif. Shell valves small, fragile, greenish, the left one having a coiled apex. Uncommon on *Caulerpa* seaweed.

Caribbean Berthelinia. *B. caribbea* Edmunds (not illus.), Fla. Keys(?)–Caribbean, is 0.2 in. long; color is greenish with yellowish rays. Uncommon on *C. verticillata* weed.

2. **CALIFORNIAN NAVANAX.** *Navanax inermis* (Cooper). S. Calif.–Mexico. Shell has velvety dark-brown background with either white or yellow spots. Overall is a sheen of blue-violet seen at certain angles only. Edge of foot and head has a line of orange-yellow. Common; on mud in grass flats and in tidepools.

3. **SPOTTED SEA HARE.** *Aplysia dactylomela* (Rang). S. Fla.–Caribbean. Animal large, pale yellow to yellowish green, marked by irregular, violet-black circles. Common; in shallow grassy areas.

The Sooty Sea Hare. *A. floridensis* (Pilsbry) (not illus.). Fla.–Texas. Animal purple-black with lighter irregular spots.

Willcox's Sea Hare. *A. willcoxi* (Heilprin) (not illus.). Cape Cod–Texas. Skin usually dark brownish and has an open tube above the interior shell.

The Californian Sea Hare. *A. californica* Cooper (not illus.). Common; about 5 in. long, shell blotched olive-green.

4. **RAGGED SEA HARE.** *Bursatella leachi plei* Rang. West coast Fla.–Caribbean. Animal large and plump, sometimes soft and flabby. Its surface is covered with numerous ragged filaments. Color greenish gray to olive-green, usually with many white flecks. Adults have no shell. Eggs are laid in long, jellylike strands, sometimes in tangled clumps. Commonly found in grassy areas of shallow, protected waters during the spring and early summer. Gives off a deep-purple ink.

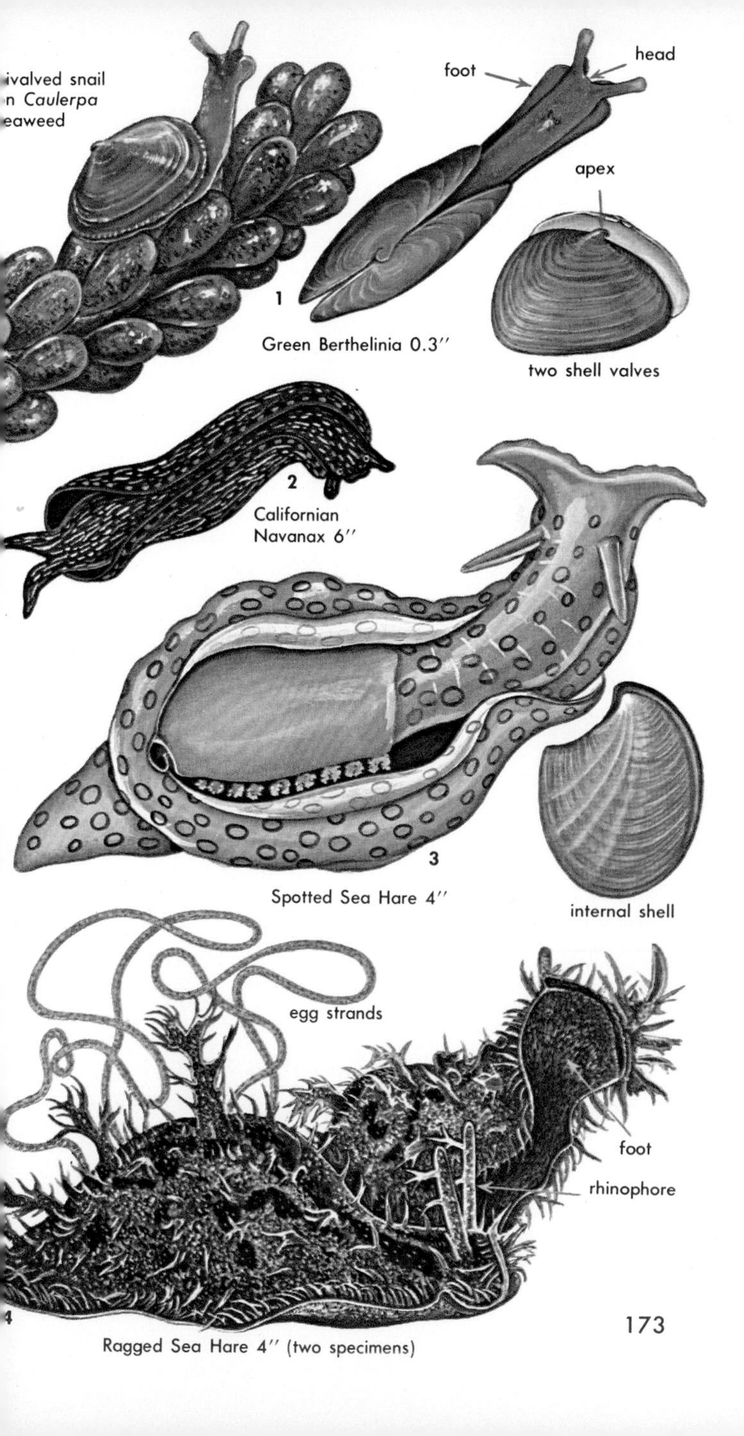

Green Berthelinia 0.3″

Californian Navanax 6″

Spotted Sea Hare 4″

Ragged Sea Hare 4″ (two specimens)

PYRAM SHELLS

The many hundreds of species of pyram shells, family Pyramidellidae, are in the subclass Opisthobranchia (p. 8). The genera and species are difficult to identify. They resemble such prosobranchs as the ceriths in having a high-spired shell, a horny operculum, and a jelly mass covering the eggs. However, the radulae are absent. Some are characterized by a single columellar ridge. Many species are part-time parasites, feeding on the juices of worms, starfish, bivalves, or *Crepidula* snails. The long, tubular proboscis pierces the host's tissues and sucks in the blood. *Odostomia* pyrams live and feed on bivalves and periwinkles. They are hermaphrodites. Sperm are placed in a balloon-like pouch with a hook at one end. The egg jelly masses are usually laid on the host's shells during the warmer months of the year and usually hatch into free-swimming veligers.

1. **GIANT ATLANTIC PYRAM.** *Pyramidella dolabrata* Lamarck. Fla. Keys—Caribbean. Shell quite large, solid, high-spired; color glossy and opaque cream-white with 3 or 4 brown spiral lines on the body whorl. Columella large, with 2 or 3 strong folds. Outer lip sometimes toothed. Uncommon; in sand, 3–90 ft.

2. **CRENATE PYRAM.** *Pyramidella crenulata* (Holmes). S.E. United States. Shell thin, glossy but frequently eroded. Whorls flat-sided with deeply channeled sutures. Color pale tan, clouded with white, and sometimes with 2 or 3 brown spiral lines on the body whorl. Common; in shallow bays on sand, mud, or grass.

3. **ADAMS' PYRAM.** *Pyramidella adamsi* Carpenter. S. Calif.–Mexico. Shell smooth and polished. Whorls moderately rounded. Color white to dark brown, sometimes spotted and banded. Sutures channeled. Columella has strongly developed folds. Aperture oval; outer lip thin. Uncommon; 6–300 ft., on sand.

4. **BROWN PYRAM.** *Sayella fusca* (C. B. Adams). Canada–Caribbean. Shell small, rather long, light brown. Columella has 3 folds. Common; in grassy areas, 6–36 ft.

5. **DALL'S TURBONILLE.** *Turbonilla dalli* Bush. N.C.–Gulf of Mexico. Shell long and slender, thin, glassy, with about 18 sturdy, rounded axial ribs on the body whorl that terminate abruptly at the periphery. There are no spiral striations. Very common; in shallow bays.

Conrad's Turbonille. *T. conradi* Bush (not illus.). West coast of Fla. Shell very similar, but surface is waxy, not glassy, with faint orange-brown spiral band near midpoint of each whorl. Upper whorls bear narrow ribs between which 5 or 6 incised spiral striations show clearly. Base rounded, smooth, spirally striate. Common; in shallow bays, in sandy mud.

6. **INTERRUPTED TURBONILLE.** *Turbonilla interrupta* (Totten). Maine–Caribbean. Shell slender, with 11–14 incised spiral lines between 20–24 smooth axial ribs. Base rounded, with spiral lines. Common; 1–20 ft.

7. **HALF-SMOOTH ODOSTOME.** *Odostomia seminuda* C. B. Adams. E. Canada–West Indies. Shell whitish, with upper whorls and upper body whorl crisscrossed. Base has spiral lines. Common; 3–30 ft.

8. **IMPRESSED ODOSTOME.** *Odostomia impressa* (Say). Mass.–Gulf of Mexico. Shell is milky white, conical, long, with 6 or 7 slightly flattened whorls. Upper whorls have 3 or 4 strongly raised, smooth, spiral ribs, with microscopic axial striations between. Common; in shallow bays, 3-60 ft.

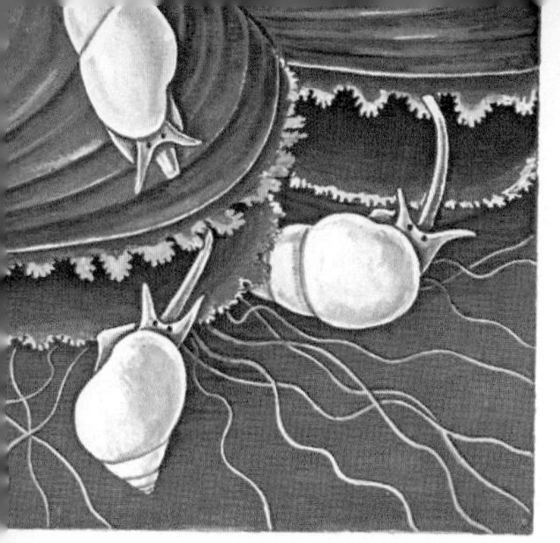

Odostomia feeding on Blue Mussels

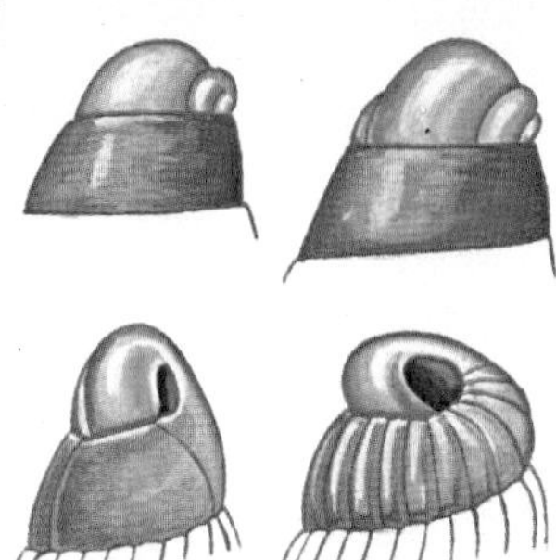

Various types of nuclear whorls in pyramidellids (greatly magnified)

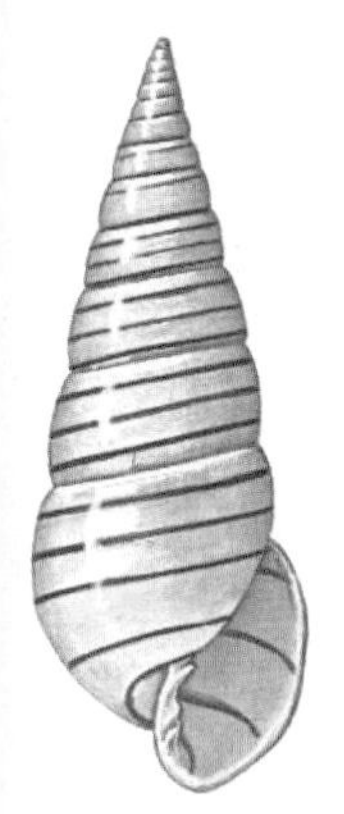

1

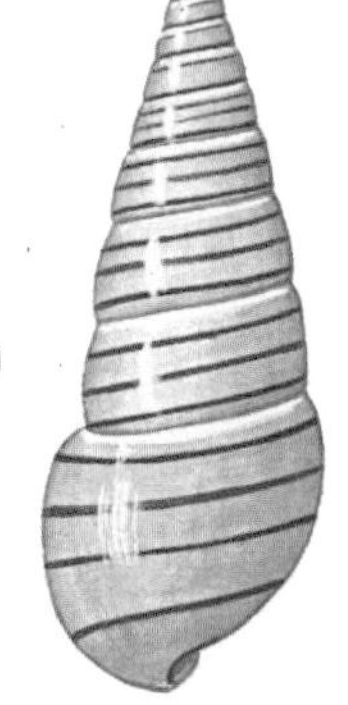

Giant Atlantic Pyram 1.5″

2

Crenate Pyram 0.5″

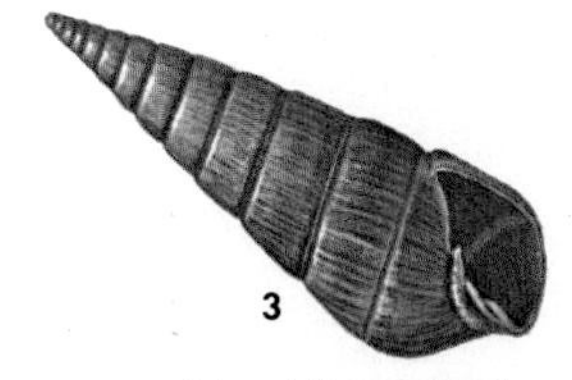

3

Adams' Pyram 0.6″

4

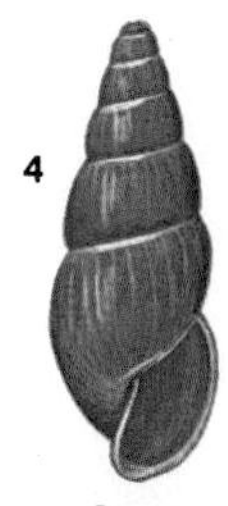

Brown Pyram 5 mm.

5

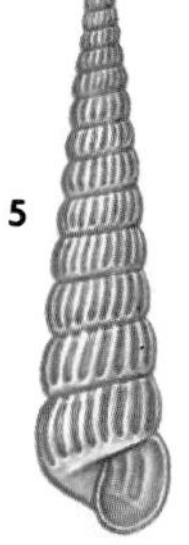

Dall's Turbonille 12 mm.

6

Interrupted Turbonille 6 mm.

7

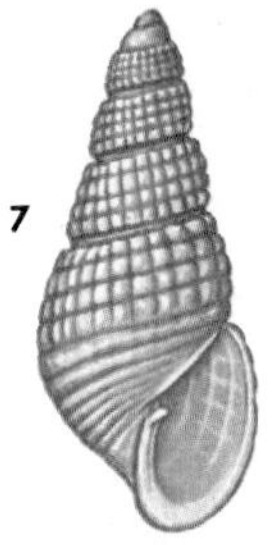

Half-smooth Odostome 4 mm.

8

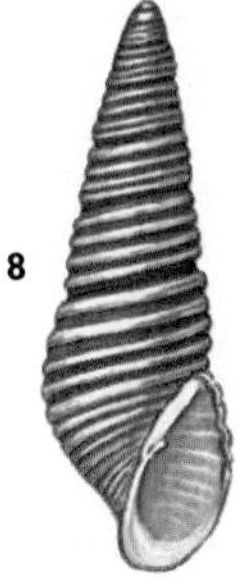

Impressed Odostome 5 mm.

PTEROPODS

Vast communities of small sea butterflies, or flying snails, as the pelagic pteropods are frequently called, form a great part of the planktonic hordes upon which most life in the seas depends, directly or indirectly. Although derived from typical gastropods, with asymmetrical internal organs, pteropods differ in several critical anatomical features and in being hermaphroditic. They have a simple radula. The order Pteropoda, containing about 15 genera, is one of the most curiously developed of all the opisthobranchs.

Some pteropods in the adult form are naked or shell-less, although the embryo does have a coiled, limy shell, complete with an operculum. The more familiar forms have delicate, glassy, translucent shells, which are sometimes brightly colored. The foot is rudimentary, with a pair of winglike flaps that enable the mollusk to swim and dart about quite rapidly. During the daylight hours, pteropods descend, swarming upwards at night to feed on microscopic animals. Pteropods are a major food for some whales and oceanic fishes.

1. **GIBBOSE CAVOLINE.** *Cavolina gibbosa* Rang. Worldwide; pelagic in warm seas. Dorsal lip thinly margined. Ventral lip without lateral points. Ventral surface transversely keeled anteriorly. Common; near surface.

2. **LONG-SNOUT CAVOLINE.** *Cavolina longirostris* Lesueur. Worldwide; pelagic in temperate and tropic seas. Dorsal lip thinly margined. Ventral lip has strong, lateral projections. Common; near surface.

3. **FOUR-TOOTHED CAVOLINE.** *Cavolina quadridentata* Lesueur. Worldwide; pelagic in temperate and tropic seas. Dorsal lip thickened into a pad. Aperture well developed. No prominent lateral spines. Common.

4. **THREE-TOOTHED CAVOLINE.** *Cavolina tridentata* (Forskäl). Worldwide; pelagic in all seas. Tan-colored. Dorsal lip thinly margined. Ventral lip without lateral points. Ventral surface not keeled. Common.

5. **CIGAR PTEROPOD.** *Cuvierina columnella* (Rang). Worldwide; pelagic in cool and warm seas. Shell cylindrical, but somewhat variably shaped and slightly constricted behind aperture. Surface smooth. Common.

6. **STRIATE CLIO.** *Hyalocylis striata* Rang. Worldwide; pelagic in warm seas. Shell conical, slightly compressed, oval in cross section. Apex slightly recurved dorsally. Surface transversely grooved. Common.

7. **STRAIGHT NEEDLE PTEROPOD.** *Creseis acicula* Rang. Atlantic and Pacific; pelagic. Shell long, straight, and slender, evenly tapered to a fine, needlelike point; almost circular in cross section. Common.

Curved Needle Pteropod. *Creseis virgula* Rang (not illus.). Closely related to the Straight Needle and has almost the same range. It is smaller, 0.5 in., and stubbier, less finely tapered, with narrow end recurved to one side. Common.

8. **CUSPIDATE CLIO.** *Clio cuspidata* Bosc. Atlantic and Indo-Pacific; pelagic. Shell somewhat angular. Lateral spines very long. No lateral keels on the posterior portion. Common; found mainly in warm waters, although appears in temperate seas.

9. **PYRAMID CLIO.** *Clio pyramidata* Linné. Worldwide; pelagic. Shell angular but somewhat variable in form; colorless, compressed, with no lateral keels posteriorly. No lateral spines. Common.

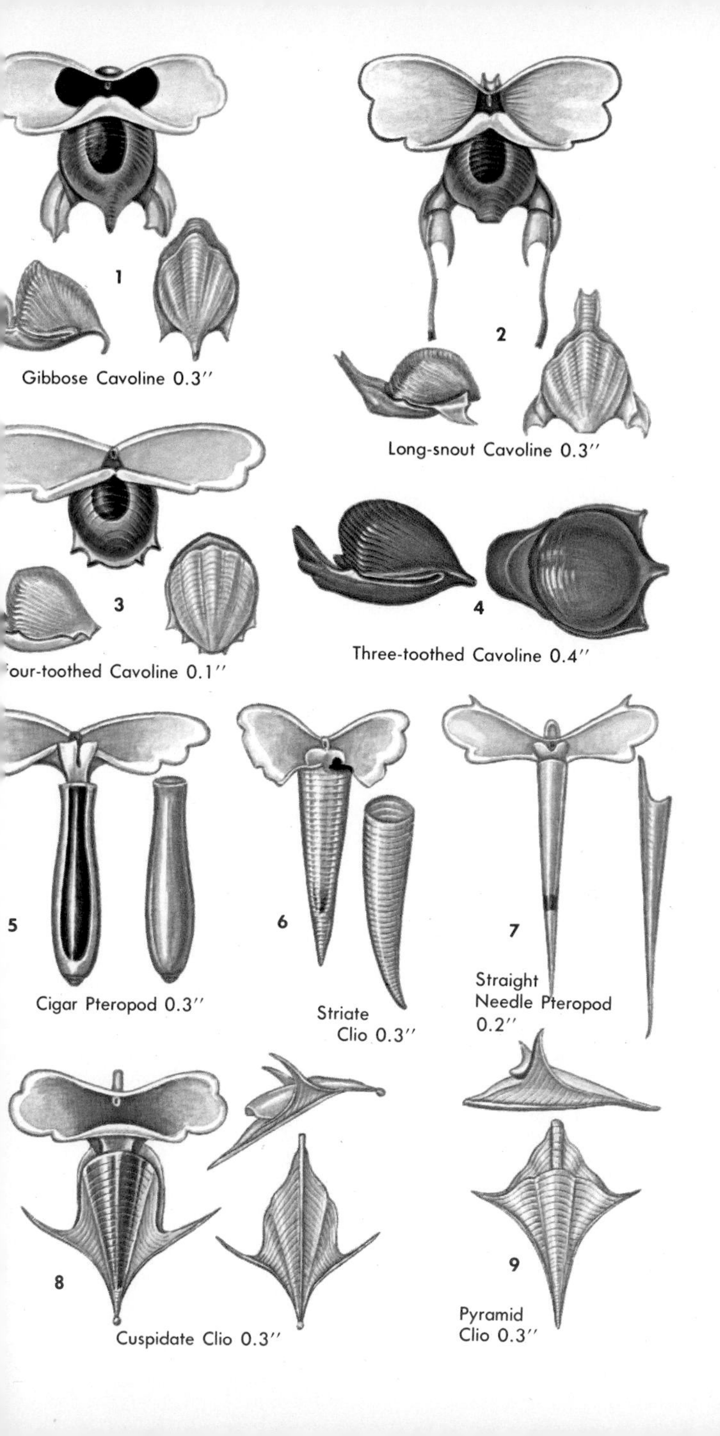
1
Gibbose Cavoline 0.3″
2
Long-snout Cavoline 0.3″
3
Four-toothed Cavoline 0.1″
4
Three-toothed Cavoline 0.4″
5
Cigar Pteropod 0.3″
6
Striate
Clio 0.3″
7
Straight
Needle Pteropod
0.2″
8
Cuspidate Clio 0.3″
9
Pyramid
Clio 0.3″

ATLANTIC NUDIBRANCHS

The highly colorful, shell-less sea slugs without true gills belong to the large and diversified order Nudibranchia. They are found in all shallow seas, and our Pacific Coast is particularly blessed with a large number of beautiful species. Unfortunately, there is no satisfactory way of preserving the colors and shapes of these exquisite little animals.

The embryonic shell in the nudibranch veliger is cast off. True gills and a mantle cavity are wholly lost. Respiration takes place through the general body surface, or through secondary gills. Pairs of club-shaped organs (rhinophores) may be used for detecting odors.

The dorsal surface of many nudibranchs carries spectacular outgrowths called cerata. These usually have some camouflaging value. One species has silvery-gray, bulbous cerata that closely resemble the fish eggs upon which it feeds.

1. **DWARF BALLOON EOLIS.** *Eubranchus exiguus* (Alder and Hancock). Arctic Seas–Mass. Small, barely visible. Body is slender, with 4 simple, pink-banded tentacles, the oral pair rather short. Has numerous spindle-shaped green- and pink-banded, inflated cerata set in 2 rows on each side of the back. Whitish, with dark olive-green markings. This species also occurs in Europe. Seasonal but uncommon; in tide pools.

2. **PAINTED BALLOON EOLIS.** *Eubranchus pallidus* (Alder and Hancock). Arctic Seas–Mass. A small animal with 4 rows of cerata on each side of the back. Color whitish, speckled with golden tan to red-brown. Uncommon; in tide pools.

3. **ORANGE–TIPPED EOLIS.** *Catriona aurantia* (Alder and Hancock). Arctic Seas–Conn. Another small snail with 80 or more slender cerata set in 10–14 rows. Cerata are pink with yellow-orange tips. Has 4 simple, equal-sized tentacles, the front pair white. Uncommon; 2–60 ft.

4. **FROND EOLIS.** *Dendronotus frondosus* (Ascanius). Arctic Seas–R.I.; Alaska–Vancouver. Body somewhat compressed, whitish with red-brown markings. Oral tentacles have leafy sheaths; rhinophores have 5 or 6 leafy processes surrounding the club-shaped head. Cerata very leafy and branching. Common; from intertidal zone to 360 ft. Often found on weed-covered wharf pilings, in summer.

5. **RED–FINGERED EOLIS.** *Coryphella rufibranchialis* Johnston. Arctic Seas–N.Y. Small, with narrow foot bearing pronounced, angular projections anteriorly. Cerata numerous, long, and set in clusters. Tentacles simple, equal-sized. Common; 2–30 ft.

6. **PILOSE DORIS.** *Acanthodoris pilosa* Abildgard. Arctic Seas–Conn.; Alaska. Body widely oval, semitransparent, variably colored from white to yellowish brown and black. Back covered with soft, slender, pointed papillae, which give it a hairy appearance. Rhinophores are long and its club has 19 or 20 leaves. Has 7–9 large, 3-branched branchial plumes. Moderately common; intertidally.

7. **HUMM'S POLYCERA.** *Polycera hummi* Abbott. N.C.–W. Fla. The anus is at the dorsal hump of the animal. In front of it is a 7–9-branched plume. At the sides of the anus are 3 or 4 tentacle-like gills, which are striped with gray, blue, and yellow. The animal is gray with black speckles. The club-shaped rhinophores on the head have 14 leaves. Common; in shallow, grassy areas.

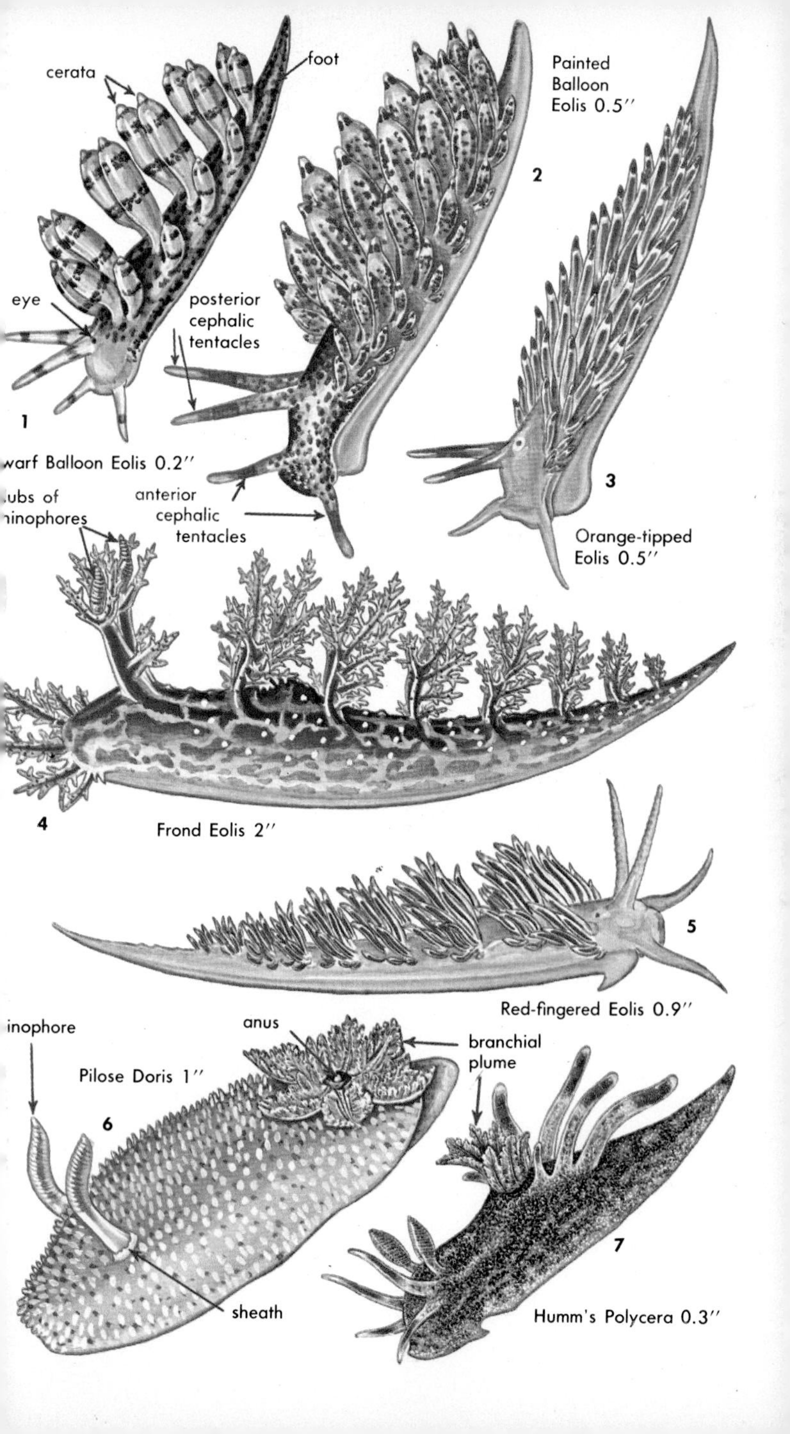
cerata
foot
Painted
Balloon
Eolis 0.5″
2
eye
posterior
cephalic
tentacles
1
warf Balloon Eolis 0.2″
3
ubs of
inophores
anterior
cephalic
tentacles
Orange-tipped
Eolis 0.5″
4
Frond Eolis 2″
5
Red-fingered Eolis 0.9″
inophore
anus
branchial
plume
Pilose Doris 1″
6
7
sheath
Humm's Polycera 0.3″

PACIFIC NUDIBRANCHS

Some kinds of nudibranchs feed on hydroids, on sea anemones, and, rarely, on other small mollusks. They have the remarkable ability to nip off small pieces of hydroids and sea anemones without causing the stinging cells, called nematocysts, in the tentacles of their prey to discharge. The stinging cells are passed from the digestive system of the nudibranch to the tips of the cerata, where they remain undischarged until disturbed by an enemy of the nudibranch.

Some species of nudibranchs, such as the Blue Glaucus, are pelagic. This one floats in warm, open seas with the aid of gas-filled tubes in its gut. Some other species are studded with tiny light organs, which glow at night.

1. **HOPKINS' DORIS.** *Hopkinsia rosacea* MacFarland. S. Calif. Body broadly sluglike, rose-colored, with numerous elongate, rose-colored papillae. Pair of rhinophores long, tapering, smooth on anterior side. Posterior side has about 20 pairs of oblique plates. Branchial plumes, 7–14, are narrow and naked. Egg ribbon rose-colored. Moderately common; in intertidal tide pools.

2. **CARPENTER'S DORIS.** *Triopha carpenteri* (Stearns). Central Calif. Body cream or white, flecked with numerous raised, bright red-orange granules and tubercles. Rhinophores are short clubs, red-orange and with 20–30 leaves. Has 5 large branchial plumes, each with 3 frilled branches. They do not retract when animal is alarmed. Common; in rock pools on green or brown algae.

3. **NOBLE PACIFIC DORIS.** *Montereina nobilis* (MacFarland). Calif. Large. Mantle deep yellow, with numerous minute red speckles and dark brown central markings. Rhinophores stout, clublike, with about 24 leaves. Sheath around base of rhinophores bumpy. Has 6 large, spreading branchial plumes, divided 3 or 4 times, and joined by a membrane. Moderately common; on wharf pilings.

4. **MACULATED DORIS.** *Triopha maculata* MacFarland. Central Calif. Body broadly sluglike. Mantle dark brown or black, edged with vermilion, with numerous rounded white spots and a few red tubercles. Underside of foot golden tan. Rhinophores retractible. Club vermilion, with 18 leaves. Has 5 branchial plumes, 3-branched and not retractible. Very common; in tide pools on brown kelp.

5. **HEATH'S DORIS.** *Discodoris heathi* MacFarland. Calif. Mantle thick, light tan, darkly speckled, with numerous minute, raised spicules. Body rather soft, somewhat oval. Rhinophores stout, conical, retractible; the club has 10–15 leaves. Has 8–10 tripinnate branchial plumes. Uncommon; in tide pools in summer.

6. **LAILA DORIS.** *Laila cockerelli* MacFarland. S. Calif. Small animals. Body sluglike, bluish white, bordered with numerous clublike, orange-tipped processes. Rhinophores stout; club red, longer than the whitish stalk; 13 leaves. Five branchial plumes are tripinnate and nonretractible. Uncommon; under rocks in tide pools.

7. **BLUE GLAUCUS.** *Glaucus marina* (DuPont). Worldwide in warm seas. Body elongate; tentacles and rhinophores small. Each side of body bears 4 vivid blue, fringed clumps. Back smooth, bright blue with long white stripes. Underside pale grayish blue. Has strong jaws and a radula with a center row of about 10 teeth. Not uncommon; seasonally, pelagic. Often washed ashore with Purple Sea Snail (p. 93).

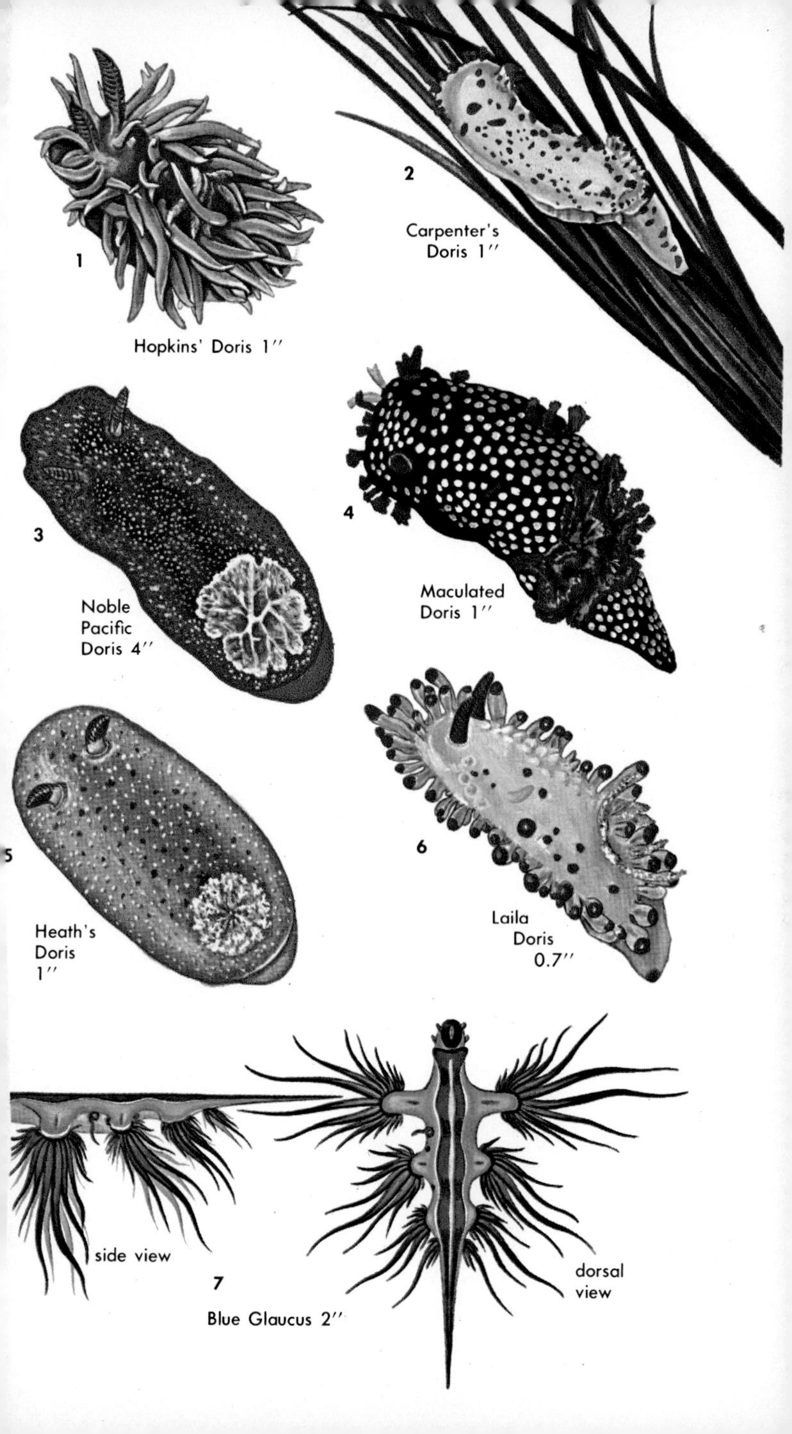
1
Hopkins' Doris 1″
2
Carpenter's
Doris 1″
3
Noble
Pacific
Doris 4″
4
Maculated
Doris 1″
5
Heath's
Doris
1″
6
Laila
Doris
0.7″
side view
7
Blue Glaucus 2″
dorsal
view

CLASS AMPHINEURA (Chitons)

The chitons are a group of primitive mollusks resembling limpet-like snails (p. 6) but lacking tentacles and cephalic eyes. The shell is secondarily divided into eight hard valves, which are embedded in the mantle or girdle of the animal. Chitons are bilaterally symmetrical, with a large central foot. The anus is at the posterior end, and there are gills on either side. The mouth contains a strong set of radular teeth.

The chiton's shelly plates are variously shaped and sculptured. In some species the plates may be partially or totally covered by the mantle. The mantle margin or girdle may be smooth or may bear bristles, spikes, or overlapping scales resembling dried split peas. In some species there are small, light-sensitive eyes on the dorsal surface. These eyes have a cornea, iris, lens, retina, and optic nerve.

About 600 living species of chitons have been described, and many more are known from past geologic eras as far back as the Paleozoic. The Pacific Coast is particularly rich in numbers and species (probably 75, all told) of chitons. Their identification is difficult and may require that the valves be removed for careful observations. Most chitons live in shallow water in the intertidal zone on rocks, but one species of *Lepidochiton* has been dredged from a depth of 13,800 feet.

Chitons are usually nocturnal, browsing over the surface of intertidal rocks and feeding on algae, bryozoans, diatoms, and, rarely, small shrimps. Some species have an ability to return to their resting areas after a night of foraging. The eggs are either free-floating single cells or embedded in jellylike strings. The males liberate sperm freely into the water. A few species are ovoviviparous, giving birth to live young. Chitons have an amazing ability to cling to rocks, and collectors must use a deft blow of a sharp knife to dislodge them. In the West Indies the meat is used as fish bait and as food.

1. **GIANT PACIFIC CHITON.** *Amicula stelleri* (Middendorff). Alaska–Calif. This largest known chiton reaches 1 ft. in length. The leathery, firm, gritty, red-brown to yellow-brown girdle completely covers the butterfly-shaped, rose-tinted valves. Common; on intertidal rocks. Formerly in genus *Cryptochiton*.

2. **BLACK KATY CHITON.** *Katharina tunicata* (Wood). Alaska–S. Calif. One of the most abundant Pacific Coast chitons. A shiny black, naked girdle covers most of the 8 gray, eroded valves. Females shed eggs in the water during the summer. Common; on rocks in tidal area.

3. **VEILED PACIFIC CHITON.** *Placiphorella velata* Dall. W. Canada–Mexico. This chiton is recognized by its flat oval shape and wide girdle, which is broad anteriorally and studded with scaly hairs. Valves are a streaked olive-brown; interior is white. This unusual chiton has the anterior end of its girdle modified into a flap, which can be raised and quickly snapped down to capture small shrimp, crabs, and amphipods. Eight tentacles assist in guiding the captured prey to the animal's mouth. This chiton also feeds on diatoms and seaweeds. Fairly common; in rock crevices near shallow-water kelp beds.

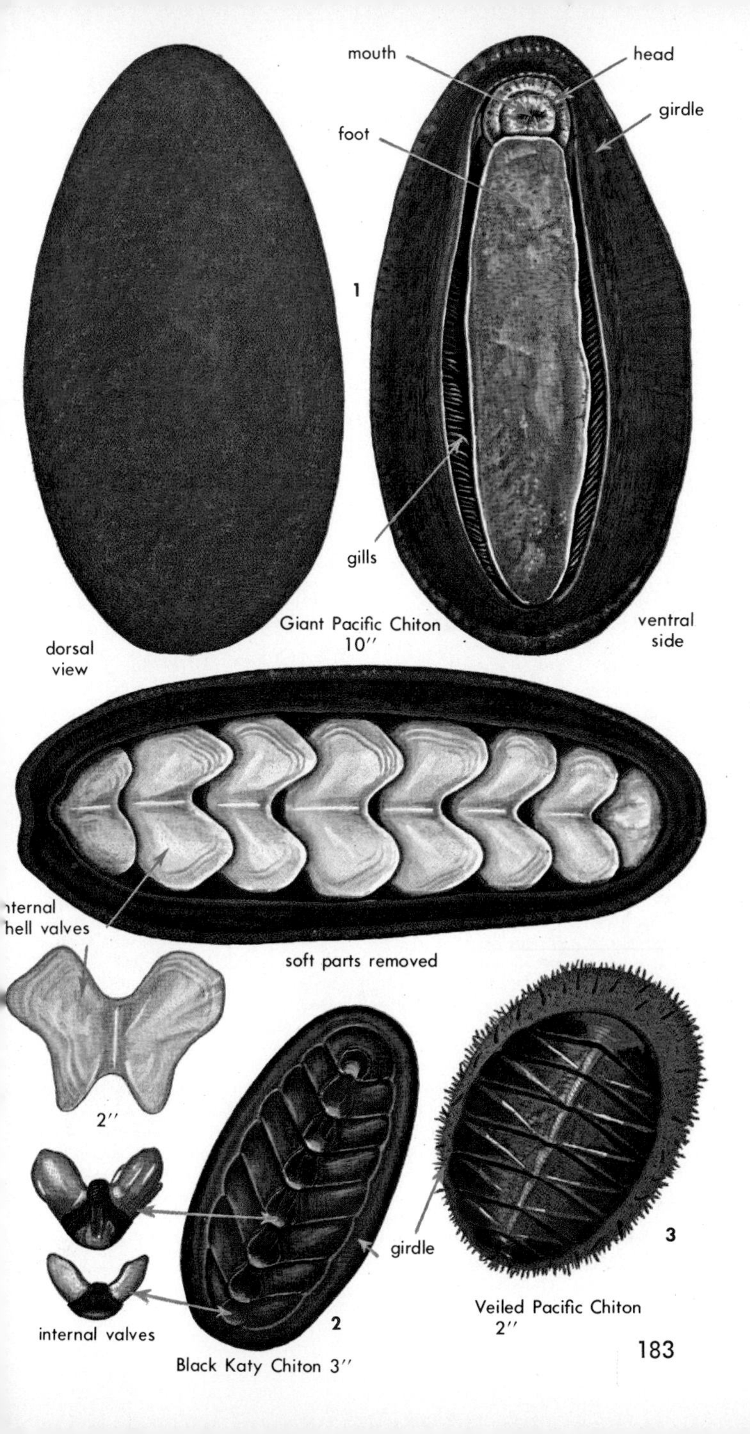
mouth
head
girdle
foot
1
gills
Giant Pacific Chiton
10″
ventral
side
dorsal
view
ıternal
hell valves
soft parts removed
2″
girdle
3
Veiled Pacific Chiton
2″
2
internal valves
Black Katy Chiton 3″

CHITONS—continued

About two dozen common species of chitons are found on the Pacific Coast, with perhaps twice this number of rarer, deep-water forms. Florida and adjacent areas also have many species. The study of chitons is fascinating, but it requires the use of advanced monographs. Specimens should be preserved in a flattened position and the girdle retained, for the latter has characters useful in identification. At least one specimen should be "disarticulated" so that the minute characters of the shelly valves can be observed. Each chiton possesses three types of valves: (1) the anterior valve at the head end, (2) six intermediate valves, and (3) the posterior valve at the hind end. The notches and slits along the valve edges are used in identification.

1. **LINED RED CHITON.** *Tonicella lineata* (Wood). Alaska–S. Calif. Shell is an elongated oval; surface smooth and shiny; brightly colored with brown or black-brown lines bordered with white. Girdle leathery, naked. In Alaska, common on shores. In Calif., uncommon.

2. **MOTTLED RED CHITON.** *Tonicella marmoreus* (Fabricius). Arctic Seas–Mass., and in S. Alaska. Valves arch rather sharply. Upper surface minutely granular; looks smooth. Interior of valves rose-colored. Posterior valve has 8 or 9 slits; others toothed. Girdle leathery, naked. Common; offshore, 6–300 ft.

3. **NORTHERN RED CHITON.** *Ischnochiton ruber* (Linné). Arctic Seas–Conn. and central Calif. A light-red chiton with irregular brown markings. Valves arch moderately and are rather rounded. Upper surface has growth wrinkles; interior valves bright rose; posterior valve has 7–11 weak slits. Girdle granulated, with long scales. Common; 6–500 ft.

4. **COMMON WEST INDIAN CHITON.** *Chiton tuberculatus* Linné. Fla.–Texas and West Indies. A variable brown to gray chiton, sometimes marked with green, black, or white. Girdle granular, with small, "split-pea" scales in alternating color zones. Valves smooth at top and with 7 or 8 longitudinal, wavy riblets on the sides. End valves have irregular wavy cords. Common; intertidally.

5. **COMMON EASTERN CHITON.** *Chaetopleura apiculata* (Say). Mass.–Fla. Small, oblong to oval. Valves slightly arched. Upper surface has 15–20 rows of tiny beads. Interior of valves whitish. Girdle narrow, minutely granular, and has a few short hairs. Common; 6–100 ft. This species is found in northern tide pools, usually in well-protected places.

6. **MOSSY MOPALIA.** *Mopalia muscosa* (Gould). Alaska–Mexico. Color usually brown to dull olive-gray. Interior of valves blue-green, rarely pinkish. Girdle has stiff hairs resembling a fringe of moss. Posterior end of girdle may have a weak indentation. Common; intertidally. There are several closely related species.

7. **FUZZY CHITON.** *Acanthopleura granulata* (Gmelin). S. Fla.–Brazil. Valves usually eroded, but perfect ones are brown with granular surface. Girdle thick, gray, with bands of black and with a thick, coarse covering of low, hairlike spines. Common; on rocks, intertidally. Used as food and bait in the West Indies.

8. **ROUGH GIRDLED CHITON.** *Ceratozona squalida* (C. B. Adams). S.E. Fla.–Caribbean. Similar to Fuzzy Chiton. Valves also badly eroded and lightly colored. Anterior valve has 10 or 11 radiating ribs. Girdle yellowish brown and has clusters of hairs. Formerly named *C. rugosa* Sowerby. Abundant; found in the intertidal zone.

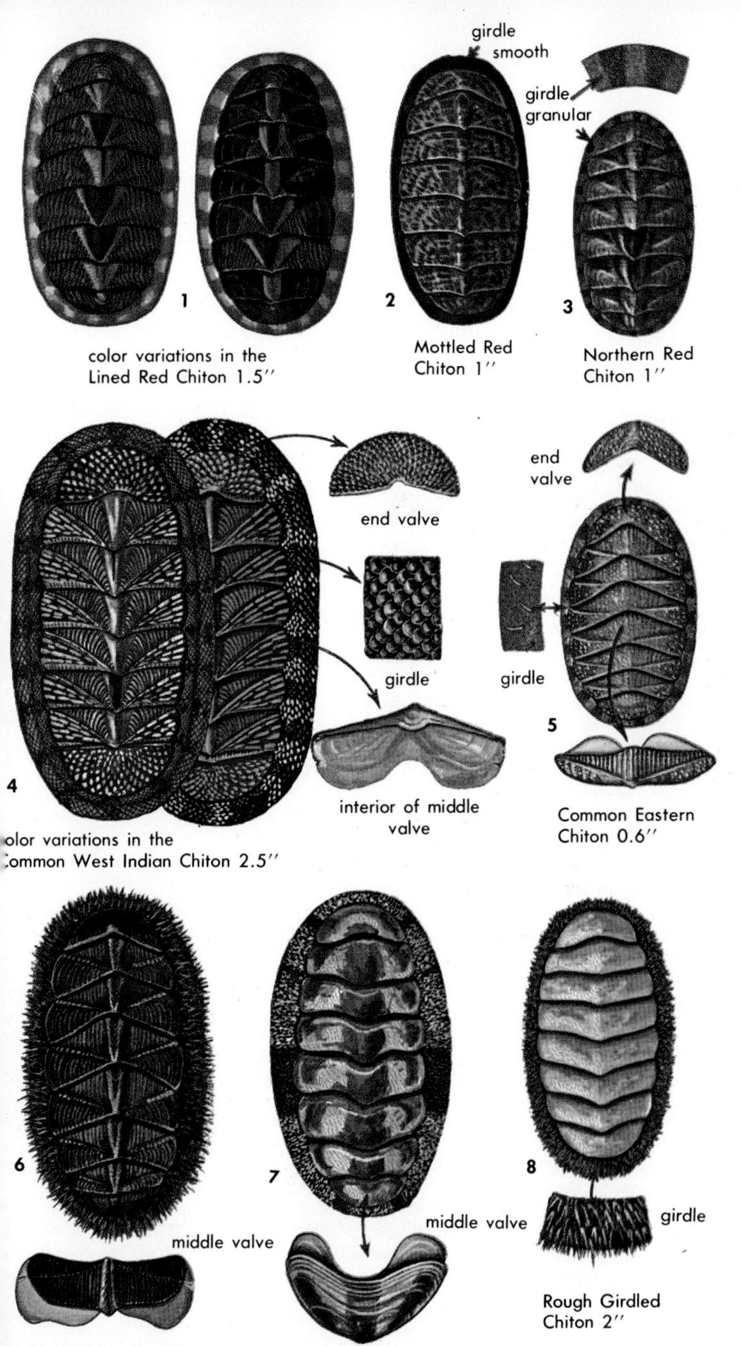

color variations in the Lined Red Chiton 1.5″

Mottled Red Chiton 1″

Northern Red Chiton 1″

olor variations in the :ommon West Indian Chiton 2.5″

Common Eastern Chiton 0.6″

Mossy Mopalia 3″

Fuzzy Chiton 3″

Rough Girdled Chiton 2″

CLASS SCAPHOPODA (Tusk Shells)

Less than 1,000 species of living tusk shells constitute the class Scaphopoda. Worldwide in distribution, and exclusively marine, they live in mud or sand from the low-tide line to great depths. The shell and mantle completely surround the body, which is open at both ends. The embryonic shell (prodissoconch) is cup-shaped and consists of two valves that later fuse along the ventral margin to form the tube. At the narrow end, which protrudes above the sandy bottom, there is an orifice through which water is drawn and expelled at intervals. At the large end, the conic-shaped foot and dozens of ciliated, prehensile captacula, or threads, project through the sand. The main food of tusks are single-celled foraminifera, which are captured by the captacula and drawn to the mouth. They are crushed by a strong set of radular teeth. The sexes are separate. The circulating system is so efficient that no true heart is present. There are no gills.

Of about 200 American species, most live at depths below 500 feet. A Pacific Northwest species was used as Indian wampum. In the Southeast some are found cast up on beaches. The tiny posterior slits or notches at the narrow end of *Dentalium* are used to distinguish species in this genus.

Tusk shells are fed upon by many species of fish. Hermit crabs sometimes use empty tusks as homes. One species of crab has modified its two semicircular front claws to fit perfectly inside the circular entrance to the tusk shell.

1. **MERIDIAN TUSK.** *Dentalium meridionale* Pilsbry and Sharp. Mass.–Caribbean. Shell mouse-gray, smoothish, with 16 riblets at small end, about 90 near middle, and none at large end. Common; deep water.

2. **FLORIDA TUSK.** *Dentalium floridense* Henderson. S.E. Fla.–Caribbean. Shell yellowish white. Apex hexagonal, with narrow slit. Has 24 riblets at large end. Uncommon.

3. **PANELLED TUSK.** *Dentalium laqueatum* Verrill. N.C.–Caribbean. Shell dull white, with 9–12 main ribs with fine lines between them. Anterior end smoothish. Common; in water 30–600 ft. deep.

4. **INDIAN MONEY TUSK.** *Dentalium pretiosum* Sowerby. Alaska–Mexico. Shell opaque white, with dirty buff growth rings. Apex has short notch on convex side. Common; offshore. Formerly used by Indians on the Pacific Coast as wampum.

5. **STIMPSON'S TUSK.** *Dentalium entale* subspecies *stimpsoni* Henderson. Nova Scotia–Mass. An American form of a European species. Color ivory-white; apex chalky and eroded. Cross section round. Common offshore.

6. **IVORY TUSK.** *Dentalium eboreum* Conrad. N.C.–Caribbean. Shell glossy ivory-white to pinkish. Apical slit narrow and on the convex side. Narrow end has about 20 fine scratches. Common; intertidally.

7. **CAROLINA CADULUS.** *Cadulus carolinensis* Bush. N.C.–Texas. Glossy clear to white. Shell swollen in middle. Apex has 4 shallow slits. Commonly dredged in sand from 18–600 ft.

Four-toothed Cadulus. *C. quadridentatus* Dall (not illus.). N.C.–Caribbean, 0.2–0.4 in. long. Similar, but the swelling is just behind the aperture end. The 4 apical slits are deep. Common; 18–300 ft.

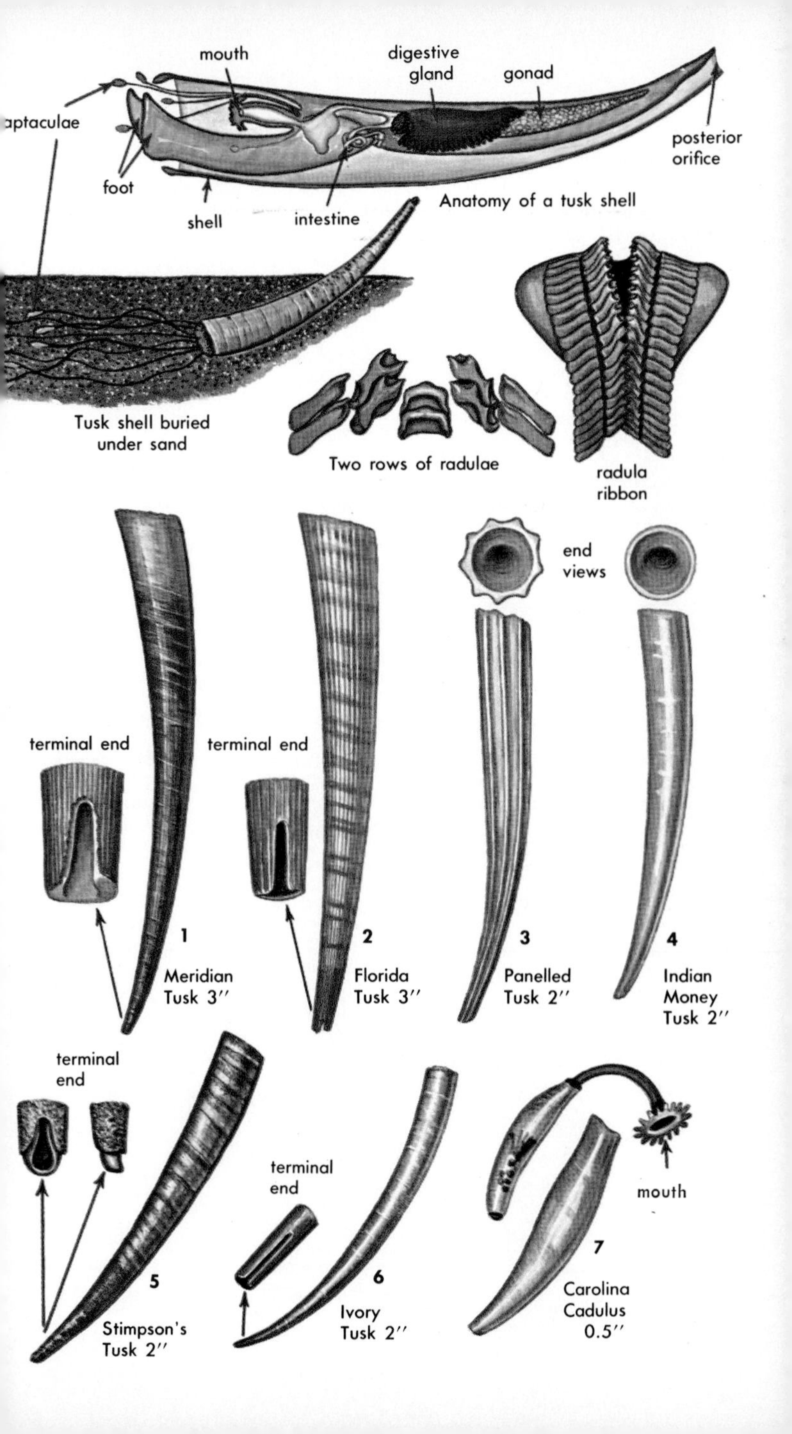
mouth
digestive gland
gonad
aptaculae
posterior orifice
foot
shell
intestine
Anatomy of a tusk shell
Tusk shell buried under sand
Two rows of radulae
radula ribbon
end views
terminal end
terminal end
1
Meridian Tusk 3″
2
Florida Tusk 3″
3
Panelled Tusk 2″
4
Indian Money Tusk 2″
terminal end
terminal end
mouth
5
Stimpson's Tusk 2″
6
Ivory Tusk 2″
7
Carolina Cadulus 0.5″

CLASS PELECYPODA (Bivalves)

The clam, or bivalve, class includes mollusks with two shelly valves hinged at the top, without a head and radula, and usually with a hatchet-shaped foot. For more details see page 10.

There is no general agreement on the higher classification of the bivalves. Some systems are based on shell alone, others on soft anatomy. The outline of major American groups given below is usually accepted. For details on orders see page 8.

ORDER Palaeoconcha
Family: Solemyacidae
ORDER Protobranchia
Superfamily: Nuculacea
ORDER Filibranchia
Superfamilies: Arcacea, Pteriacea, Pinnacea, Mytilacea, Ostreacea, Pectinacea, Anomiacea
ORDER Eulamellibranchia
Superfamilies: Chamacea, Astartacea, Carditacea, Lucinacea, Cardiacea, Veneracea, Mactracea, Tellinacea, Solenacea, Myacea, Pholadacea, Pandoracea (the latter family constitutes the order Anomalodesmata of other classifications)
ORDER Septibranchia
Superfamily: Poromyacea

GLOSSARY OF PELECYPOD SHELL TERMS

Accessory plate. Extra, small, shelly or horny plate over the hinge area or on the siphons.

Adductor muscle. One or two large muscles inside the shell that close the two valves.

Anterior end. Front end; where the foot usually protrudes; opposite the posterior end where the siphons protrude.

Gross Anatomy of a Clam

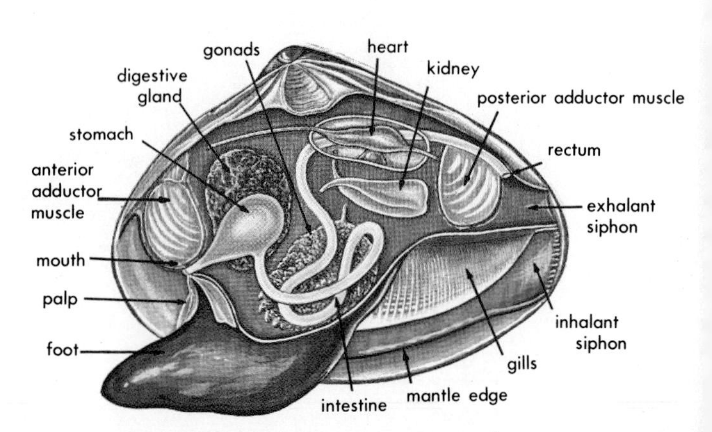

Apophysis. Shelly, finger-like projection inside each valve under the umbo, especially in pholad clams (p. 258).
Beak (or **umbo**). The first formed part of a valve, usually above the hinge.
Byssus. Clump of horny threads spun by foot and used for attachment to the substrate, as in pen shells (p. 204).
Chondrophore. Spoon-shaped shelf in the hinge, holding the padlike resilium, or cartilage.
Concentric. Sculpturing of ridges, ribs, or threads, or color marking, running parallel to the margins of the shell valve.
Equilateral. Front and back halves of the valve are the same size and shape; the umbones are at the center, as in bittersweets (p. 196).
Equivalve. Each valve is the same in shape and size.
Escutcheon. A smooth, long surface on the upper margin of the valve behind the ligament.
Gape. Opening between the margins of the valves when the shell is shut.
Hinge. Top margin of a bivalve, where the shelly teeth, if present, interlock.
Inequilateral. Front and back halves of the valve are unequal in shape and size.
Inequivalve. One valve is larger or fatter.
Ligament. An external or internal horny band, usually behind the beaks, holding the valves together or ajar.
Lunule. A long or heart-shaped impression on the upper margin of the valves in front of the beaks, one half being on each valve.
Pallial line. A scar line on the inside of the valve, where the mantle muscles are attached.
Pallial sinus. An embayment in the pallial line indicating where the siphon-retracting muscles are attached.
Prodissoconch. Tiny, first-formed shell on the beaks.
Resilium. A horny, padlike cushion located on the chondrophore.
Radial. Sculpturing or color rays running from the beaks to the ventral margins of the valves.
Tooth. Shelly ridge in the upper, hinged portion of the valve. Cardinal teeth are the largest 2 or 3 just under the beak; lateral teeth are off to the side and are usually narrow and long.
Valve. One of the main shelly halves of a bivalve.

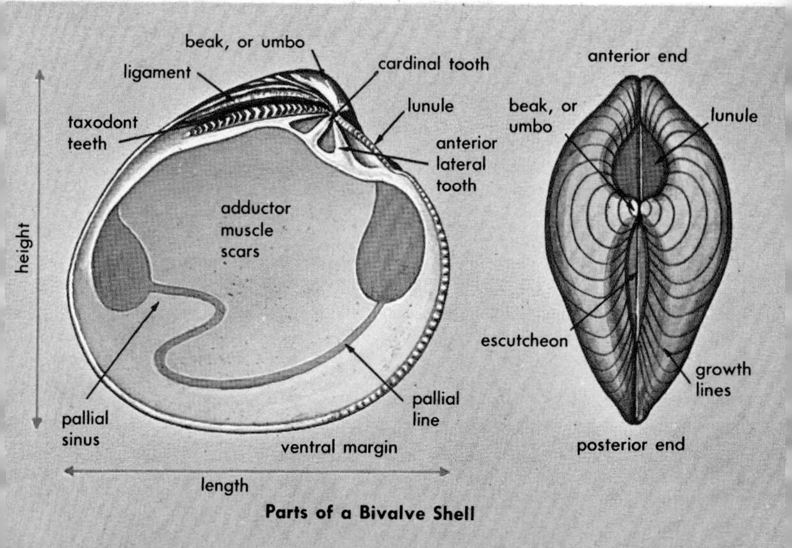

Parts of a Bivalve Shell

AWNING AND NUT CLAMS

Awning clams (family Solemyacidae) are believed to be very primitive. Their thin, cigar-shaped shells are without hinge teeth and are covered with a glossy, brown periostracum. The foot is disc-shaped and has serrated edges. The protective siphonal appendages are used to distinguish the four known American species. They are not uncommonly dredged in muddy areas. They move with ease up and down long burrows, which are U-shaped and open at both ends.

Nut clams belong to two families. Both are characterized by their small, generally thin shells, which bear very distinct, sharp interlocking teeth along the hinge line. Shells of the family Nuculidae are pearly inside and usually oval in shape. Those of Nuculanidae are whitish within, usually elongate, and the hinge bears a much larger, spoon-like chondrophore, the depression on the hinge into which the horny cushion or resilium fits. Most of the several hundred members of these sand-dwelling families live in cool, temperate seas and are a source of food for bottom-feeding fish.

1. **BOREAL AWNING CLAM.** *Solemya borealis* Totten. Nova Scotia–Conn. Interior of strong shell is bluish gray, not white. Periostracum is tan-brown. About 40 siphonal appendages. Common; 20–600 ft.

The Common Awning Clam. *Solemya velum* (Say) (not illus.). Nova Scotia–Fla. Shell is similar but more fragile and with lighter radial bands. Interior whitish. About 16 siphonal appendages. Delicate brown periostracum. Common; intertidal to 30 ft.

2. **ATLANTIC NUT CLAM.** *Nucula proxima* Say. Nova Scotia–Fla. and Texas. Exterior smooth but may have fine gray axial lines. Ventral edge minutely scalloped. Common; in mud, 3–100 ft.

3. **DIVARICATE NUT CLAM.** *Acila castrensis* (Hinds). Alaska–Baja Calif. Exterior has strong diverging riblets. Anterior edge somewhat squarish. Common; in sand, 20–600 ft.

4. **POINTED NUT CLAM.** *Nuculana acuta* Conrad. Mass.–Caribbean. A small clam with a long beak on the shell's posterior, over which run concentric threads. Periostracum thin, pale yellow. Common; in sandy mud, 5–1,200 ft.

The Concentric Nut Clam. *Nuculana concentrica* Say (not illus.). N.W. Fla.–Texas. Shell is 0.5 in. long, fat, with a smooth, pointed end and with many fine concentric lines, but not over beak. Common; in sand, 6–20 ft.

5. **MÜLLER'S NUT CLAM.** *Nuculana pernula* (Müller). Arctic Ocean–Mass. Long, with concentric growth lines. Interior shiny white with a strong, low radial rib. Lunule long, prominent. Common; 50–1,000 ft.

6. **FILE YOLDIA.** *Yoldia limatula* (Say). Maine–N.J. and N. Alaska. Greenish tan-brown; long, and narrowing at posterior end. Subspecies *Y. l. gardneri* Oldroyd, Alaska–Calif., has a small depression on anterior ventral margin. Both common; beyond low water.

7. **SHORT YOLDIA.** *Yoldia sapotilla* Gould. Arctic Ocean–N.C. Shell thin, translucent, smooth. Chondrophore small. About 100 teeth along hinge line. Periostracum yellowish. Common; in muddy shallows.

8. **BROAD YOLDIA.** *Yoldia thraciaeformis* Storer. Arctic Ocean–N.C. and Puget Sound. Has squarish, upturned posterior end. Chondrophore large, prominent. Periostracum dull. Moderately common; 50–1,000 ft.

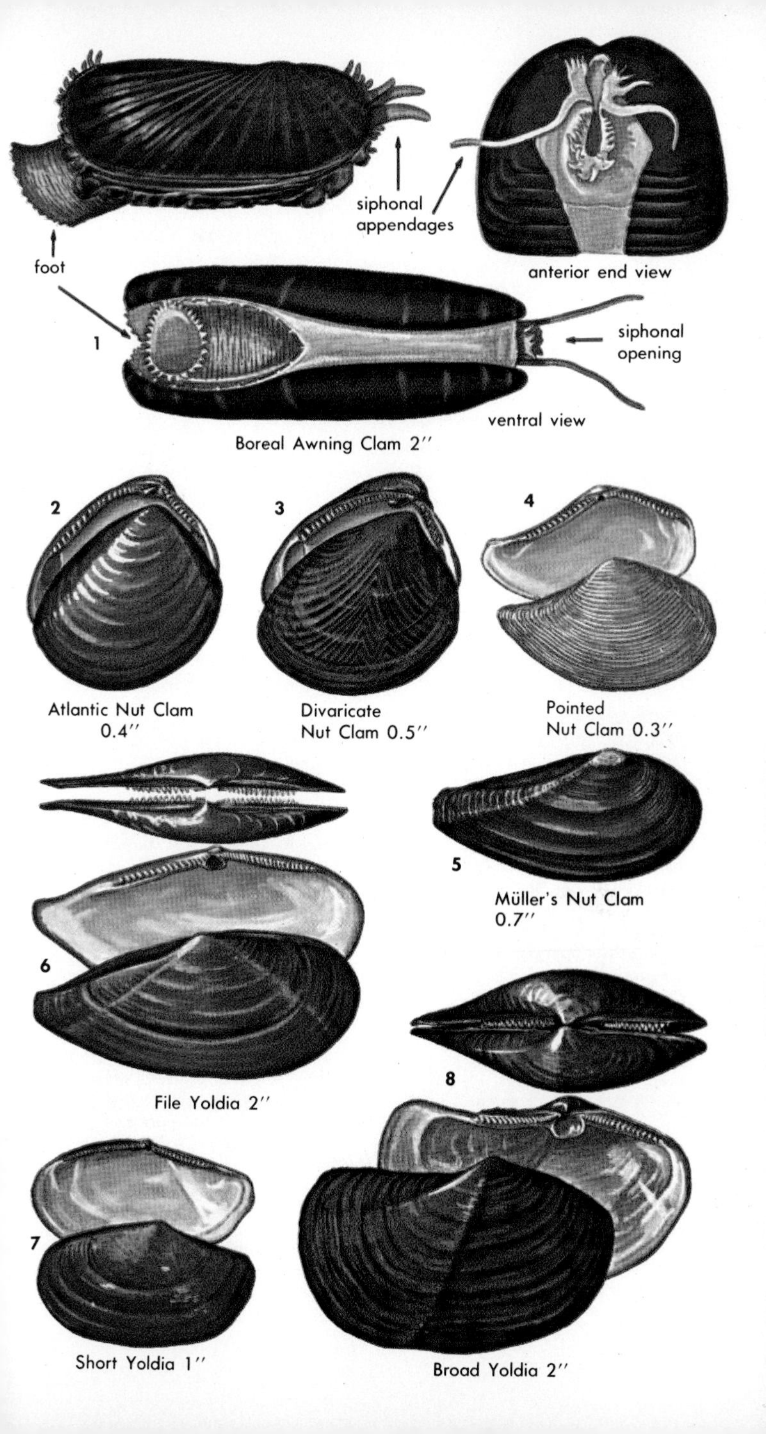

Boreal Awning Clam 2″

Atlantic Nut Clam 0.4″

Divaricate Nut Clam 0.5″

Pointed Nut Clam 0.3″

Müller's Nut Clam 0.7″

File Yoldia 2″

Short Yoldia 1″

Broad Yoldia 2″

ARK SHELLS

Arks of the family Arcidae are common, worldwide clams, most living in warm, shallow seas, although a few, like the Blood Ark, are found in cold waters as far north as New England. The strong, boxlike shells are usually heavily ribbed, without bright coloration, and covered with a feltlike, blackish periostracum. The hinge is straight and bears many small taxodont teeth. The number of teeth varies considerably, even within a species. Between the beaks and on the upper surface or cardinal area is a blackish, horny ligament, which is helpful in identification.

Of the approximately 200 species, about 16 are fairly common along the Atlantic Coast, and 4 on the Pacific. The true arks, of the tropical genus *Arca,* have a very long, straight hinge and a large, wide ventral gape in the shell for the massive byssus. True arks and many members of the genus *Barbatia* attach themselves to the undersides of rocks. Other genera, such as *Anadara,* live buried in sandy mud, and their shells have no ventral gape. The young spin a single byssus for temporary attachment to small pebbles.

The siphons are reduced and are bordered with fine, protective filaments. Poorly developed, simple eyes are present along the mantle edge in some species, usually in small clumps, like a compound eye. Many members of the Arcidae family are unique in having a "duplicated" heart—that is, in having two ventricles widely separated because of the intervening massive muscles employed in retracting the foot and large byssus.

1. **TURKEY WING.** *Arca zebra* Swainson. N.C.–Brazil. Also called Zebra Ark; has a strong shell with red-brown zebra-like stripes. Twice as long as deep, or more. Numerous smooth, rounded ribs of irregular size. Shiny olive-green byssus, short and strong, comes from a moderately large byssal opening. The animal has a series of eyespots along the middle fold of the mantle edge. Turkey Wings grow attached to rocks but are often overlooked because of encrusting marine growths. Washed ashore after storms. Do not confuse with Mossy Ark. Very common; 3–25 ft.

2. **MOSSY ARK.** *Arca imbricata* Bruguière. N.C.–Caribbean. Surface crisscrossed, except for 6–8 finely beaded posterior ribs. Periostracum dark brown, heavy, and shaggy, particularly on posterior ridge. Byssal opening very large. Common; in moderately shallow water.

3. **BLOOD ARK.** *Anadara ovalis* (Bruguière). Mass.–Caribbean and Texas. Shell sturdy, with 26–35 squarish, smooth ribs separated by narrow grooves. Beaks almost touch. Periostracum is dark brown. Ligament very narrow. This unusual species has red blood. Very abundant; at 6–100 ft.

4. **EARED ARK.** *Anadara notabilis* (Röding). S.C.–Brazil. Shell quite sturdy, with 25–27 heavy, grooved ribs. Fine concentric lines cross ribs and interspaces. Young specimens have prominent dorsal wing, or "ear." Common; on mud and grassy bottoms, in shallow water.

5. **CUT-RIBBED ARK.** *Anadara lienosa floridana* (Conrad). N.C.–Texas. It has 30–38 grooved ribs. Fine concentric lines cross both ribs and interspaces. Common; in shallow water.

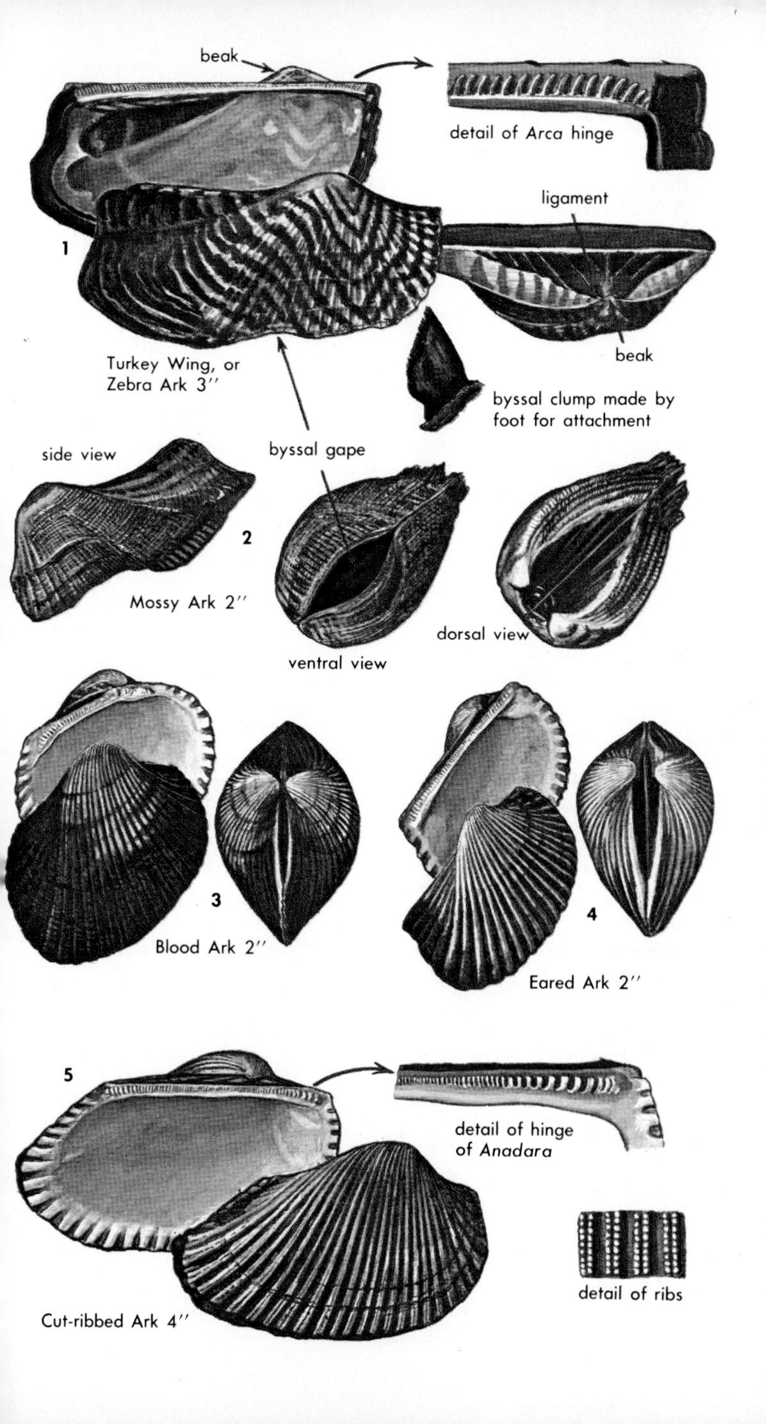
beak
detail of *Arca* hinge
ligament
1
beak
Turkey Wing, or
Zebra Ark 3″
byssal clump made by
foot for attachment
side view
byssal gape
2
Mossy Ark 2″
dorsal view
ventral view
3
Blood Ark 2″
4
Eared Ark 2″
5
detail of hinge
of *Anadara*
detail of ribs
Cut-ribbed Ark 4″

ARK SHELLS—continued

Arks are an ancient stock that appeared in the Jurassic. Despite the superficial similarity of the taxodont teeth in both the ark clams and the Nuculidae nut clams, the two groups are quite unrelated. By Tertiary times, the arks had developed into several fairly distinctive groups, each having characteristic ligaments, ribbing, and beak formation.

In the genus *Noetia,* such species as the ponderous ark have beaks that point posteriorly, rather than toward each other, as in members of the genus *Anadara*. The ligament of *Noetia* is striated crosswise from beak to beak, rather than lengthwise or in chevrons, as in *Barbatia* and *Anadara*. In some arks, such as the Incongruous Ark, the left valve is so large that it overlaps the edges of the right valve. Other genera, such as *Arcopsis,* are characterized by a very short, small, black ligament located between the two beaks.

Ark clams are edible but are not usually eaten in the United States because of their somewhat bitter taste and probably because in some species the blood's hemoglobin gives the flesh a red color.

1. **INCONGRUOUS ARK.** *Anadara brasiliana* (Lamarck). N.C.–Texas and Brazil. Shell is about as long as high, with 26–28 ribs, each with strong barlike beads, particularly on the larger (left) valve. Posterior ribs on right valve generally quite smooth. The left valve overlaps the right valve toward the posterior end of this inflated species. Beaks centrally located. Ligament sometimes transversely striate. Hinge teeth noticeably smaller toward center. Periostracum light brown and rather thin. Quite common; in sand, in shallow water.

2. **TRANSVERSE ARK.** *Anadara transversa* (Say). Mass.–Texas and Caribbean. Smallest of the genus, with 30–35 ribs, usually only beaded on left valve. Left valve larger and overlaps right. Periostracum grayish brown. Common; below low tide on rocks, in sandy mud. Interior rarely pinkish.

3. **WHITE MINIATURE ARK.** *Barbatia domingensis* (Lamarck). N.C.–Caribbean. Shell small and white; surface has a coarse network. Posterior end dips downward. Ligament long and very narrow. Periostracum thin, yellow-brown. Common; under rocks and sponges, below tide line.

4. **WHITE-BEARDED ARK.** *Barbatia candida* (Helbling). N.C.–Brazil. Ribs numerous, crossed by growth lines. Interior of shell white. Periostracum shaggy, yellow-brown. Byssal opening small. Common; on rocks in shallow water.

5. **RED-BROWN ARK.** *Barbatia cancellaria* (Lamarck). S. Fla.–Brazil. Shell somewhat compressed. Readily identified by its red-brown coloring and crisscrossed surface. Interior of shell brownish; exterior darker. Common; attached to the undersides of rocks in shallow water.

6. **ADAMS' MINIATURE ARK.** *Arcopsis adamsi* (E. A. Smith). N.C.–Brazil. Shell dark, very small, somewhat inflated. Surface finely crisscrossed. Easily distinguished by the small black, triangular to oblong ligament between the umbones. Common; under dead coral in shallow water.

7. **PONDEROUS ARK.** *Noetia ponderosa* (Say). Va.–Fla. Keys and Texas. Shell very sturdy, covered with velvety, black periostracum that is absent on beaks. Has 27–31 ribs, each divided by a fine-cut line. No byssus in the adult. Common; in sand in shallow water.

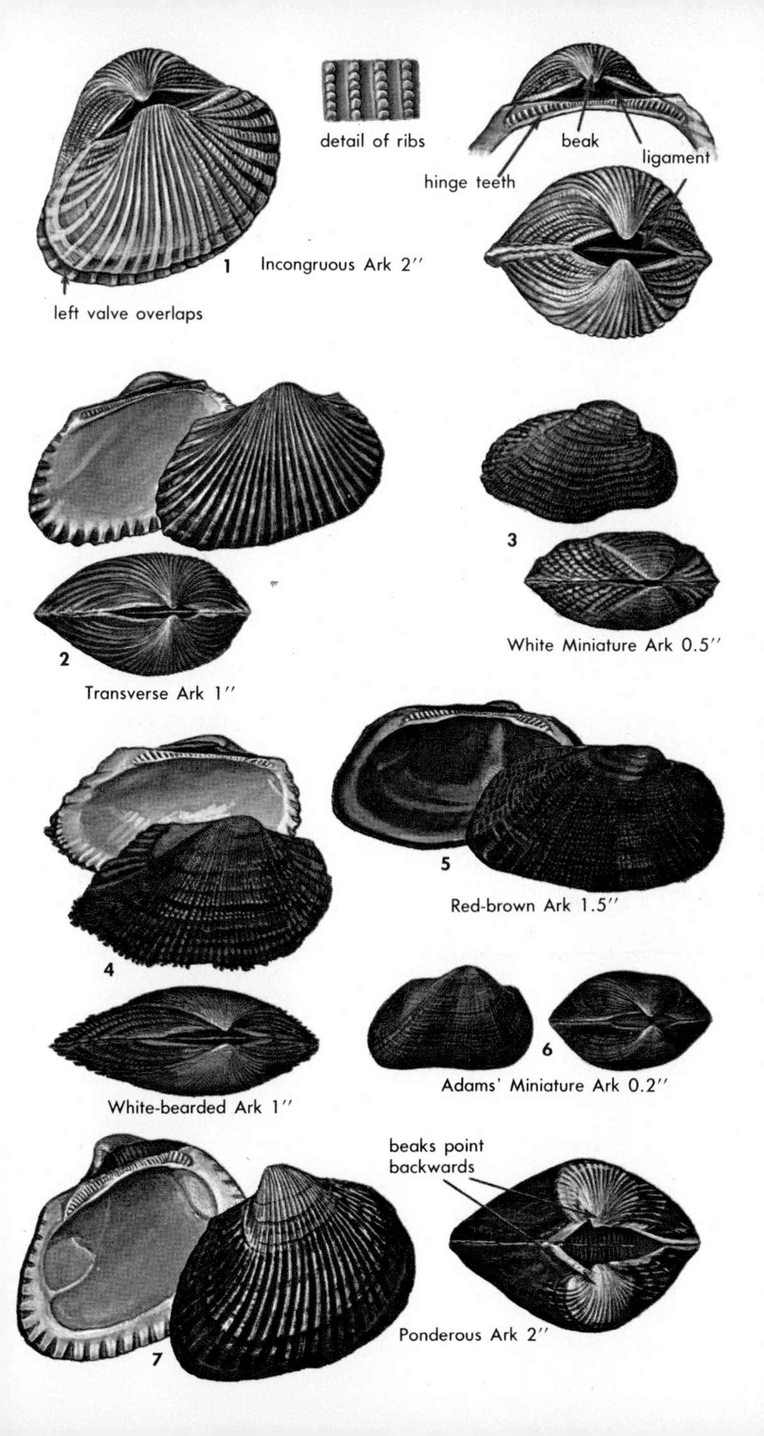
detail of ribs
beak
ligament
hinge teeth
1
Incongruous Ark 2″
left valve overlaps
3
White Miniature Ark 0.5″
2
Transverse Ark 1″
5
Red-brown Ark 1.5″
4
White-bearded Ark 1″
6
Adams' Miniature Ark 0.2″
beaks point
backwards
Ponderous Ark 2″
7

BITTERSWEET CLAMS

Bittersweet, or dog-cockle, clams, family Glycymeridae, are related to the arks but differ in having colorful, compressed, oval shells with relatively fewer teeth in the hinge. The ligament is external, and the strongly curved hinge has numerous strong teeth. There is no pallial sinus scar on the inside of the valves. The posterior muscle scar usually has a built-up, calcareous ridge. The smoother, anterior muscle scar is larger in area. Living specimens are covered with a thin velvety periostracum, usually worn away near the center of the valves. Species are distinguished by the direction in which the beaks turn, their distance apart, and by the number and size of the radial ribs. There are less than a dozen living species in American waters, and most of them are tropical.

Glycymeris generally are found in sandy, shallow areas, although several species are abundant on the continental shelf at depths of 100 to 300 feet. The mantle edge is simple and bordered by numerous light-receptive eyes. The muscular, curved foot is powerful enough to permit the clam to leap off the bottom.

The bittersweet clams are an ancient group, having evolved probably during late Jurassic times, some 150 million years ago. The family originated in southern European waters and had a tendency to produce heavier, smoother, and more rounded shells during later times.

1. **DECUSSATE BITTERSWEET.** *Glycymeris decussata* (Linné). S.E. Fla.–Brazil. Shell is inequilateral. Ribs smoothish, with fairly strong radial scratches. The beaks point posteriorly. Nearly all the ligament is on anterior portion of shell. Moderately common; on sandy bottoms, 6–200 ft.

2. **COMB BITTERSWEET.** *Glycymeris pectinata* (Gmelin). N.C.–Fla. and Caribbean. Shell gray, splotched with brown; equilateral. Has 20–40 well-rounded, raised ribs that lack radial striae or scratches. Ligamental area evenly placed on both sides of the beak. Common; on sandy bottoms in shallow water.

3. **SPECTRAL BITTERSWEET.** *Glycymeris spectralis* Nicol. N.C. and S.C. Shell slightly inequilateral, somewhat oval. The beaks point slightly toward the posterior end. Color almost uniform light brown. Do not confuse with small specimens of Atlantic Bittersweet. Moderately common; 20–200 ft.

4. **PACIFIC COAST BITTERSWEET.** *Glycymeris subobsoleta* (Carpenter). Aleutian Is.–Baja Calif. Shell inequilateral, chalky textured, usually white but occasionally with brownish marking. Ligament area short. Ribs flat with narrow interspaces. Fairly common; shallow to deep water.

5. **GIANT AMERICAN BITTERSWEET.** *Glycymeris americana* (DeFrance). N.C.–Caribbean. A dull gray or tan shell, considerably larger and more compressed than other bittersweets. Beaks at midpoint of hinge and point toward each other. See dorsal view and that of Atlantic Bittersweet. Rather rare; 50–300 ft.

6. **ATLANTIC BITTERSWEET.** *Glycymeris undata* (Linné). N.C.–S.E. Fla. and Brazil. Valves somewhat rounded, with numerous weak ribs. Fine radial and concentric scratches give the shell a silky look. Beak at middle of the ligament. Common; in sand, 3–80 ft. This species was formerly called *G. lineata* Reeve.

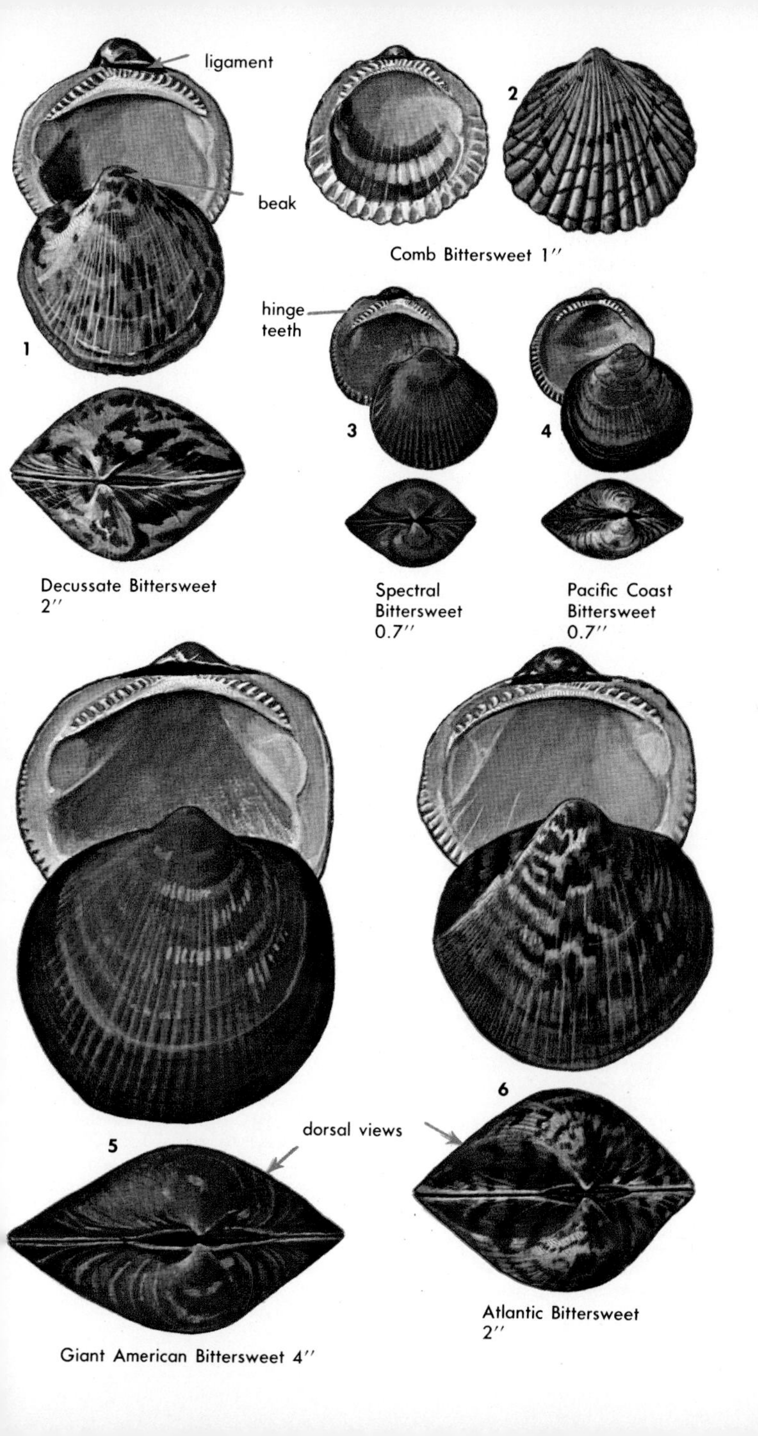

Comb Bittersweet 1″

Decussate Bittersweet 2″

Spectral Bittersweet 0.7″

Pacific Coast Bittersweet 0.7″

Giant American Bittersweet 4″

Atlantic Bittersweet 2″

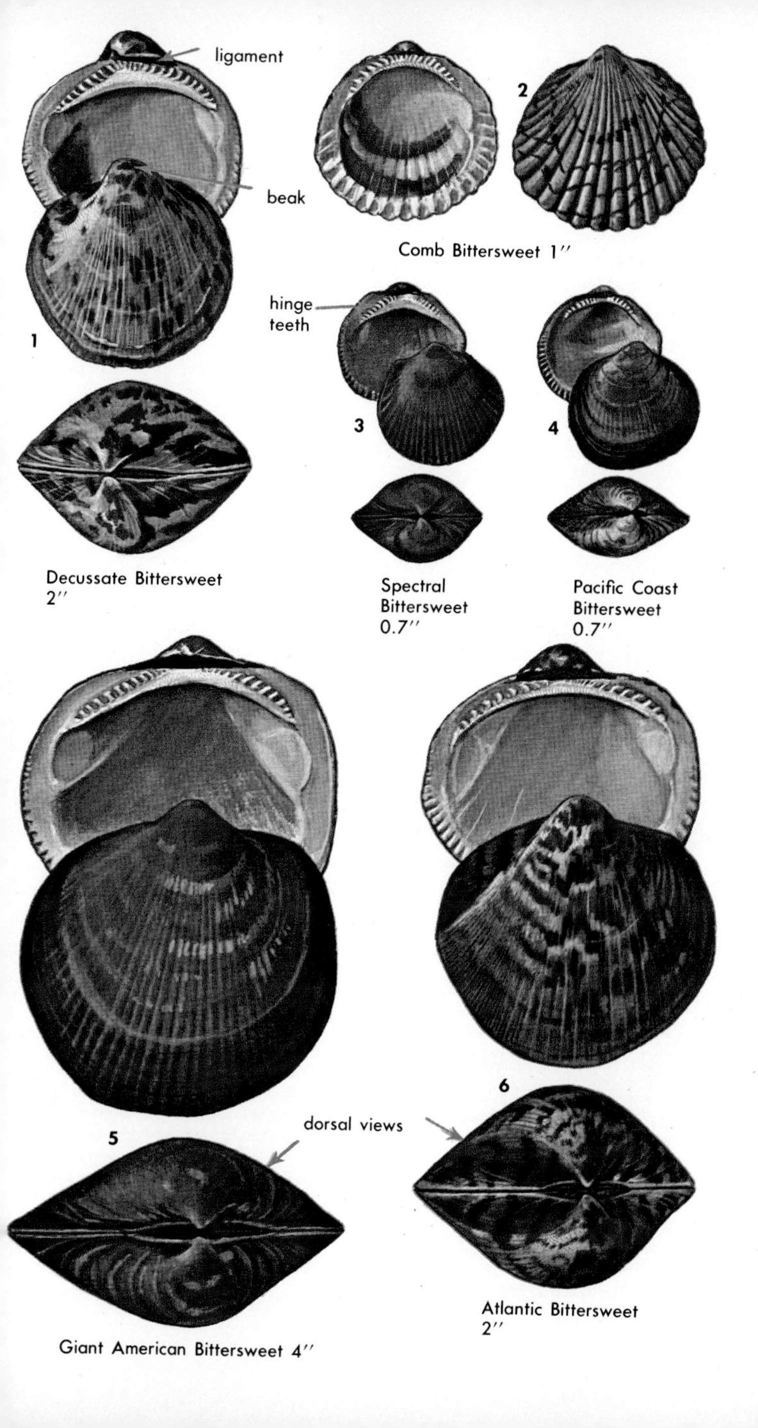

Comb Bittersweet 1″

Decussate Bittersweet 2″

Spectral Bittersweet 0.7″

Pacific Coast Bittersweet 0.7″

Atlantic Bittersweet 2″

Giant American Bittersweet 4″

TRUE MUSSELS

Mussels of the family Mytilidae are found in all seas, but are most prolific in cool waters. Most species have thin, strong, more or less pear-shaped shells with somewhat iridescent interiors. Some genera have several small, weak hinge teeth, while others have none at all. Shells have valves of equal size and are much longer than they are wide. All have shiny, polished interiors.

Although sedentary, mussels are not permanently anchored by the byssus. Some species live attached to rocks, exposed to strong surf. Others prefer more protected locations, clustering by the millions to underwater objects—to such an extent that they help form a barrier against wave action. Some species of mussels seek shelter in the burrows of various rock borers, attaching themselves to the walls with byssal threads. There are about 40 members of this family in North American waters.

1. **COMMON BLUE MUSSEL.** *Mytilus edulis* Linné. Arctic Ocean–S.C. and Calif. Shell bluish black, often with purplish eroded areas. Some specimens show brownish radial rays under the shiny, varnish-like periostracum. Ventral margin straight or somewhat curved. No ribs are present but coarse, prominent growth lines are often seen. Interior pearly white or grayish, much darker or purple along the border. Four small teeth are found in the margin, under the beak at the apex of the shell. The ligament is external. Very common; in quiet, shallow waters, attached to rocks and pilings. The Blue Mussel often occurs in crowded colonies on intertidal rocks. Great quantities of this species are sold for food in northern Europe.

2. **CALIFORNIAN MUSSEL.** *Mytilus californianus* Conrad. Alaska–Calif. Shell tan or brownish, strong, inflated, with very coarse growth lines. Some radial ribs are always present. Ventral margin almost straight. Apex usually eroded. Abundant; on rocks between tides.

3. **ATLANTIC RIBBED MUSSEL.** *Modiolus demissus* (Dillwyn). Nova Scotia–Fla. Introduced to San Francisco Bay about 1894. Shell thin, strong, yellowish brown, with numerous strong radiating ribs. Interior bluish white. Beaks near apex. No hinge teeth. Common; embedded in mud-sand flats at low-tide mark; prefers brackish water. The subspecies *M. d. granosissima* Sowerby, Fla.–Yucatan, has many more finer, beaded ribs.

4. **CAPAX HORSE MUSSEL.** *Modiolus capax* Conrad. Calif.–Peru. Shell bright orange-brown under its thick, hairy periostracum. Has coarse growth lines and no ribs. Interior half bluish white, half purple. Fairly common; on rocks from tide line to 200 ft. A small, soft-bodied peacrab lives inside.

The Northern Horse Mussel. *Modiolus modiolus* (Linné) (not illus.). N.E. United States, 2–6 in.; similar to Capax Horse Mussel, but with shell mauve-white under a thick, brown periostracum. Largest and commonest mussel of New England.

5. **TULIP MUSSEL.** *Modiolus americanus* (Leach). N.C.–Caribbean. Shell is smooth, thin but strong; usually light brown but sometimes has fine rose or light purple rays. Ventral area has a large splotch of brown. Periostracum brown and often hairy. The dull white interior is sometimes stained with blue, rose, or light brown. Very common; in clumps on rocks, 6–100 ft. This species attaches itself to broken shells and rocks. Washed ashore after storms.

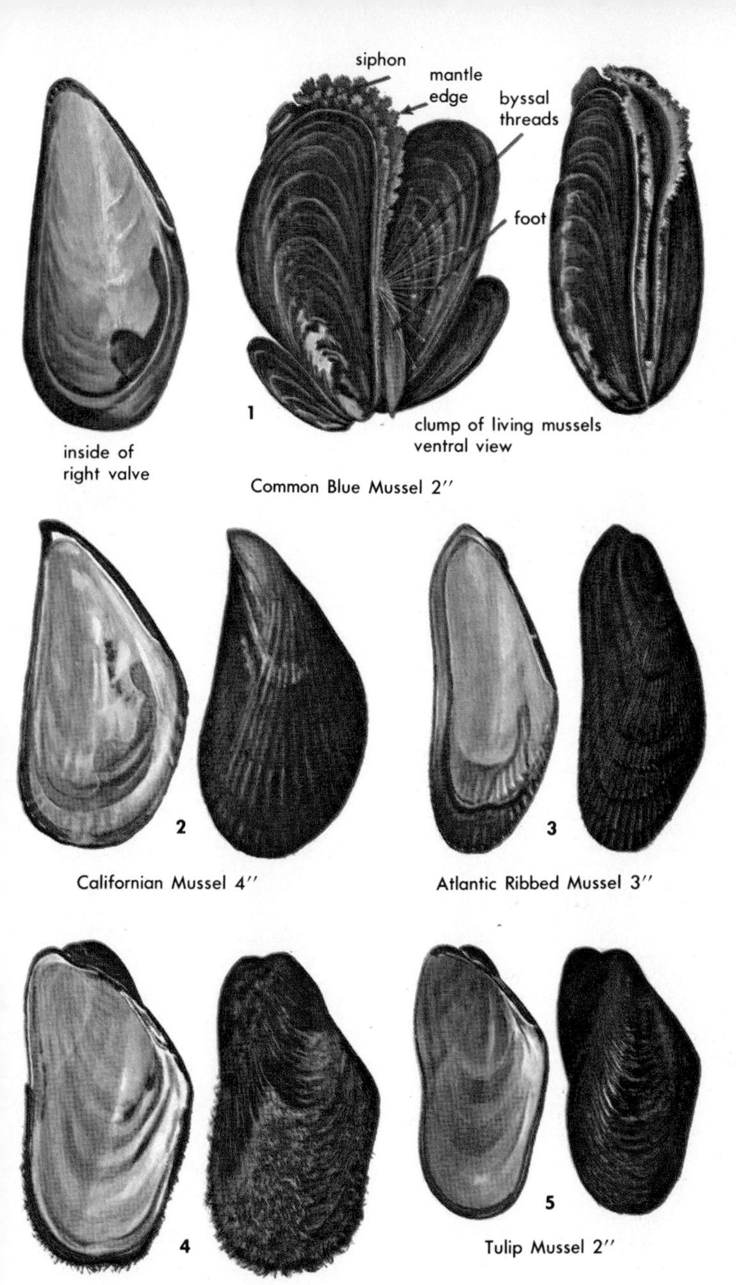

Common Blue Mussel 2″

Californian Mussel 4″

Atlantic Ribbed Mussel 3″

Tulip Mussel 2″

Capax Horse Mussel 4″

MUSSELS—continued

In some parts of the world, mussel poisoning is not uncommon. This serious, sometimes fatal, illness is caused by a toxic substance released by microscopic algae (dinoflagellates). At certain times, these organisms multiply enormously and become concentrated in the gill cavities of mussels and other mollusks that feed by filtering sea water. Cooking does not destroy the poison. In California and Nova Scotia, state laws forbid the collection and eating of mussels during the summer months.

Date mussels (*Lithophaga* and *Botula*) burrow in rock, shells, or coral by rotating constantly until a cavity, which exactly fits the shell, has been formed. The heavy periostracum prevents damage to the shell during the burrowing process. The shapes of the siphonal holes in the rocks are usually characteristic for each species.

Many parasites are associated with mussels, and these include ciliates, sporozoans, larval worms, blood-sucking snails, and small, adult *Pinnotheres* crabs. The latter live in the mantle chamber.

1. **YELLOW MUSSEL.** *Brachidontes citrinus* (Röding). S. Fla.–Caribbean. Shell long, thin but strong, with numerous fine radial ribs. Anterior end has 4 minute white teeth. Ligament bordered by about 30 tiny, equal-sized teeth along the edge of shell. Periostracum thin and a conspicuous light brownish yellow. Common; 6–30 ft., often with Scorched Mussel. Small clusters are washed ashore after winter storms.

2. **HOOKED MUSSEL.** *Brachidontes recurvus* (Rafinesque). Mass.–Caribbean. Shell solid, flat, wide, with numerous strongly curved radiating ribs. At umbonal end, 3 or 4 small teeth mark the edge of the shell. Common; on pilings.

3. **SCORCHED MUSSEL.** *Brachidontes exustus* (Linné). N.C.–Caribbean. Smaller, wider than Yellow Mussel, with only 2 tiny purplish teeth at the anterior end. Beyond the ligament, the edge of the shell has teeth. Common; intertidal.

4. **FALCATE DATE MUSSEL.** *Botula falcata* (Gould). Ore.–Calif. Shell thin, long, and slightly curved. Marked angular ridge from beak. Periostracum dark brown, wrinkled. Bores deeply into hard rock. Common in shallow water.

5. **CALIFORNIAN DATE MUSSEL.** *Botula californiensis* (Philippi). W. Canada–Calif. Shell not so long, curved, smooth. Periostracum velvety, hairy at posterior end. Moderately common; boring in rocks.

6. **BLACK DATE MUSSEL.** *Lithophaga nigra* (Orbigny). S.E. Fla.–Caribbean. Shell long and cylindrical, with strong vertical ribs on anterior part of each valve. Remainder of shell smooth. Common; boring in coral.

7. **SCISSOR DATE MUSSEL.** *Lithophaga aristata* Dillwyn. S. Fla.–Brazil., also S. Calif.–Peru. Pointed tips at posterior end look like crossed, yellowish brown fingers. Fairly common; boring in soft rock.

8. **GIANT DATE MUSSEL.** *Lithophaga antillarum* (Orbigny). S. Fla.–Caribbean. Long, cylindrical, with numerous irregular, vertical lines. Exterior light yellow-brown; interior cream. Fairly common; in soft coral rock, in shallow water.

9. **MAHOGANY DATE MUSSEL.** *Lithophaga bisulcata* (Orbigny). N.C.–Caribbean. Each valve divided by a sharp, oblique line; usually lime encrusted, especially posterior end. Fairly common; in soft rock and corals, in shallow water.

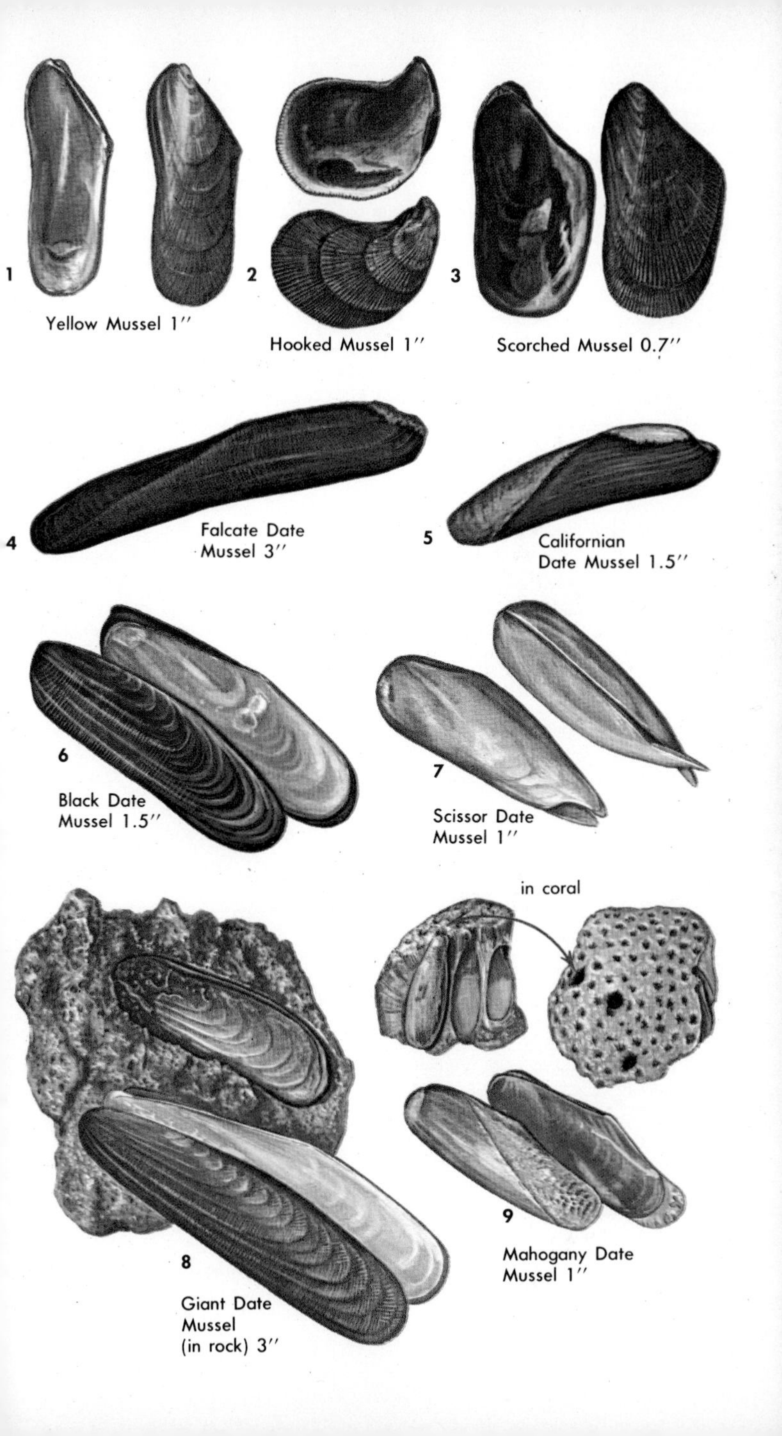

1
Yellow Mussel 1″
2
Hooked Mussel 1″
3
Scorched Mussel 0.7″
4
Falcate Date Mussel 3″
5
Californian Date Mussel 1.5″
6
Black Date Mussel 1.5″
7
Scissor Date Mussel 1″
in coral
8
Giant Date Mussel (in rock) 3″
9
Mahogany Date Mussel 1″

TREE AND PEARL OYSTERS

Two families, the tree oysters (Isognomonidae) and the pearl oysters (Pteriidae), are found in the warmer waters of the tropics. Only distantly related to the edible oysters, these families are of great economic importance because six tropical species are producers of precious pearls. All species produce nacre, or mother-of-pearl. Layers of nacre, when coated around a grain of sand or some other small irritant in the mantle, become the only gem of animal origin. Our small native pearl oysters produce free pearls only on rare occasions. Edible oysters can produce a valueless pearl.

Only about 25 species are in these two closely related families. Most are found in shallow water, from the intertidal zone to about 50 feet; a few are found in deeper water. The family Isognomonidae has many equal-sized teeth in the hinge between which are small, square pads of horny material.

The pearls from specimens of the Atlantic Pearl Oyster that grow in Florida waters are usually too small to be of any great value. Those growing in the southern Caribbean are larger and of commercial value. Pearls of gem quality are usually judged by the perfection in shape, the weight, size, color, and orient. The latter refers to the luster, or reflection of light, and to the iridescence, or interference of the light waves. Most of the pearl is made up of calcium carbonate, the remaining parts being water and organic matter.

1. **FLAT TREE OYSTER**. *Isognomon alatus* (Gmelin). S. Fla.–Brazil. Valves extremely flat, with rough to smooth growth lines. Interior somewhat pearly, stained or mottled with brown, black, or purple. Hinge has 8–12 long grooves. Common; in large clumps attached by byssi to mangrove roots and wharf pilings.

2. **BICOLOR TREE OYSTER.** *Isognomon bicolor* (C. B. Adams). Fla.–Texas–Caribbean. Shell fairly heavy, somewhat oval. Exterior often has strong plates representing periodic growth. Both interior and exterior of shell strongly splashed with purple. Pallial line raised. Common; in intertidal waters.

3. **LISTER'S TREE OYSTER.** *Isognomon radiatus* (Anton). S.E. Fla.–Caribbean. Long but irregular in shape, often quite twisted. Exterior rough, with weak plates or ridges. Shell yellowish, with a few very light radial rays. Hinge has 4–8 widely spaced grooves. Common; on rocks in shallow water.

4. **ATLANTIC WING OYSTER.** *Pteria colymbus* (Röding). N.C.–S. Fla. and Brazil. The long posterior wing identifies this species. Left valve inflated; right valve flattened. Interior pearly with a wide nonpearly margin. Periostracum brown, matted. Common; attached to sea whips, 12–100 ft.

5. **WESTERN WING OYSTER.** *Pteria sterna* (Gould). S. Calif.–Panama. Similar to Atlantic species but with longer posterior wing and preferring sandy mud in shallow, offshore waters as a habitat. Deep purplish brown with occasional paler rays. Once gathered commercially for the mother-of-pearl. Fairly common.

6. **ATLANTIC PEARL OYSTER.** *Pinctada radiata* (Leach). S. Fla.–Brazil. Usually flattened, rather thin-shelled, and brittle. Color variable. Interior pearly. Posterior wing very short. Long, delicate spines on the tan-brown periostracum sometimes occur on specimens from quiet waters. Common; on rock and sea fans, 6–60 ft.

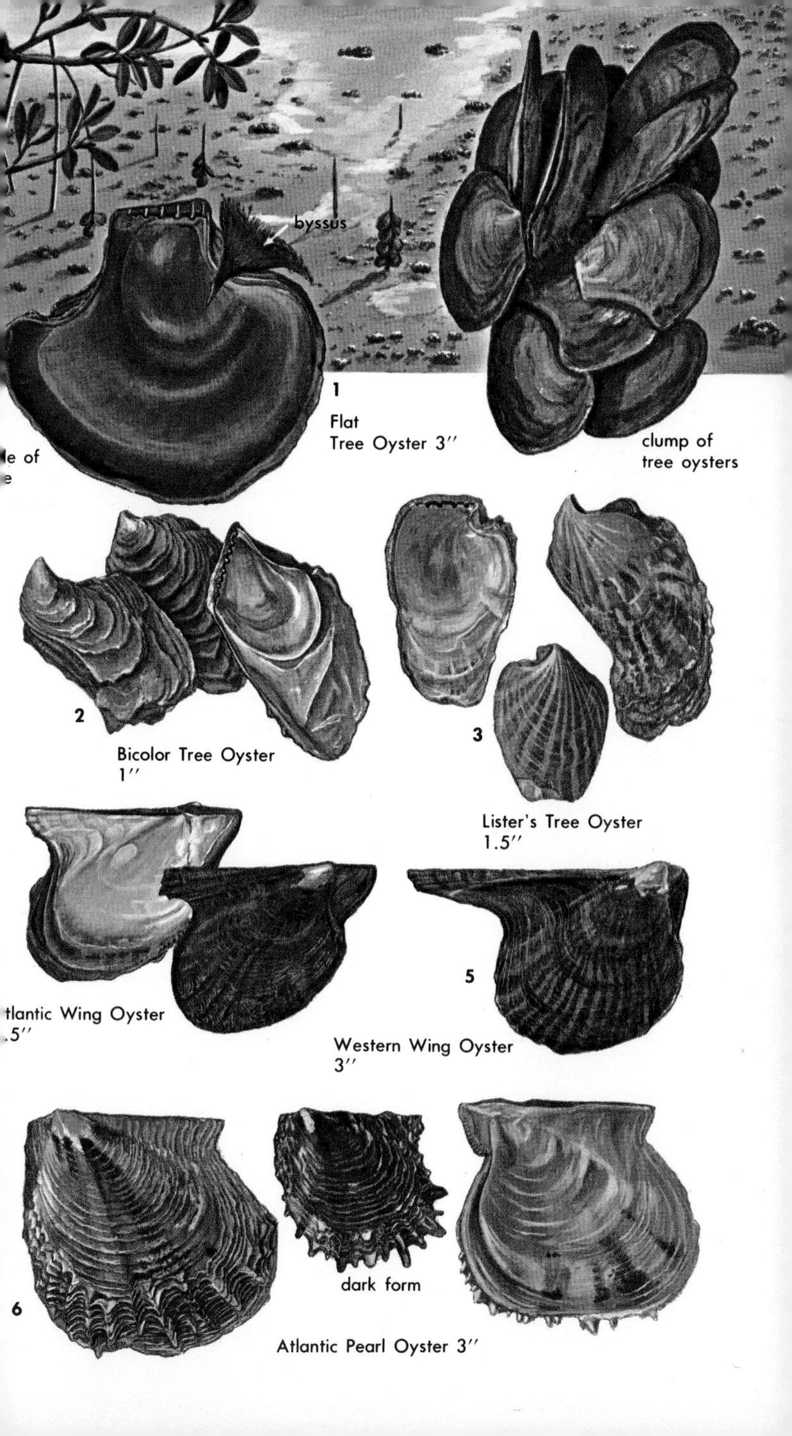
byssus
1
Flat
Tree Oyster 3″
clump of
tree oysters
le of
e
2
Bicolor Tree Oyster
1″
3
Lister's Tree Oyster
1.5″
tlantic Wing Oyster
.5″
5
Western Wing Oyster
3″
dark form
6
Atlantic Pearl Oyster 3″

PEN SHELLS

The family Pinnidae of thin, brittle, fan-shaped shells includes only four species and two genera in American waters—*Pinna* and *Atrina*. The shell of the latter lacks the central groove (or sulcus) which runs down the middle of the inner surface of *Pinna*. Pen shells live in soft, sandy mud with the narrow, umbonal tip downward. The small foot spins clumps of thin, chitinous threads that are attached to buried stones and broken shells. If uprooted, the pen shells succumb to enemies or soon become broken. The brittleness of the shell is due in part to the very large size of the prismatic crystals, which can be easily seen with the aid of a hand lens.

The soft parts of these clams are unique in possessing a finger-like pallial organ and a gutter-like waste canal. These organs remove pieces of broken shell, rejected food, and other debris from the mantle cavity. The anterior muscle is small and located at the narrow end of the shell valves. The posterior muscle is very large and centrally located. It is used as seafood in some countries.

The rough, spiny exterior of the valves helps anchor the pen shell in the mud and also serves as an excellent holdfast for barnacles, oysters, tube worms, and algae. Within the mantle cavity of this bivalve a small crab, *Pinnotheres*, receives protection and feeds upon surplus food particles.

1. **SAW-TOOTHED PEN SHELL.** *Atrina serrata* Sowerby. N.C.–Texas; Caribbean. Shell very thin, with over 30 ribs bearing numerous small scales. Large muscle scar in nacreous area. Moderately common; low-tide line to 20 ft.

2. **RIGID PEN SHELL.** *Atrina rigida* (Lightfoot). N.C.–Fla. and Caribbean. Shell wide, usually dark olive-brown, with 15 or more radial rows of tube-like spines. Large muscle scar on border of nacreous area. Mantle bright orange. Common; in shallow bays.

Half-naked Pen Shell. *Atrina seminuda* (Lamarck) (not illus.). N.C.–Texas–Argentina. Exterior like Rigid Pen, but muscle scar is within nacreous area. Mantle pale yellowish. One half of each valve smoothish. Common; 1–70 ft.

3. **AMBER PEN SHELL.** *Pinna carnea* Gmelin. S.E. Fla.–Caribbean. Shell pale orange to amber, thin, and very fragile; rather narrow, with about 10 radial ridges, which may be spiny or smooth. Uncommon in Fla.; rare on buoys. Usually found deeply buried in fine coral sand.

Muscle Scars on Inner Surface on Pen Shells

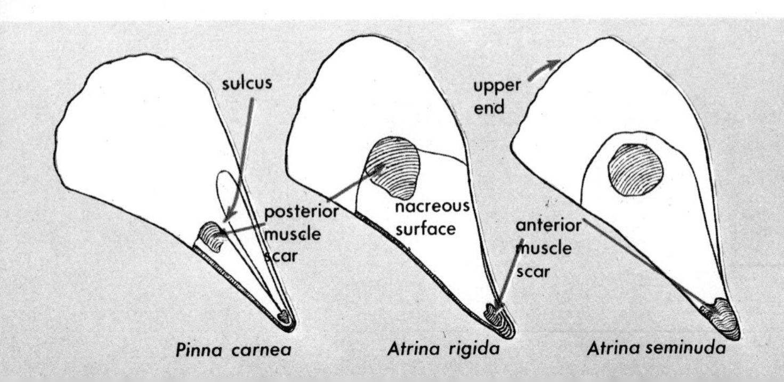

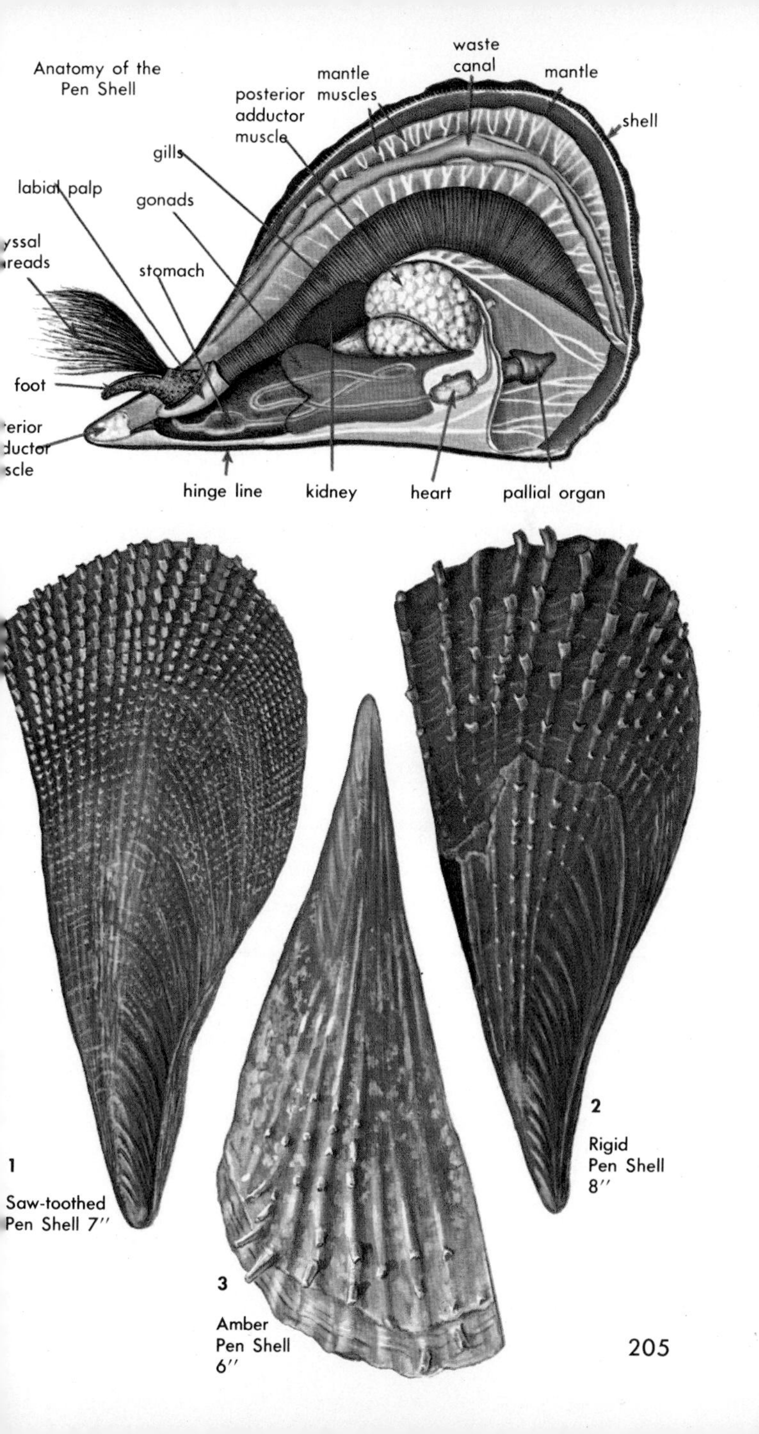

1 Saw-toothed Pen Shell 7″

2 Rigid Pen Shell 8″

3 Amber Pen Shell 6″

SCALLOPS

This large and diverse group of bivalves has a single, fused adductor muscle, and a hinge bearing a socket-like arrangement (see Kitten's Paw). The foot is greatly reduced and no siphons have developed along the mantle edge. The true scallops (family Pectinidae) include several genera of edible, commercially important bivalves. Two closely related families (or subfamilies) are the Plicatulidae, or kitten's paw, which have heavy, strongly hinged valves, and the Propeamussiidae, or glass scallops, which have delicate, translucent valves with internal ribs. Some glass scallops feed on crustaceans.

Over 50 genera of worldwide scallops have been proposed, but the differences are so slight that it appears that most of our 50 American species may be placed in about 7 genera. In true *Pecten,* the "ears" on the hinge line are about equal in size and the right or bottom valve is very deeply convex, while the left or upper valve is almost flat. In *Chlamys,* one "ear" is much larger than the other. *Hinnites,* the sessile rock scallop, has the characters of *Chlamys* when young, but it later grows heavy and irregular. The *Lyropecten* Lion's Paw Scallop usually has large knobs on its strong ribs. These are filled with an aqueas solution akin to seawater. The cavities dry out in dead specimens in collections. The knobs and heavy ribs in most kinds of scallops serve as strengthening structures against marauding fish.

1. **KITTEN'S PAW.** *Plicatula gibbosa* (Linné). N.C.–Caribbean. Shell thick. Several high, rounded ribs give both valves a strong, zigzag margin. Two strong, equal-sized teeth in the upper valve fit into 2 sockets in the lower. Common; intertidal to 200 ft., attached to shells and rocks.

2. **POURTALES' GLASS SCALLOP.** *Propeamussium pourtalesianum* (Dall). S.E. Fla.–Caribbean. Shell very thin, transparent. Each valve has 8 or 9 opaque, white reinforcing ribs internally. One valve quite smooth externally, the other has numerous extremely fine concentric lines. Common; dredged at 50–500 ft.

3. **GIANT ROCK SCALLOP.** *Hinnites multirugosus* Gale. Alaska–Baja Calif. Free-swimming, young specimens resemble *Chlamys* except for the purple splotch along the inside of the hinge line. The massive, oyster-like adult is attached by the lower valve to rocks. Fairly common; 1–150 ft., on wharf pilings.

4. **SAN DIEGO SCALLOP.** *Pecten diegensis* Dall. Calif. Has 22 or 23 strong, flattened, grooved ribs on the convex valve. The flatter, upper valve has 21 or 22 more widely spaced, narrower, rounded ribs. Uncommon; 50–400 ft. A rare Capshell (p. 100) may live on upper valve.

5. **ZIGZAG SCALLOP.** *Pecten ziczac* (Linné). N.C.–Caribbean. Has 18–20 broad, widely spaced, low ribs on the convex, mottled, reddish brown lower valve, and 35 or more compressed ribs on the upper valve. Interior of upper valve white, margins tinted. Common; 10–100 ft. Rare albinos occur.

6. **RAVENEL'S SCALLOP.** *Pecten raveneli* Dall. N.C.–Caribbean. Similar to Zigzag Scallop. Lower valve convex. About 25 whitish ribs with wide grooves in between. Interior margin lightly tinted. Ribs on flat upper valve are rounded. Color variable. Uncommon; 20–200 ft. This is a popular collector's item.

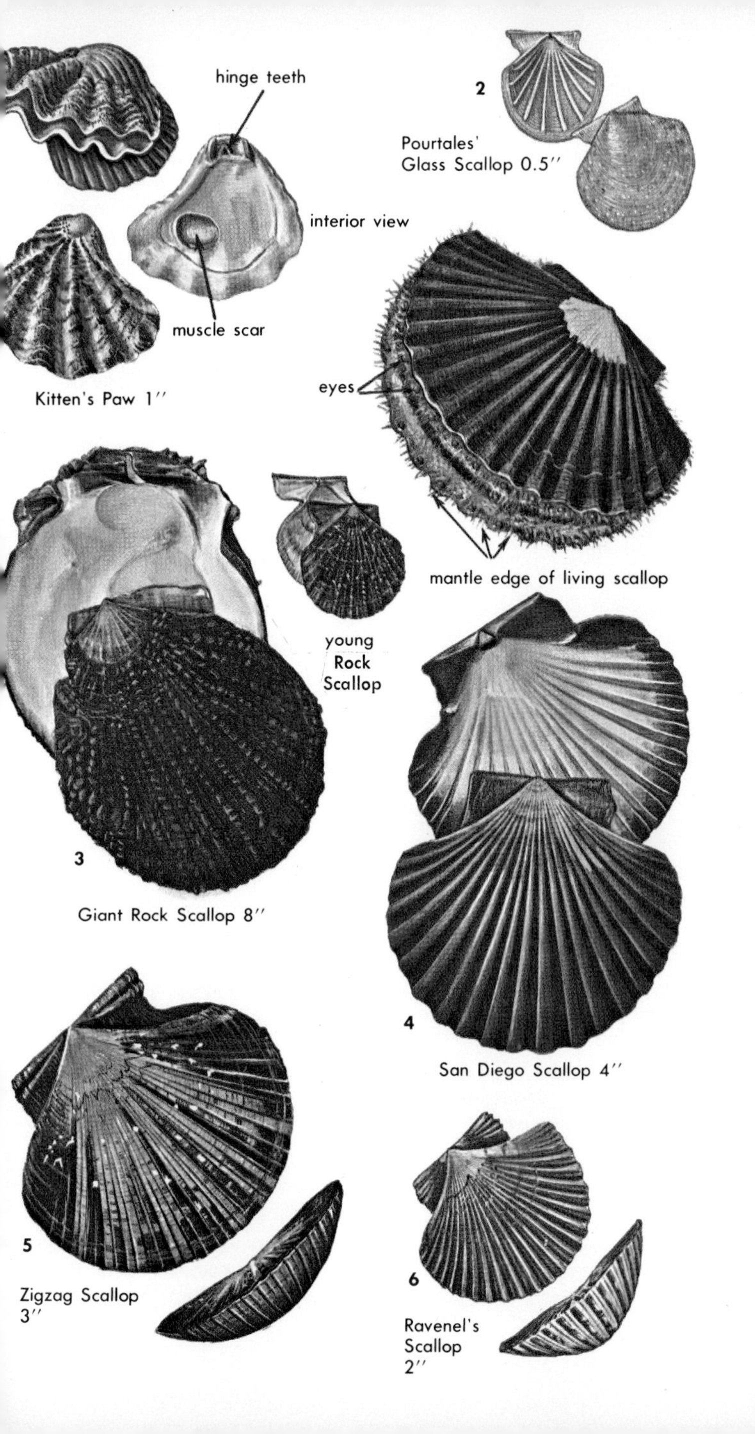
hinge teeth
interior view
muscle scar
Kitten's Paw 1″
2
Pourtales'
Glass Scallop 0.5″
eyes
mantle edge of living scallop
young
Rock
Scallop
3
Giant Rock Scallop 8″
4
San Diego Scallop 4″
5
Zigzag Scallop
3″
6
Ravenel's
Scallop
2″

SCALLOPS—continued

Scallops are gregarious in nature, and some free-swimming species are capable of short-range migrations. Species in the genus *Chlamys* are sedentary and live attached by a byssus under rocks in intertidal areas. Scallops are renowned for their ability to swim. The opening and sudden snapping shut of the valves creates strong jets of water that propel the scallop. By varying the position of escaping water by means of the muscular mantle edge, the scallop may swim in any direction. When swimming normally, the scallop claps its valves as though it were taking a series of bites out of the water. Water escaping behind at either side of the hinge pushes the scallop forward. The swimming ability prevents burial in muddy bottoms and often enables the scallop to escape predatory starfish and fish.

All scallops possess well-developed eyes set along the edge of the fleshy mantle. Some species may have up to 100 jewel-like eyes, each of which has a lens, retina, and optic nerve. Scallops respond instantly to changes in light intensity or nearby moving objects. Small tentacles along the edge of the mantle are extremely sensitive to odors and changes in water pressure and serve as an advance guard against potential enemies. Members of the genus *Chlamys* anchor themselves with a tough byssus.

1. **SENTIS SCALLOP.** *Chlamys sentis* (Reeve). N.C.–Caribbean. Valves quite flat, with about 50 various-sized, finely scaled ribs. Colors variable. Common; under rocks, 1–20 ft.

2. **PACIFIC SPEAR SCALLOP.** *Chlamys hastata* (Sowerby). Alaska–Calif. 18–21 weakly scaled primary ribs on right valve. Weak ribs in between. Common; 20–200 ft.

3. **ICELAND SCALLOP.** *Chlamys islandica* (Müller). Arctic Ocean–Mass.; Alaska–Wash. Valves somewhat convex, with 50 or more irregular, weakly scaled ribs, which are occasionally arranged in groups. Color variable, usually gray or cream but occasionally yellow, reddish, or purplish. Very common; 10–300 ft., on coarse sand. Fished commercially in a few areas.

4. **HINDS' SCALLOP.** *Chlamys hindsi* (Dall). Alaska–Calif. 25 or more primary ribs, each with 3 rows of spines. There are microscopic crisscross markings between the main ribs. Right valve has fewer, smoothish, rounded ribs. Color variable: yellow, lavender, rose, white, or orange. Rather common; in offshore waters, 10–2,000 ft.

5. **KELP-WEED SCALLOP.** *Leptopecten latiauratus* Conrad. Central Calif.–Baja Calif. Thin shelled, with 12–16 low, squarish ribs. Usually yellowish to reddish brown, with white zigzag markings. Common; on kelp and stones in shallow water. Subspecies *monotimeris* Conrad has rounded ribs and less sharply pointed ears.

6. **GIANT PACIFIC SCALLOP.** *Pecten caurinus* Gould. Alaska–Calif. Upper valve almost flat, with about 17 very low, rounded ribs. The lower valve is more convex, with wider ribs, and whitish. Offshore to 200 ft.; fished commercially.

7. **ATLANTIC DEEP-SEA SCALLOP.** *Placopecten magellanicus* (Gmelin). Labrador–N.C. Large; valves flattened. Exterior rough, with numerous small, raised, threadlike ribs. Rarely, has color rays. Common; 20–300 ft.; fished commercially.

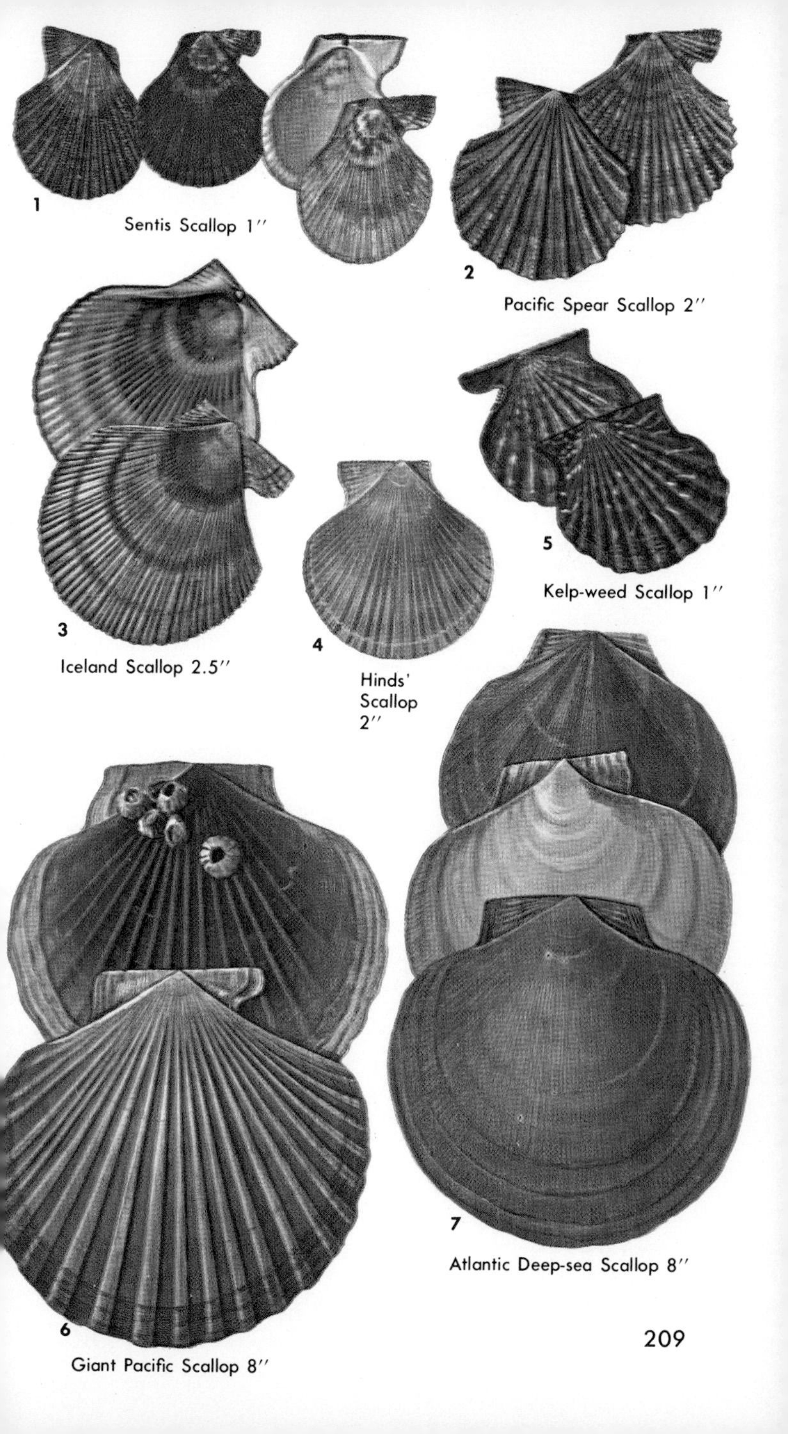

1 Sentis Scallop 1″

2 Pacific Spear Scallop 2″

3 Iceland Scallop 2.5″

4 Hinds' Scallop 2″

5 Kelp-weed Scallop 1″

6 Giant Pacific Scallop 8″

7 Atlantic Deep-sea Scallop 8″

SCALLOPS—continued

In some species, such as the Atlantic Deep-sea Scallop, the sexes are in separate individuals, but many other species are hermaphroditic. Eggs and sperm are shed freely into the ocean water in early spring. The larval scallop is free-swimming for several days to a few weeks. Young "spat" scallops attach themselves by means of delicate threads spun by the foot. Sexual maturity is reached at the second year, and some scallops may live to an age of 5 or 6 years. The growth rings in the shell can sometimes be used to determine age.

Only the round, disc-shaped muscle of the scallop is used for food purposes. The Atlantic Deep-sea Scallop, *Placopecten,* is dredged year-round along the Atlantic coast. Most of the annual, 30-million-pound catch is made off Massachusetts. The scallops are shucked at sea and the meats bagged, iced, and stored below deck until ready for auctioning at such ports as New Bedford. There is also a large fishery in Alaska. A new scallop industry is growing in southeastern United States, based upon the Calico Scallop, *Aequipecten gibbus.* Large schools were recently discovered off the Carolinas. The live scallops, brought to shore on the decks of trawlers, are shucked by hand and the meat sent iced or frozen to local markets.

1. **ATLANTIC BAY SCALLOP.** *Aequipecten irradians* (Lamarck). Has three subspecies. 1a. The typical subspecies, *irradians,* ranges from Nova Scotia to N.Y. Has 17 or 18 low, rounded ribs. Drab gray-brown. Less convex than the two following subspecies. These are the common, edible bay scallops, found in eel-grass.

1b. The subspecies *concentricus* (Say) ranges from N.J. to Georgia and from Tampa Bay to La. Has 19–21 somewhat squarish ribs. Lower valve light-colored, commonly pure white, and more convex than the grayish to brownish upper valve, which is sometimes white-spotted.

1c. The subspecies *amplicostatus* (Dall) ranges from central Texas to Colombia, S.A. Has 12–17 well-developed, squarish or slightly rounded ribs. Lower valve usually white. Shell very inflated. Common in Texas.

2. **ROUGH SCALLOP.** *Aequipecten muscosus* (Wood). N.C.–Caribbean. The 18–20 main ribs are strongly scaled. Hinge ears very wide. Rarely lemon-yellow specimens are found. Moderately common; in shallow water to 200 ft.

3. **CALICO SCALLOP.** *Aequipecten gibbus* (Linné). N.C.–Brazil. 19–21 squarish ribs. Lower valve commonly whitish with flecks of color. Upper valve has colorful, mixed shades. Common; 6–30 ft.

4. **LION'S PAW.** *Lyropecten nodosus* (Linné). N.C.–Brazil. Has 7–9 strong ribs with large hollow nodes and numerous smaller, distinct riblets. Usually reddish; rarely orange or yellow. Fairly common; 20–100 ft.

The Pacific Lion's Paw, *L. subnodosus* (Sowerby) from southern Calif.–Ecuador is similar, about 6 inches, and lacks the strong nodes.

Front Views of Subspecies of the Bay Scallop

typical subspecies *irradians*

subspecies *concentricus*

subspecies *amplicostatus*

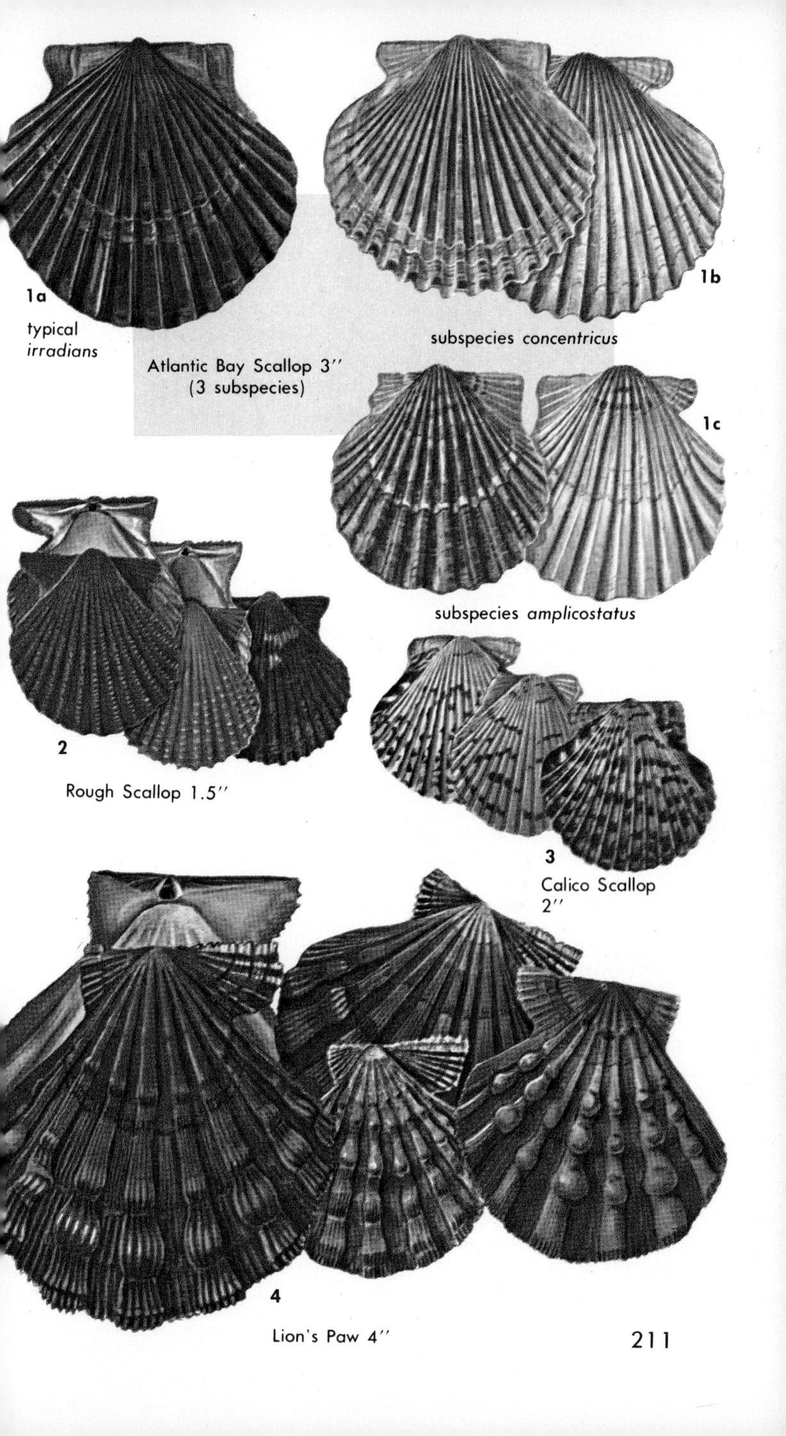
1a
typical *irradians*
Atlantic Bay Scallop 3″
(3 subspecies)
1b
subspecies *concentricus*
1c
subspecies *amplicostatus*
2
Rough Scallop 1.5″
3
Calico Scallop 2″
4
Lion's Paw 4″

THORNY OYSTERS, LIMAS, AND JINGLES

Thorny oysters (family Spondylidae) are related to the scallops but differ in having long spines and a ball-and-socket type of hinge. They have eyes along the mantle edge. Most *Spondylus* live attached to rocks at depths of 10 to 100 feet. In quiet waters, large thorny oysters often develop long spines. They are usually well camouflaged with heavy growths of sponge, bryozoans, and algae. There are two North American species.

Jingle shells (family Anomiidae) are common, translucent shells that are attached to rocks, other shells, and wood. The lower, attached valve has a large hole through which the sturdy byssal clump extends. There are two species on each of the North American coasts.

Lima shells (family Limidae) have long, colorful tentacles and oblique, eared shells that gape open along one side. Although able to swim, they spend most of their time in nests constructed of byssal threads and debris held together by mucus. The sticky tentacles are harmless to touch. There are about 100 worldwide species.

1. **PACIFIC THORNY OYSTER.** *Spondylus princeps* Broderip. Baja Calif.–Panama. Spines usually less than 1½ in. long, somewhat bent. Uncommon; in fairly deep water. Formerly called *S. pictorum.*

2. **ATLANTIC THORNY OYSTER.** *Spondylus americanus* Hermann. N.C. –Caribbean. Usually white, variously tinted at umbones, but sometimes colored vivid purple, red, or yellow throughout. Spines fairly erect, 2 or 3 in. long. Common; in moderately deep water (50–150 ft.).

3. **ATLANTIC JINGLE.** *Anomia simplex* Orbigny. N.Y.–Caribbean. Top valve convex, strong, and either yellow-orange or silvery black. Lower valve flat and fragile, with a slotlike hole near the hinge line. Common; shore to 30 ft. Dead valves common on beaches.

Peruvian Jingle. *A. peruviana* Orbigny (not illus.). S. Calif.–Peru. Almost identical to Atlantic Jingle, sometimes greenish. Common.

Prickly Jingle. *A. aculeata* (Gmelin) (not illus.). Canada–N.C. Also similar to Atlantic Jingle, but much smaller, brown to gray, with minute prickles. Common; on rocks, 6–100 ft.

4. **FALSE PACIFIC JINGLE.** *Pododesmus macroschisma* (Deshayes). Alaska–Baja Calif. Shell quite heavy, strong, irregularly rounded, with many coarse, irregular, radiating ribs. Exterior yellow to greenish white; interior greenish and pearly. Lower attachment valve flat and has a large byssal opening. Very common; attached to stones and pilings from low tide to 200 ft.

False Atlantic Jingle. *P. rudis* (Broderip) (not illus.). Georgia–Caribbean. Very similar to False Pacific Jingle. Exterior is brown to purple. Uncommon; in crevices of coral rocks and on old wrecks, 3–100 ft.

5. **SPINY LIMA.** *Lima lima* (Linné). S.E. Fla.–Caribbean. Shell white with many radial, sharply spined ribs. Byssal gap small; posterior ear much smaller than anterior ear. Fairly common under stones in shallow water.

6. **ROUGH LIMA.** *Lima scabra* (Born). S. Car.–Caribbean. Coarsely sculptured with many irregular, barlike ribs (fig. 6b). Byssal gap fairly large. Ears similar in size. The smooth form *tenera* Sowerby (fig. 6a) is smaller, with more numerous, finer ribs. Common; 1–30 ft.

1
Pacific Thorny
Oyster 5″
2
Atlantic Thorny
Oyster 5″
bottom
valve
young spat
in clam shell
3
Atlantic Jingle
1″
False Pacific
Jingle
2″
4
5
Spiny Lima 2″
living Rough Lima
smooth
phase
Rough Lima 2″
6
b

EDIBLE OYSTERS

The several dozen kinds of oysters of the family Ostreidae are well-known edible shellfish. The porcelaneous shells vary greatly in shape, depending upon environmental conditions. Pearls produced by edible oysters are not nacreous and are of little value.

Two genera occur in American waters. *Crassostrea* has a deep left, or lower, valve and a purplish muscle scar and sheds eggs directly into the ocean. *Ostrea* has two almost equal-sized valves and a nonpurple muscle scar, and the eggs are incubated within the mantle cavity. There are about eight North American species.

Sexes in oysters are separate, but sex reversal is common. An oyster may function as a female for several weeks and then, within a few days, change into a sperm-producing male. The production of eggs occurs in the summer and is usually triggered by a rise in water temperature or by hormones liberated by free-swimming sperm. Eggs develop into tiny, ciliated larvae, which swim about for one to three weeks. In its later stages the larval oyster crawls about by means of a small foot until it locates a suitable hard surface to which to cement itself. Oysters take about three years to reach an edible size, unless they are raised in special food-stocked ponds.

The enemies and diseases of oysters are numerous. Of the approximately 50 million eggs laid by an oyster during one spawning period, only about a dozen reach maturity. Starfish, crabs, and oyster drills do great damage to adult and young "spat" oysters. A parasitic fungus, *Dermocystidium,* kills many oysters but is harmless to man.

1. **EASTERN OYSTER.** *Crassostrea virginica* (Gmelin). New Brunswick–Gulf of Mexico. Generally narrow and elongate, the rough, heavy grayish shell of the common edible oyster is extremely variable in shape. Upper valve smaller and flatter than the lower. Interior white except for purple muscle scar and purple edging.

2. **GIANT PACIFIC OYSTER.** *Crassostrea gigas* (Thurnberg). W. Canada–Calif.; also native to Japan. This large marketable oyster is introduced yearly on the Pacific Coast. Shell very coarsely sculptured and variable in shape. Interior white; muscle scar purplish.

3. **NATIVE PACIFIC OYSTER.** *Ostrea lurida* Carpenter. Alaska–Baja Calif. Shell small, variable, generally rough with coarse growth lines. Interior stained olive-green. This small oyster is much prized as food. Common in the intertidal zone.

4. **CRESTED OYSTER.** *Ostrea equestris* Say. Va.–Texas, and West Indies. Shell usually oval; lower valve has a high, vertical, crenulated margin; muscle scar almost central. Interior pearly green to gray. Edge of upper valve has a row of denticles. Common; subtidal to 300 ft.

5. **COON OYSTER.** *Ostrea frons* Linné. Fla.–Caribbean. Reddish to purple-brown. Variable in shape: long and with claspers when attached to stems of sea whips; rounded in outline when on flat rocks or wood pilings. Inner margins dotted with tiny pimples. Common; 6–60 ft.

Sponge Oyster. *Ostrea permollis* Sowerby (not illus.). N.C.–Caribbean. This is 2 in. long, flat, oval in outline, yellow-tan, and with a silky smooth surface. Beak twisted into a spiral. Inner margins have denticles. Interior white. Lives embedded in the sponge *Stellata,* with only margins of valves showing. Common; 6–60 ft.

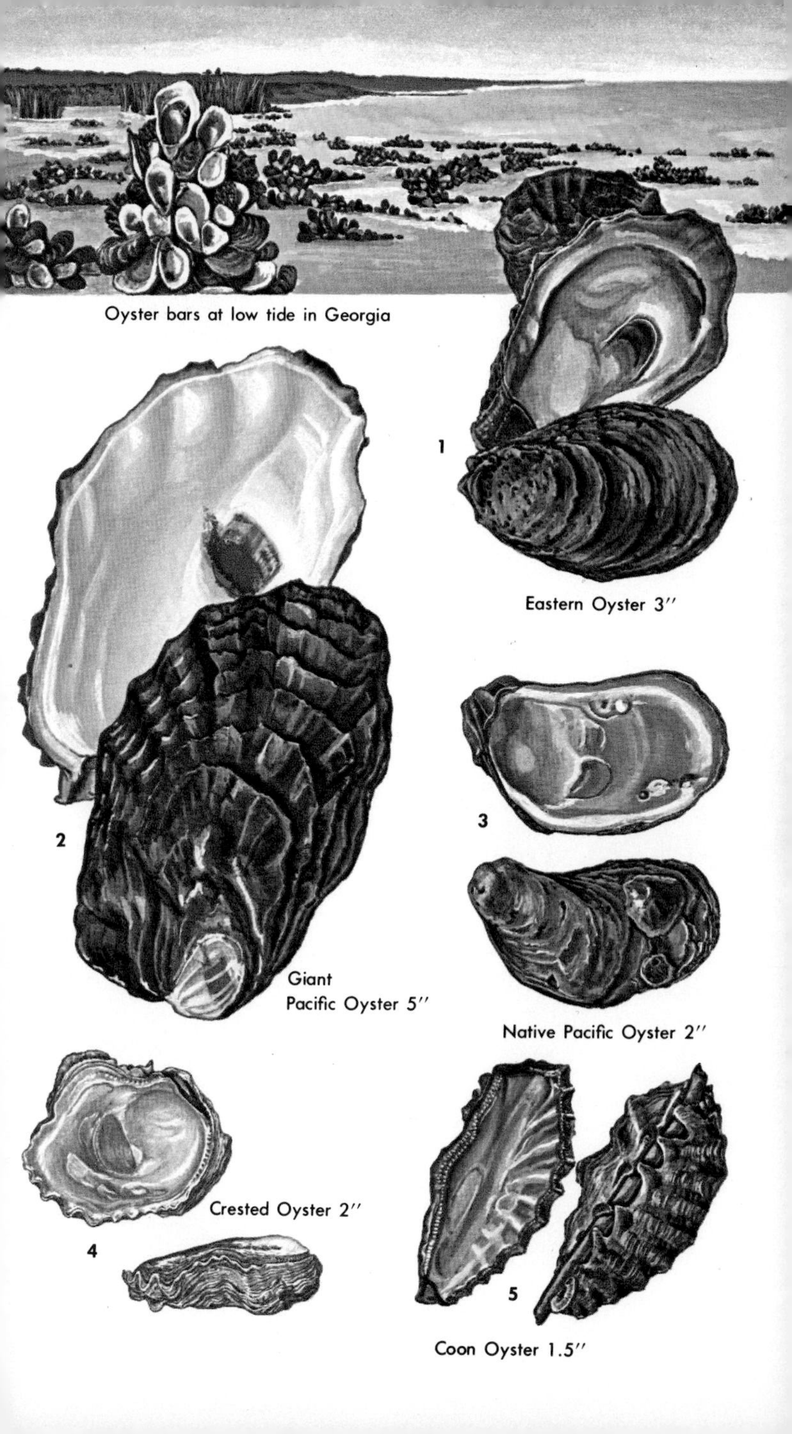

Oyster bars at low tide in Georgia

Eastern Oyster 3″

Giant Pacific Oyster 5″

Native Pacific Oyster 2″

Crested Oyster 2″

Coon Oyster 1.5″

ASTARTES AND CRASSINELLAS

Both of these families of bivalves are characterized by strong, heavy, whitish to pinkish shells, covered by a brown or black periostracum. The flesh is bright orange or reddish. Sexes are separate. These clams have no siphons, and hence the pallial line on the interior of the shell lacks a sinus or an indentation. In Astartidae, of which there are about two dozen North American species, the ligament and resilium are outside the strongly toothed hinge. In Crassatellidae they are internal, and the large resilium sits in a concave pit behind the cardinal or center teeth, as shown in the drawings below. This family has about eight North American species.

Concentric sculpture dominates most species in both families, and the margin of the valves may be minutely serrated. The northern species have a thick periostracum for protection against boring sponges.

1. **BOREAL ASTARTE.** *Astarte borealis* Schumacher. Arctic Seas–Mass. Shell egg-shaped, somewhat compressed, with a large, external ligament. Concentric ridges distinct only near beak; absent near margins of shell. Inner margins smooth. Common; in shallow water to 500 ft.

2. **SMOOTH ASTARTE.** *Astarte castanea* Say. E. Canada–Cape Cod, Mass. Shell thick, solid, and a glossy light brown; almost smooth except for weak concentric lines. External ligament small. Beaks pointed and hooked. Inner margins fine and slightly wavy. Common; in mud, in fairly shallow water to 100 ft.

3. **WAVED ASTARTE.** *Astarte undata* Gould. E. Canada–Mass. Shell solid with about 10 strong concentric ridges. Beaks curve inward. Interior shining white. Periostracum reddish brown. Extremely common; in mud below low-tide mark.

4. **LENTIL ASTARTE.** *Astarte subequilatera* Sowerby. Labrador–Fla. Similar to Waved Astarte but with more numerous concentric, evenly spaced ridges. Beaks often eroded and slightly turned forward. Inner margins finely wavy. Periostracum light yellowish brown, occasionally darker. Common; in shallow water.

5. **LUNATE CRASSINELLA.** *Crassinella lunulata* Conrad. N.C.–Caribbean. Shell compressed, interior brown. Beaks tiny, centrally located. Dorsal margins straight. Common; to 1,800 ft.

6. **LINDSLEY'S CRASSINELLA.** *Crassinella mactracea* Lindsley. Mass.–N.Y. Almost identical with Lunate Crassinella, but texture usually chalky, lunule more oval, ribs weaker. Common; to 150 ft.

7. **GIBB'S CLAM.** *Eucrassatella speciosa* (A. Adams). N.C.–Caribbean. Shell thick and heavy. Sculpture consists of many neat, closely packed concentric ridges. Exterior covered with thin brown periostracum. Interior glossy tan with orange stain. Lunule and escutcheon sunken and elongated. Moderately common; in sand from 12–100 ft. It serves as a food for manta rays.

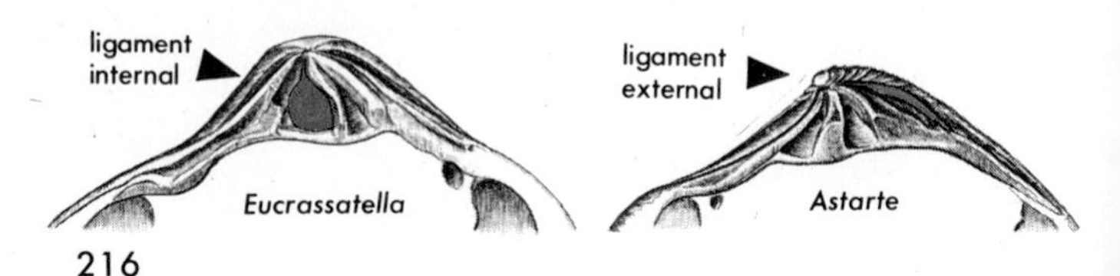

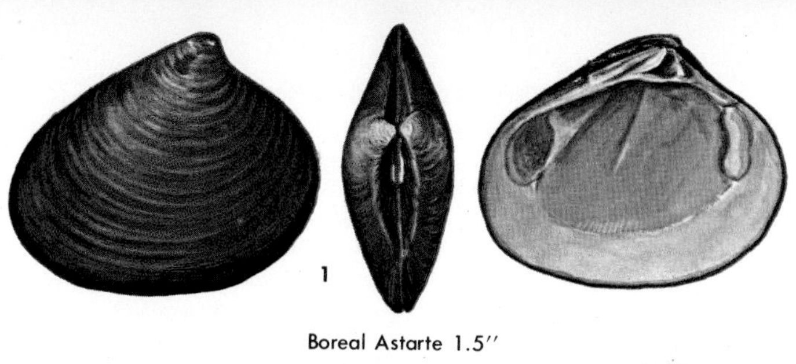

Boreal Astarte 1.5″

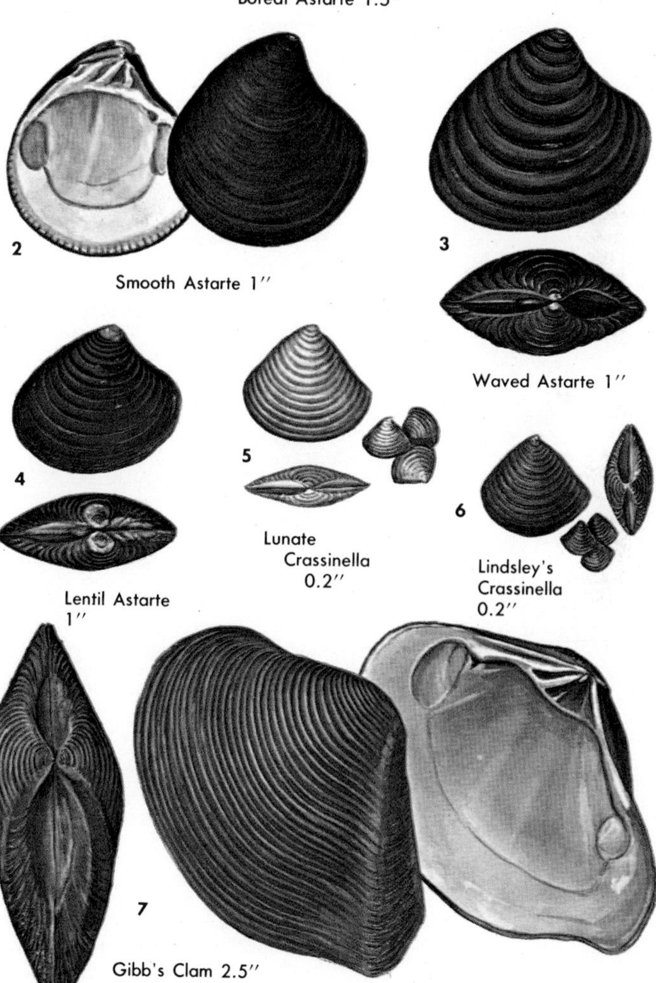

Smooth Astarte 1″

Waved Astarte 1″

Lentil Astarte 1″

Lunate Crassinella 0.2″

Lindsley's Crassinella 0.2″

Gibb's Clam 2.5″

CARDITAS AND MARSH CLAMS

The family Carditidae, with about two dozen species, includes the thick-shelled *Cardita*, an interesting group because some species retain eggs within a marsupium mantle cavity. The shells are heavily ribbed and usually spin a byssus. The closely related, cold-water *Venericardia* are more rounded in form, have a heavy periostracum, and do not produce a byssus. Neither has a pallial sinus.

Marsh clams of the family Corbiculidae consist of several genera of brackish and marine clams. Locally abundant, they serve as a main source of food for many kinds of waterfowl.

The family Arcticidae has only one living American species, the Ocean Quahog. It is similar to the Common Quahog (*Mercenaria*) but lacks a pallial sinus and has a thick brown to black periostracum. A small shellfishery based on this delicious species exists in New England, but the orange color of the clam's flesh is a detraction.

False mussel (family Dreissenidae) superficially resembles *Mytilus* or *Septifer*, but the shelflike plate at the beak has a tiny projecting tooth on the underside. These mussels become a nuisance when they clog the intake pipes of filtering plants. There is one American species.

1. **BROAD-RIBBED CARDITA.** *Cardita floridana* Conrad. S. Fla.–Mexico. Shell long, solid, and heavy, with about 20 raised, beaded, radial ribs. Periostracum gray. Beaks close together. Interior white, lunule small. Very common; in shallow water.

2. **NORTHERN CARDITA.** *Venericardia borealis* (Conrad). Labrador–S.C. Shell is thick, strong, fat, heart-shaped, with about 20 raised, somewhat beaded ribs. Lunule small, deeply sunken. Periostracum rust-brown, fairly thick. Common; 6–200 ft.

3. **STOUT CARDITA.** *Venericardia ventricosa* (Gould). Wash.–Calif. Shell small, somewhat oval, moderately inflated; with about 20 radiating ribs crossed by minute concentric lines. Inner margin wavy. Fairly common; 10–600 ft.

4. **THREE-TOOTHED CARDITA.** *Venericardia tridentata* (Say). N.C.–Fla. Shell gray-brown, very small, rather triangular, inflated, with 15–18 strongly beaded radial ribs. Lunule oval, impressed. Moderately common; 6–200 ft. This tiny clam is found by screening bottom sand.

5. **FLORIDA MARSH CLAM.** *Pseudocyrena floridana* (Conrad). Fla.–Texas. Shell oval or long, thin but strong. Periostracum smooth, thin, and shining. Interior usually purple or with purple margins. Common; in mud, in brackish water.

6. **CAROLINA MARSH CLAM.** *Polymesoda caroliniana* (Bosc). Va.–N. Fla.–Texas. Shell strong, inflated, usually eroded at the beaks. Periostracum glossy brown with minute scales, thin, and shining. Common; at mouths of rivers.

7. **OCEAN QUAHOG.** *Arctica islandica* (Linné). Newfoundland–N.C. Shell almost circular, strong. Periostracum quite thick, brown to black. An edible species, also called the Black or Mahogany Clam. Common; in sandy mud, 30–500 ft.

8. **CONRAD'S FALSE MUSSEL.** *Congeria leucophaeata* (Conrad). N.Y.–Fla. and Texas. Periostracum thin and rather glossy. Byssus short. Hinge has a long thin bar under the ligament. Interior not pearly. Common on rocks in brackish to fresh water near rivers.

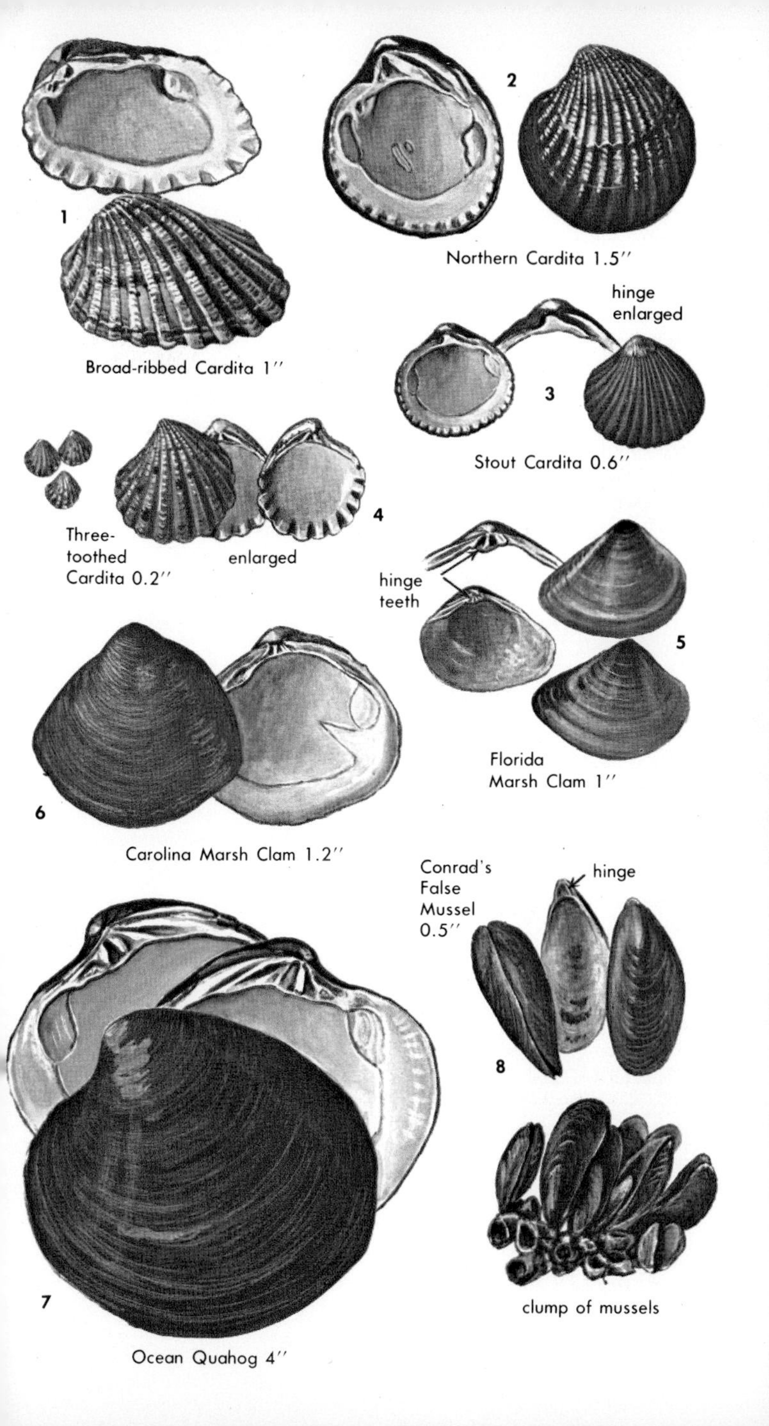
1
Broad-ribbed Cardita 1″
2
Northern Cardita 1.5″
hinge
enlarged
3
Stout Cardita 0.6″
4
Three-
toothed
Cardita 0.2″
enlarged
hinge
teeth
5
Florida
Marsh Clam 1″
6
Carolina Marsh Clam 1.2″
Conrad's
False
Mussel
0.5″
hinge
8
clump of mussels
7
Ocean Quahog 4″

DIPLODONS, CLEFT CLAMS, AND LUCINES

A number of similar species in three families, the Diplodontidae, the Thyasiridae, and the Lucinidae (p. 222), are found along the Atlantic and Pacific coasts. All have shells that are strong and equivalved, with an external ligament which is attached to a furrow lying below the margins of the valves. The pallial line is entire and without a sinus. In these families the foot is very long and narrow. The sexes are separate.

The *Diplodonta* have two cardinal teeth in each valve; the left anterior and the right posterior are prominently split. The muscle scars are of different sizes. The shell is usually smooth or sculptured with fine growth lines; however, some species have minute punctations or even coarse granulations on the surface. There are no siphons. There are five or six North American species. Some species envelop themselves in a protective covering of mucus and debris.

Shells in the family Thyasiridae have a deep posterior furrow and lack hinge teeth, two characters which make recognition relatively simple. Most of the two dozen American species are found in moderately deep water.

1. **PACIFIC ORB DIPLODON.** *Diplodonta orbella* Gould. Alaska–Panama. Shell sturdy, almost circular and smooth, but with coarse growth lines. Ligament conspicuous. Periostracum brown, fibrous. Common; from intertidal zone to 180 ft.

2. **ATLANTIC DIPLODON.** *Diplodonta punctata* (Say). N.C.–Fla. and West Indies. Shell pure white, small, quite strong, with very fine concentric lines except at beaks. Growth lines somewhat coarse, widely spaced. Fairly common; 12–200 ft.

3. **PIMPLED DIPLODON.** *Diplodonta semiaspera* Philippi. N.C.–Fla. and West Indies. Shell delicate, chalky white, with numerous concentric rows of microscopic pimples. Fairly common; from low-tide mark to 300 ft., sometimes found embedded in coral.

4. **ATLANTIC CLEFT CLAM.** *Thyasira trisinuata* (Orbigny). Nova Scotia–Caribbean. Shell small, fragile, with 2 strong radial grooves on posterior slope. Sculptured with fine concentric lines. Fairly common; in sand, 30–200 ft. The specimen shown here was bored by a nassa snail.

5. **THICK LUCINE.** *Phacoides pectinata* (Gmelin). N.C.–Caribbean. Shell strong, compressed, white or tinted with bright orange; with unequally spaced, rather sharp, concentric ridges. Lunule raised. Lateral tooth strong. Common; 6–30 ft.

6. **WOVEN LUCINE.** *Phacoides nassula* (Conrad). N.C.–Fla.; Texas and Bahamas. Shell small, almost circular, pure white. Surface has radial and concentric ribs forming a crisp crisscrossed pattern. Shell margins finely serrate and beaded. Very common, from low-tide line to 600 ft.

7. **NUTTALL'S LUCINE.** *Phacoides nuttalli* (Conrad). S. Calif.–Mexico. Shell circular and somewhat inflated, with strong, fine, even, cancellate sculpture. Dorsal area distinct. Lunule short and very deep. Fairly common in sand; tidal flats to 300 ft.

8. **CROSS-HATCHED LUCINE.** *Divaricella quadrisulcata* Orbigny. Mass.–Brazil. Shell round, inflated, glossy white with fine branching or chevron-like sculpture. Inner margins finely scalloped. Common; in sand, 6–200 ft. Often washed ashore.

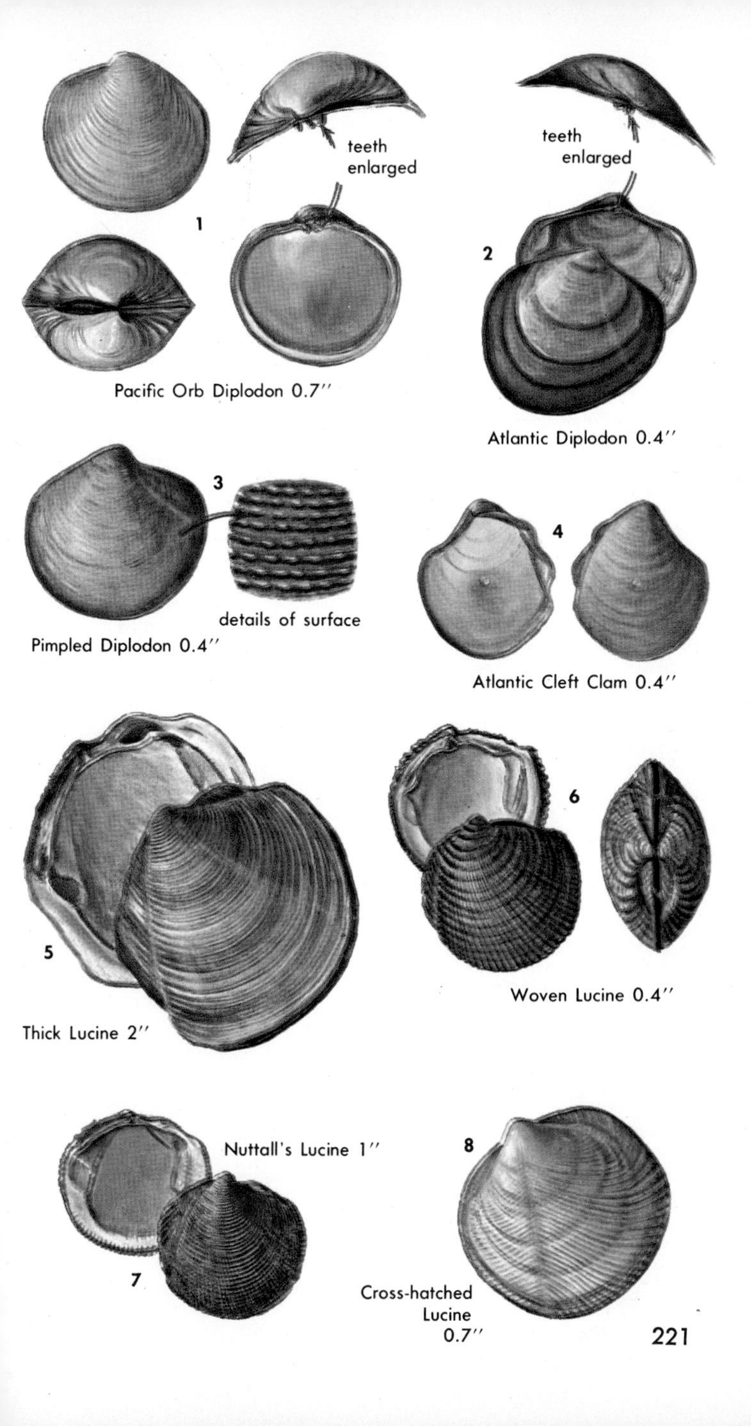

Pacific Orb Diplodon 0.7″

Atlantic Diplodon 0.4″

Pimpled Diplodon 0.4″

Atlantic Cleft Clam 0.4″

Woven Lucine 0.4″

Thick Lucine 2″

Nuttall's Lucine 1″

Cross-hatched Lucine 0.7″

LUCINES—continued

The lucines (family Lucinidae) are rather easy to recognize, because of the long, narrow, anterior muscle scar and the absence of a pallial sinus. The worm-shaped foot of lucines is six times as long as the shell. With it the clam manufactures a mucus-lined, sandy tube that serves as an inhalant water conduit. Periodically the foot pushes debris from the tube.

There is no inhalant siphon, but the exhalant siphon is very long. The latter is retracted in a fashion unique among the bivalves: the tube is contracted and is withdrawn into the mantle cavity in the manner in which the fingers of a glove are turned inside out. There are about 20 American species.

The family Lucinidae is worldwide in distribution, and, curiously, many Caribbean species have closely resembling counterparts in far-away oceans. The Tiger Lucine (fig. 7) is almost identical with the Pacific Tiger of the southwest Pacific.

1. **PENNSYLVANIAN LUCINE.** *Lucina pensylvanica* (Linné). N.C.–Fla. and Caribbean. Shell solid, somewhat inflated, sculptured with delicate, distinct concentric ridges. Pure white, with a thin, yellowish periostracum. Lunule heart-shaped and raised at the center. Moderately common; in sand, 3–50 ft.

2. **FLORIDA LUCINE.** *Lucina floridana* Conrad. West Coast of Fla.–Texas. Shell circular, compressed, and smooth, except for irregular growth lines. Lunule small and deep. Hinge teeth are poorly developed. Periostracum light tan, flaky. Common; 1–40 ft.

3. **FOUR-RIBBED LUCINE.** *Lucina leucocyma* Dall. N.C.–Fla.; Bahamas. Shell small, somewhat elongated and quite fat, with 4 large, conspicuous, rounded, radial ribs and numerous, fine concentric, squarish riblets. Inner margins finely serrate. Fairly common; from low-tide line to 200 ft.

4. **BUTTERCUP LUCINE.** *Anodontia alba* Link. N.C.–Caribbean. Shell inflated, circular in shape, with weak concentric growth lines. Interior flushed with orange; exterior dull white with orange bands. Hinge teeth very weak, anterior muscle scar long. Common; 3–30 ft.

5. **CALIFORNIAN LUCINE.** *Codakia californica* (Conrad). S. Calif. Shell generally circular, moderately inflated, and with numerous distinct, fine, concentric lines. Lunule on right valve fits into recess in left valve. Common; from low-tide line to 500 ft. in sand.

6. **COSTATE LUCINE.** *Codakia costata* (Orbigny). N.C.–Fla. and Caribbean. Shell quite small, inflated, and usually somewhat circular but variable in shape. Surface has fine radial riblets which are crossed by finer, concentric threads. Moderately common; 6–200 ft. on sandy bottoms. This is used in shell jewelry.

7. **TIGER LUCINE.** *Codakia orbicularis* (Linné). Fla.–Caribbean. With numerous coarse radial ribs crossed by finer concentric threads, producing a beaded appearance. Lunule small, deep and heart-shaped. Interior margins often pink. Common; on sand in 6–200 ft.

8. **DWARF TIGER LUCINE.** *Codakia orbiculata* (Montagu). N.C.–Caribbean. Similar to the Tiger Lucine but with a much smaller, fatter shell with ribs strong, often divided. Lunule large and long. Interior never pink. Common; in sand from the low-water line to 600 ft.

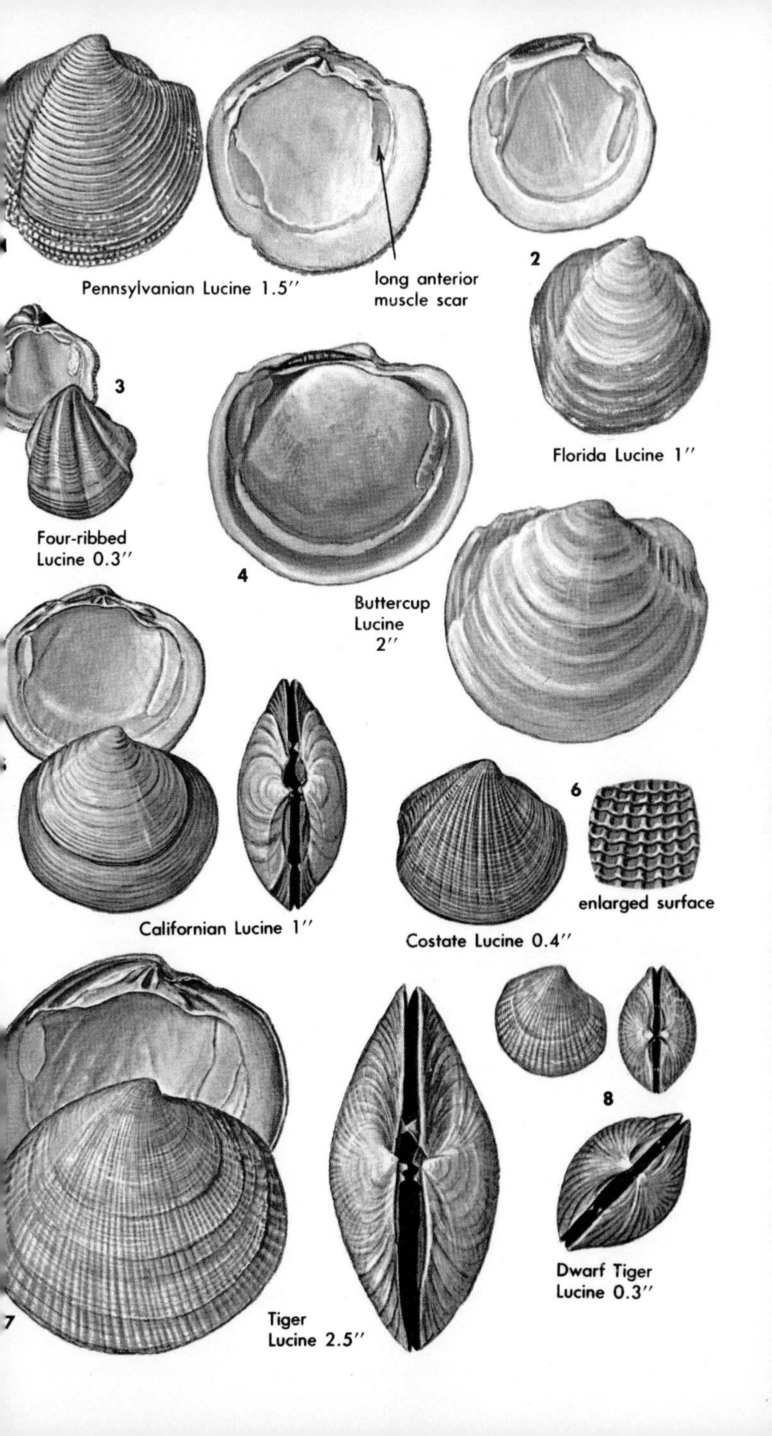
1
Pennsylvanian Lucine 1.5″
long anterior
muscle scar
2
Florida Lucine 1″
3
Four-ribbed
Lucine 0.3″
4
Buttercup
Lucine
2″
5
Californian Lucine 1″
6
enlarged surface
Costate Lucine 0.4″
7
Tiger
Lucine 2.5″
8
Dwarf Tiger
Lucine 0.3″

JEWEL BOXES

In tropical seas are found the beautiful, vividly colored jewel boxes (family Chamidae), which sometimes rival the thorny oysters (p. 212) by developing long, spinelike projections. Jewel boxes live attached to rocks, coral, and shells, from the intertidal zone to quite deep water. Three genera live in American waters. In *Chama*, the True Jewel Box, one valve is much deeper than the other. It is attached by the deep, left valve and the umbones turn from right to left. *Pseudochama* is almost a mirror image of *Chama*. The shell is attached by the deep, right valve, and the umbones turn from left to right (see the drawings below). The two genera differ considerably in hinge details. The Spiny Jewel Box, *Arcinella*, is attached by the right valve to a small pebble or piece of shell which is usually lost, leaving only the imprinted scar near the right beak. The shell has valves of equal fatness and has several radial rows of strong spines. It has a distinct lunule, which is lacking in both *Chama* and *Pseudochama*.

1. **CLEAR JEWEL BOX.** *Chama pellucida* Broderip. Ore.–Chile. Shell usually quite round with strong fluted spines and leafy folds which are waxy white. Interior chalk-white except at finely scalloped margins. Common; attached to pilings and rocks in shallow water; dredged to 150 ft. There are various color forms, but white predominates.

2. **LEAFY JEWEL BOX.** *Chama macerophylla* Gmelin. N.C.–Fla. and Caribbean. Color very variable, often brilliant. The large, scalelike projections are marked with minute, radial lines. Inner margins of the valves have a finely scalloped margin. Common in 3–100 ft. of water.

Smooth-edged Jewel Box. *Chama sinuosa* Broderip (not illus.). S. Fla.–Caribbean. This is always whitish, although the interior may be stained green. Inner edges of the valves have a smooth margin. Uncommon; on coral reefs.

3. **PACIFIC LEFT-HANDED JEWEL BOX.** *Pseudochama exogyra* (Conrad). Ore.–Panama. Shell rather coarse, opaque. Inner margins usually smooth. Beaks curve strongly toward the left. Common; in intertidal zone; often associated with the Clear Jewel Box.

4. **LITTLE CORRUGATED JEWEL BOX.** *Chama congregata* Conrad. N.C.–Texas and Caribbean. Shell usually small with low axial corrugations rather than elongated scales. Inner margins finely scalloped. Common; 3–30 ft., often attached to pen shells.

5. **FLORIDA SPINY JEWEL BOX.** *Arcinella cornuta* (Conrad). N.C.–Fla. and Texas. Shell white, pitted, with 7–9 rows of slender, tubular spines. Interior white or tinged with pink. Inner margins finely scalloped. Common; 12–60 ft., on old shells. Formerly called *Echinochama*.

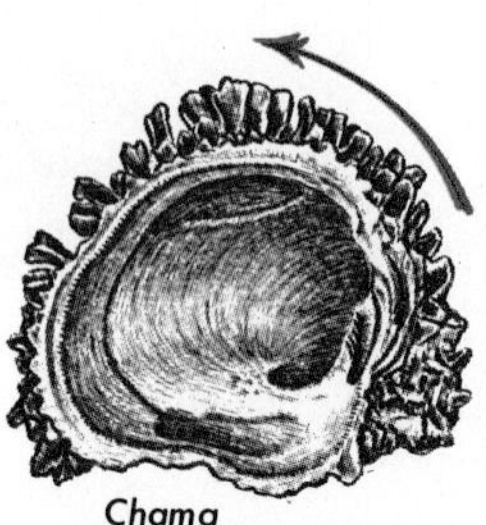

Chama
(deep attachment valve)

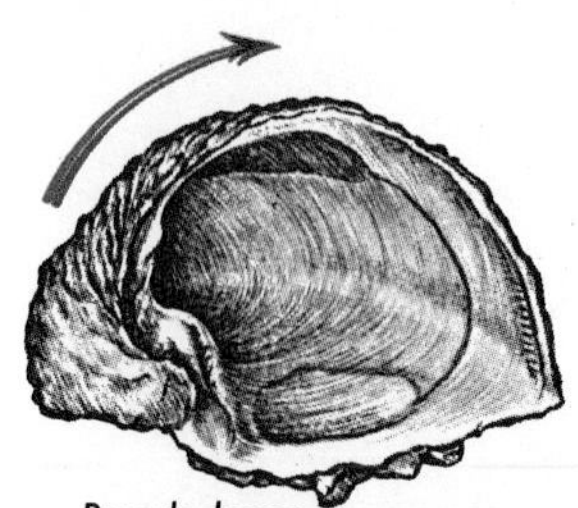

Pseudochama
(deep attachment valve)

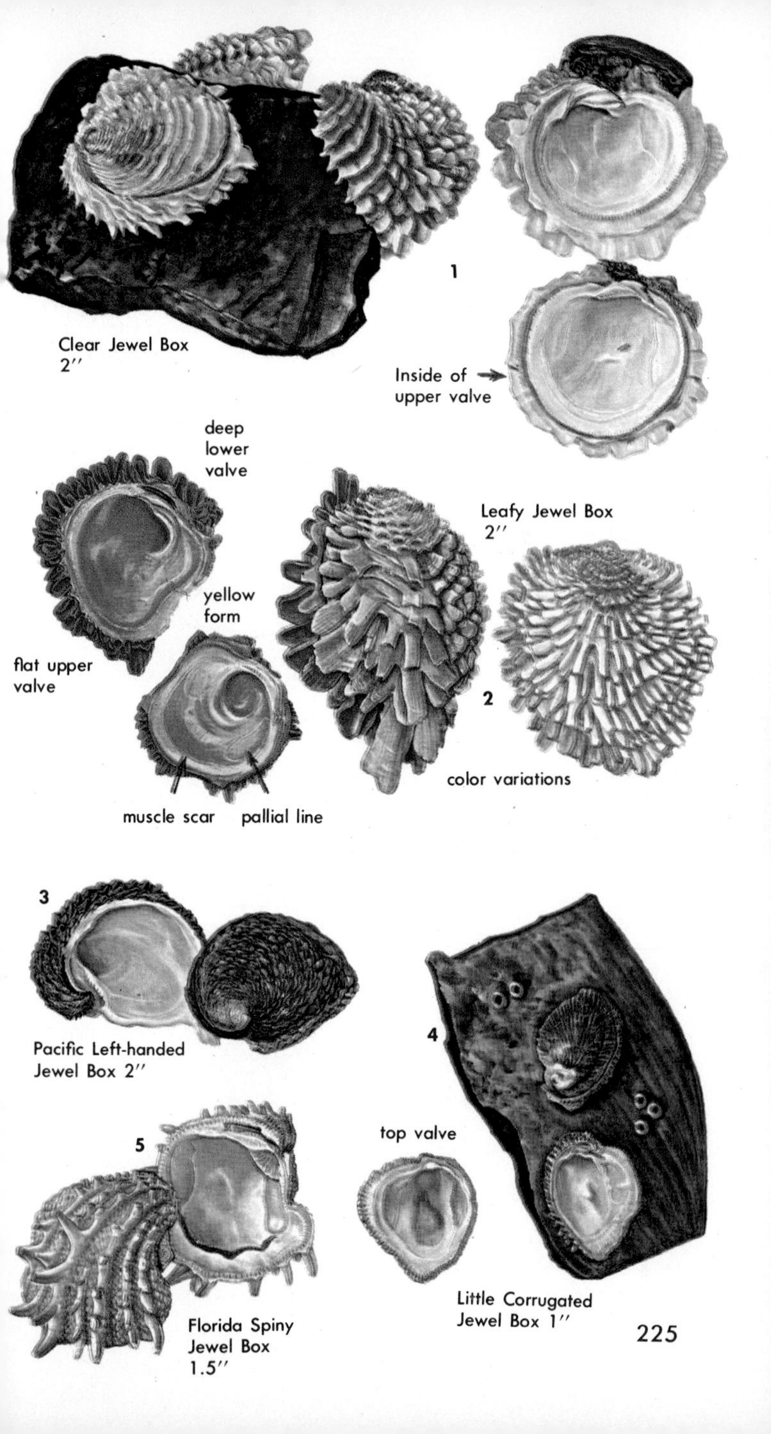
1
Clear Jewel Box
2″
Inside of upper valve
deep
lower
valve
Leafy Jewel Box
2″
yellow
form
flat upper
valve
2
color variations
muscle scar
pallial line
3
Pacific Left-handed
Jewel Box 2″
4
top valve
5
Little Corrugated
Jewel Box 1″
Florida Spiny
Jewel Box
1.5″

COCKLES

Cockles, or heart clams (family Cardiidae), are a large group of over 200 living species and many more fossil forms. They are found in all seas, generally in sand from the intertidal zone to a considerable depth. A few live in brackish water. Cockles are edible and are commonly sold in the markets of western Europe.

Despite the great variety of shapes and ornamentation, cockles are readily recognized. All are heart-shaped when viewed from either end. All have strong hinge teeth with well-developed posterior laterals and a large external ligament. The beak hinge and rib characteristics of the nine genera of cockles in North America are important in identification.

The genus *Clinocardium* (p. 228) differs in having beaks which point strongly forward. In *Trachycardium* the ribs are spined, whereas in *Laevicardium* the margins of the valves are smooth.

1. **PRICKLY COCKLE.** *Trachycardium egmontianum* (Shuttleworth). N.C.–Fla. Shell has 27–31 strong prickly ribs. Interior has vivid pinks and purples. Less oval and more colorful inside and has fewer ribs than the Yellow Cockle. Common; in sand, 1–25 ft. Rarely albino.

2. **GIANT PACIFIC COCKLE.** *Trachycardium quadragenarium* Conrad. S. Calif.–Baja Calif. Shell strong and inflated and has 41–44 strong, squarish, spiny radial ribs. Exterior cream-tan with a brown periostracum. Moderately common; shore to 200 ft.

3. **YELLOW COCKLE.** *Trachycardium muricatum* (Linné). N.C.–Texas; Brazil. Shell almost circular, with 30–40 moderately scaled, radial ribs. Exterior cream to yellow or brownish. Interior white, sometimes yellow. Very common; 6–30 ft.

4. **RAVENEL'S EGG COCKLE.** *Laevicardium pictum* (Ravenel). S.C.–Caribbean. Shell smooth, colorful, obliquely triangular, with very low umbones. Some have strong brown zigzag streaks. Fairly common; 20–300 ft.

5. **MORTON'S EGG COCKLE.** *Laevicardium mortoni* (Conrad). Mass.–Gulf of Mexico. Shell thin but strong, inflated, glossy. Interior usually vivid yellow which fades rapidly. Common; in sandy mud from low-tide mark to 20 ft.

6. **COMMON PACIFIC EGG COCKLE.** *Laevicardium substriatum* (Conrad). S. Calif.–Baja Calif. Shell tan with brownish radial lines; ribs faintly visible. Common; in sand, 6–200 ft.

7. **GIANT PACIFIC EGG COCKLE.** *Laevicardium elatum* (Sowerby). S. Calif.–Panama. Shell huge and inflated and has numerous weak radial ribs. Posterior and anterior smooth. Exterior yellowish; interior white. Common; 6–30 ft.

8. **COMMON EGG COCKLE.** *Laevicardium laevigatum* Linné. N.C.–Caribbean. Shell thin and rather smooth. Color whitish with tints of brown, orange, or purple. Quite common; 2–60 ft.

9. **ATLANTIC STRAWBERRY COCKLE.** *Americardia media* (Linné). N.C.–Caribbean. Shell thick, strong; posterior slope sharply descending; about 33–36 radial ribs. Common; in water 6–600 ft. deep.

Western Strawberry Cockle. *A. biangulata* (Sowerby) (not illus.). Calif. This is a similar species, 1.5 in., with 30 ribs and with a reddish-purple interior. Moderately common; in sand, 6–50 ft.

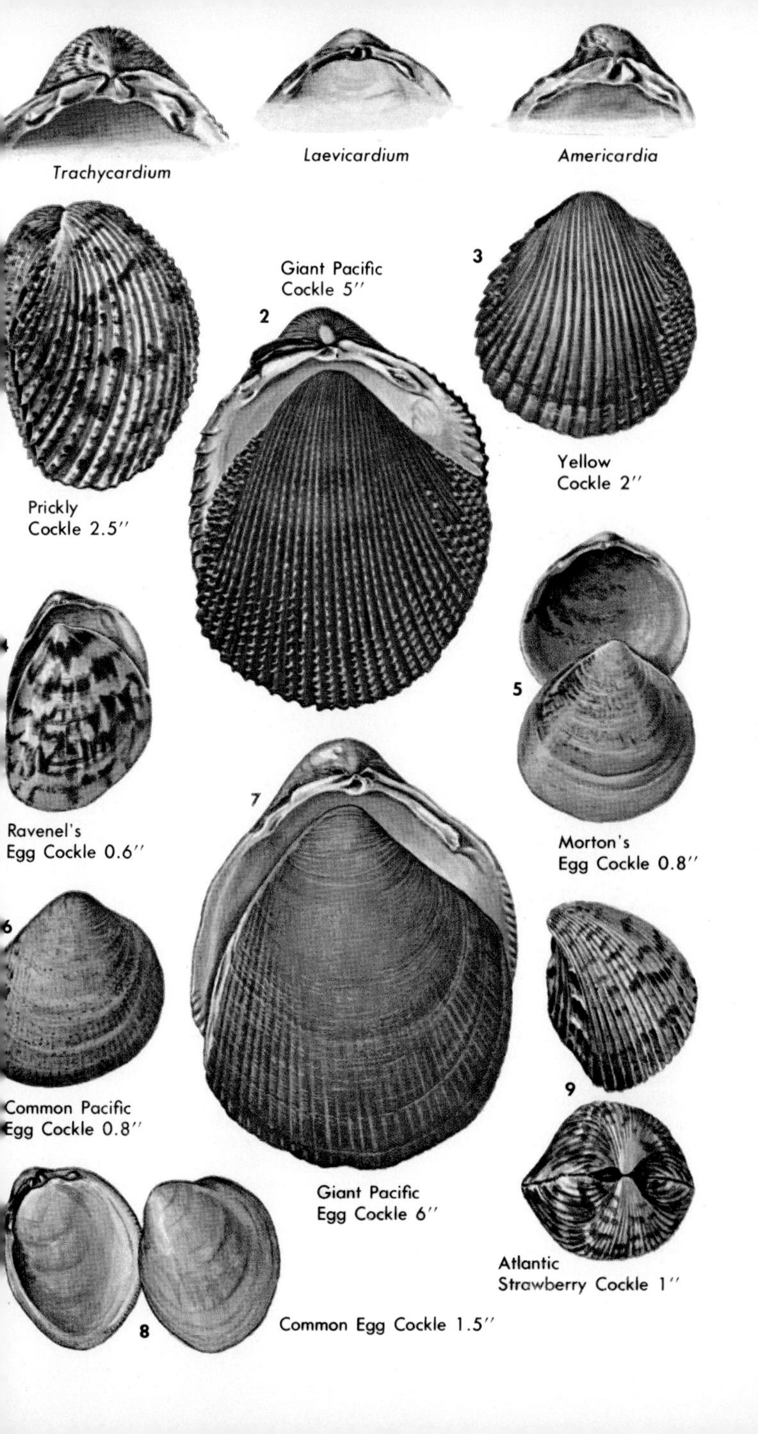
Trachycardium
Laevicardium
Americardia
Giant Pacific
Cockle 5″
2
3
Prickly
Cockle 2.5″
Yellow
Cockle 2″
Ravenel's
Egg Cockle 0.6″
5
Morton's
Egg Cockle 0.8″
7
6
Common Pacific
Egg Cockle 0.8″
9
Giant Pacific
Egg Cockle 6″
Atlantic
Strawberry Cockle 1″
8
Common Egg Cockle 1.5″

Cockles are active animals. They have a long, powerful, sickle-shaped foot that enables them to leap several inches off the bottom. In some species the foot is quite colorful. A weak byssus is spun in most species. The siphons are short and delicately fringed. The ligament is external. All have solid shells with raised, somewhat flattened radial ribs which are ornamented with scales and spines. Most species are hermaphrodites.

Because of their very short siphons, cockles must live near the surface of the substrate and consequently are affected by shifting sands and, in shallow-water species, by great temperature changes. Cockles grow steadily, except during the coldest months, when a retardation causes dark annual growth rings. Nuttall's Cockle of the Northwest grows faster and lives longer in the northern part of its range. It may reach an age of 15 years and a length of about four inches. Overcrowding in large beds of cockles causes dwarf shells. Storms and feeding fish prune populations so that ample food is available for the surviving cockles.

1. **GIANT ATLANTIC COCKLE.** *Dinocardium robustum* (Lightfoot). Va.–N. Fla.; Texas. Shell inflated, somewhat squarish, quite heavy; with 32–36 strong, rather rounded radial ribs. Interior rose; pale or white at margin. Often washed ashore. Very common; 3–100 ft.

2. **VANHYNING'S COCKLE.** *D. r. vanhyningi* (Clench and L. C. Smith). From Tampa, Fla., southward. A subspecies of the Giant Atlantic Cockle. Shell is inflated and obliquely lengthened, which gives it a triangular appearance; glossy and colorful. Common; 6–60 ft.

3. **NORTHERN DWARF COCKLE.** *Cerastoderma pinnulatum* (Conrad). Labrador–N.C. Shell creamy, thin, inflated; 22–28 wide, flattened ribs, thinly scaled except on central portion of valve. Interior brownish white. Common; 20–600 ft.

4. **FUCAN COCKLE.** *Clinocardium fucanum* (Dall). Alaska–Calif. Shell moderately inflated; 45–50 low radial ribs, crossed by microscopic concentric lines. Periostracum grayish. Usually uncommon, but common in Puget Sound, 12–100 ft.

5. **NUTTALL'S COCKLE.** *Clinocardium nuttalli* Conrad. Alaska–Calif. Shell white, strong, usually higher than long; 33–37 coarse radial ribs are crossed by crescent-shaped riblets; older specimens often worn smooth. Periostracum brownish. Common; 6–80 ft.

6. **ICELAND COCKLE.** *Clinocardium ciliatum* Fabricius. Greenland–Mass. and Alaska–Wash. Shell yellowish gray, banded, inflated, moderately thin; 32–38 strong, ridged radial ribs. Periostracum conspicuous, grayish. Very common; 20–600 ft.

7. **SPINY PAPER COCKLE.** *Papyridea soleniformis* (Bruguière). N.C.–Caribbean. Shell fragile, compressed; with about 12 low, weakly prickled ribs. Posterior margin strongly toothed. Exterior mottled; rarely, all orange. Quite common; 10–100 ft.

8. **GREENLAND COCKLE.** *Serripes groenlandicus* (Bruguière). Arctic–Mass. and to Wash. Shell brownish or gray, rather thin but strong and almost round. Radial ribs very weak; posterior end gapes slightly. No lunule. Very common; 30–200 ft. A common food of sea mammals.

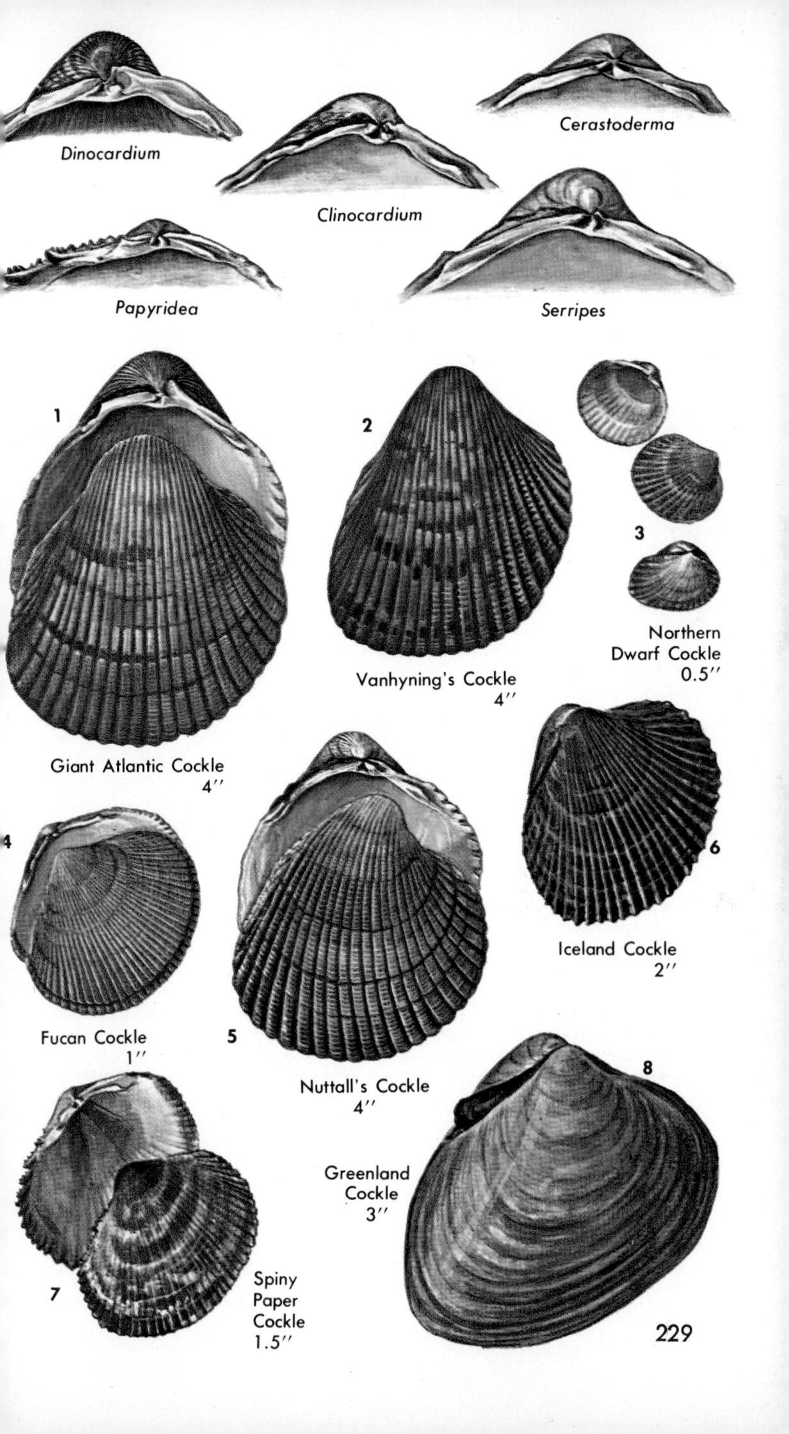
Dinocardium
Clinocardium
Cerastoderma
Papyridea
Serripes
1
2
3
Northern
Dwarf Cockle
0.5″
Vanhyning's Cockle
4″
Giant Atlantic Cockle
4″
4
6
Iceland Cockle
2″
Fucan Cockle
1″
5
Nuttall's Cockle
4″
8
Greenland
Cockle
3″
7
Spiny
Paper
Cockle
1.5″

VENUS CLAMS

There are hundreds of species in the family Veneridae, many of them very beautifully sculptured and brilliantly colored. All have common family characteristics which are conspicuous and which make initial identification fairly simple. They are equivalve and generally egg- or heart-shaped, and all have their beaks placed at and pointing toward the anterior end. The sculpture is varied. It may be concentric or radiating or a combination of both. The ligament is external. The hinge is strong. There are usually three well-developed primary cardinal teeth in each valve (see interior of Northern Quahog, opposite). In a few species these are somewhat divided. Some forms also have lateral teeth, which may be weak or well-developed. There are two muscle scars on the interior of each valve, connected by a distinct pallial line with a more or less deep sinus. The size and shape of the pallial sinus is very distinctive for each species.

The venus clams are commercially important in many areas of the world. Great numbers of the Quahog, *Mercenaria,* are hand collected and marketed in eastern United States.

1. **NORTHERN QUAHOG OR HARD-SHELL CLAM.** *Mercenaria mercenaria* (Linné). Canada–Fla. and introduced in Calif. Shell moderately inflated, heavy, quite thick; sculptured with numerous closely spaced concentric growth lines except for a smooth area near middle of the valve. Interior white, usually with a deep purple stain. Lunule is ¾ wide as long.

A color form, *M. m. notata* (Say), with same range, is externally marked with brown zigzag lines. This species is extremely important commercially, marketed as Littleneck, Cherrystone, and Hard-shell clams. Very common; in 2–40 ft.

2. **TEXAS QUAHOG.** *Mercenaria mercenaria texana* Dall. Western Gulf of Mexico. A fat subspecies with large, irregular, flat-topped concentric ridges and a glossy central area on the exterior.

3. **SOUTHERN QUAHOG.** *Mercenaria campechiensis* (Gmelin). Va.–Fla. This species is heavier and fatter than the Northern Quahog. It lacks the smooth area on the outside of the valves. Lunule is as long as it is wide. Interior almost always white; rarely, stained with purple or mottled near beaks and sides. It is more strongly flavored than *M. mercenaria* and is seldom used commercially. Common; in sand, 3–50 ft.

4. **POINTED VENUS.** *Anomalocardia cuneimeris* (Conrad). S. Fla.–Texas. Shell small with numerous rounded, concentric ribs. Color whitish to tan with brown or purple rays. Interior white, brown, or purple. Common; in sand just below low-tide line. Specimens from brackish water are smaller.

5. **PRINCESS VENUS.** *Antigona listeri* (Gray). S.E. Fla.–Caribbean. Resembles the Southern Quahog somewhat but has numerous, fine radial riblets that cross the concentric raised ribs, giving the surface a beaded or criss-crossed appearance. Lunule bounded by a deep furrow on each side. Both the exterior and the interior are cream-colored; posterior muscle scar usually stained brown or purple. Moderately common; in sand, 6–36 ft.

Empress Venus, *A. strigillina* Dall (not illus.), is a very similar species from S.C.–Fla., but it is not as long and has more distinct concentric riblets. The pallial sinus is small or absent. Dredged from 100–300 ft.

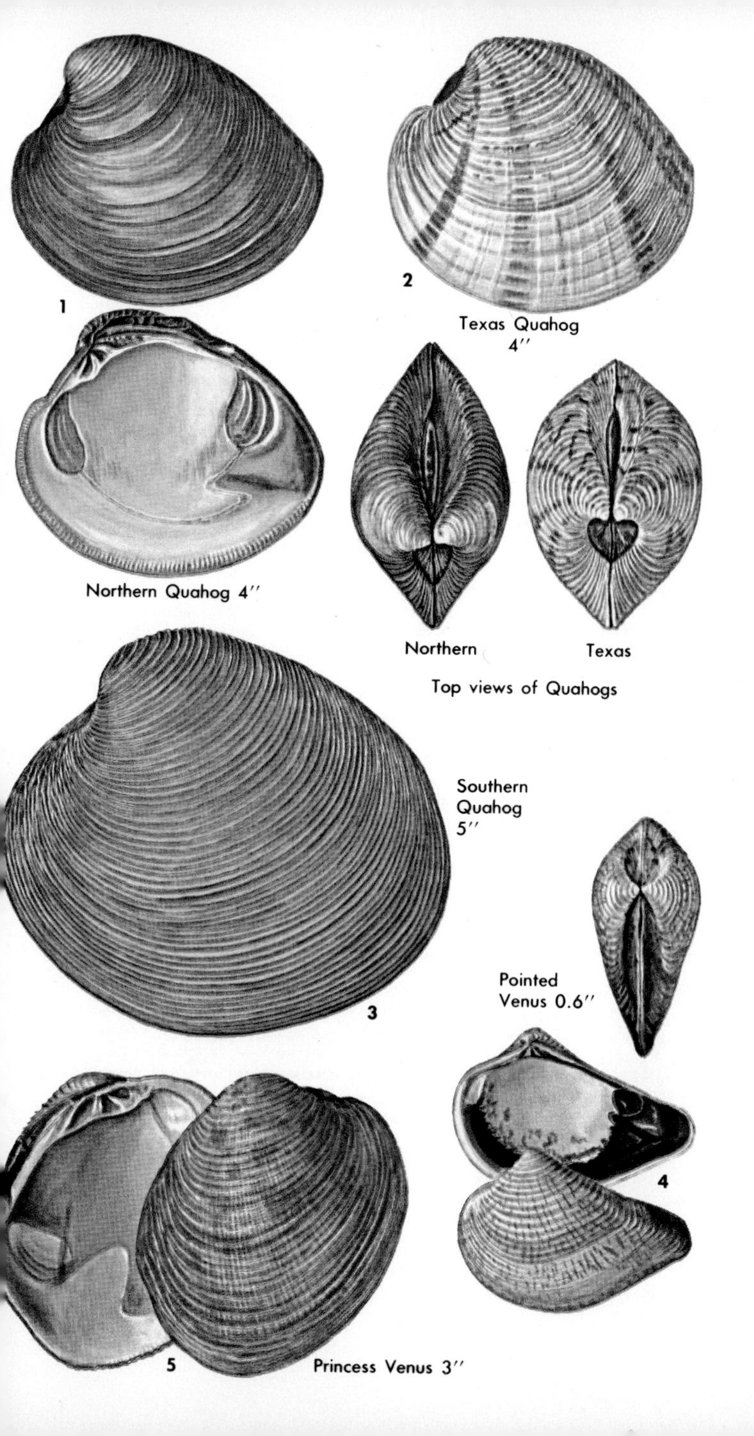
1
2
Texas Quahog
4″
Northern Quahog 4″
Northern
Texas
Top views of Quahogs
Southern
Quahog
5″
Pointed
Venus 0.6″
3
4
5
Princess Venus 3″

VENUS CLAMS—continued

The shells of the genus *Chione* are characterized by trigonal shapes, beveled escutcheons, an impressed lunule, cancellate sculpturing, fine scalloping of the inner margins of the valves, and a short, ascending pallial sinus. There are over a dozen species in American waters, most living in shallow, protected bays. The hard-shelled clams are very prolific and are adapted to survival under difficult conditions. They spawn when the tide is out and usually during a part of the month when tidal fluctuation is small. The larvae swim and crawl over the bottom until a suitable mud-covered, hard surface is found. They then secrete a byssus and remain attached for about a week until siphons develop. The larvae suffer great mortality from filter-feeding adult bivalves and small crustaceans. Young clams are the main food of both the blue and green crabs and the drilling moon snails, *Polinices*.

1. **KING VENUS.** *Chione paphia* (Linné). S.E. Fla.–West Indies. Shell solid, with 8–10 strong concentric ridges which are thin at the posterior end. Dorsal margin of lunule strongly curved. Fairly common; in sand, 6–60 ft.

Imperial Venus, *Chione latilirata* Conrad (not illus.), from N.C.–Texas, is similar to the above but has 5–7 large concentric ribs that are usually rounded and heavy, although often sharply shelved at the top; they are never thin or flattened at the ends. Lunule heart-shaped with a dorsal margin that is almost straight. Rather uncommon; 12–80 ft.

2. **GRAY PYGMY VENUS.** *Chione grus* (Holmes). N.C.–Fla. and La. Shell small, oblong with fine radial ribs crossed by concentric threads. Lunule narrow, heart-shaped, brown. Exterior whitish or dull gray. Interior white, with a purple-brown area at posterior. Hinge is colored purple at both ends. Common; 6–50 ft. Usually found in coarse gray sand or sometimes under rocks.

3. **LADY-IN-WAITING VENUS.** *Chione intapurpurea* (Conrad). N.C.—Texas; Caribbean. Shell thick, with numerous low, smooth, rounded concentric ribs, serrate along the lower edge, giving a somewhat crisscrossed appearance. Interior white, usually with violet markings. Uncommon; 12–60 ft.

4. **CROSS-BARRED VENUS.** *Chione cancellata* (Linné). N.C. Fla.—Brazil. Shell white-gray, heavy, with strong, raised concentric ridges and numerous coarse radial ribs. Lunule heart-shaped, dark, with minute vertical threads. Interior glossy white, usually marked with purple. Extremely common; 3–60 ft.

5. **COMMON CALIFORNIAN VENUS.** *Chione californiensis* (Broderip). Calif.–Panama. Shell moderately compressed, with numerous raised, concentric ribs (7 or 8 per in.), and with low, rounded radial ribs. Interior white; posterior commonly tinged with purple. Lunule heart-shaped, striated. Common; in mud, from the intertidal zone to water 200 ft. deep.

6. **SMOOTH PACIFIC VENUS.** *Chione fluctifraga* (Sowerby). Calif.–Baja Calif. Shell has numerous smooth, polished concentric ribs and irregular radial ribs which are weakest in the central area. Lunule poorly defined. Interior white, purple-stained near muscle scars or teeth. Fairly common; intertidal zone.

7. **FRILLED CALIFORNIAN VENUS.** *Chione undatella* (Sowerby). S. Calif.–Peru. Similar to Common Californian Venus but more-inflated shell, with more numerous, thinner, sharper concentric ribs (10–15 per in.). Common; in sand from intertidal zone to 30 ft. Edible; also no's 5 and 6.

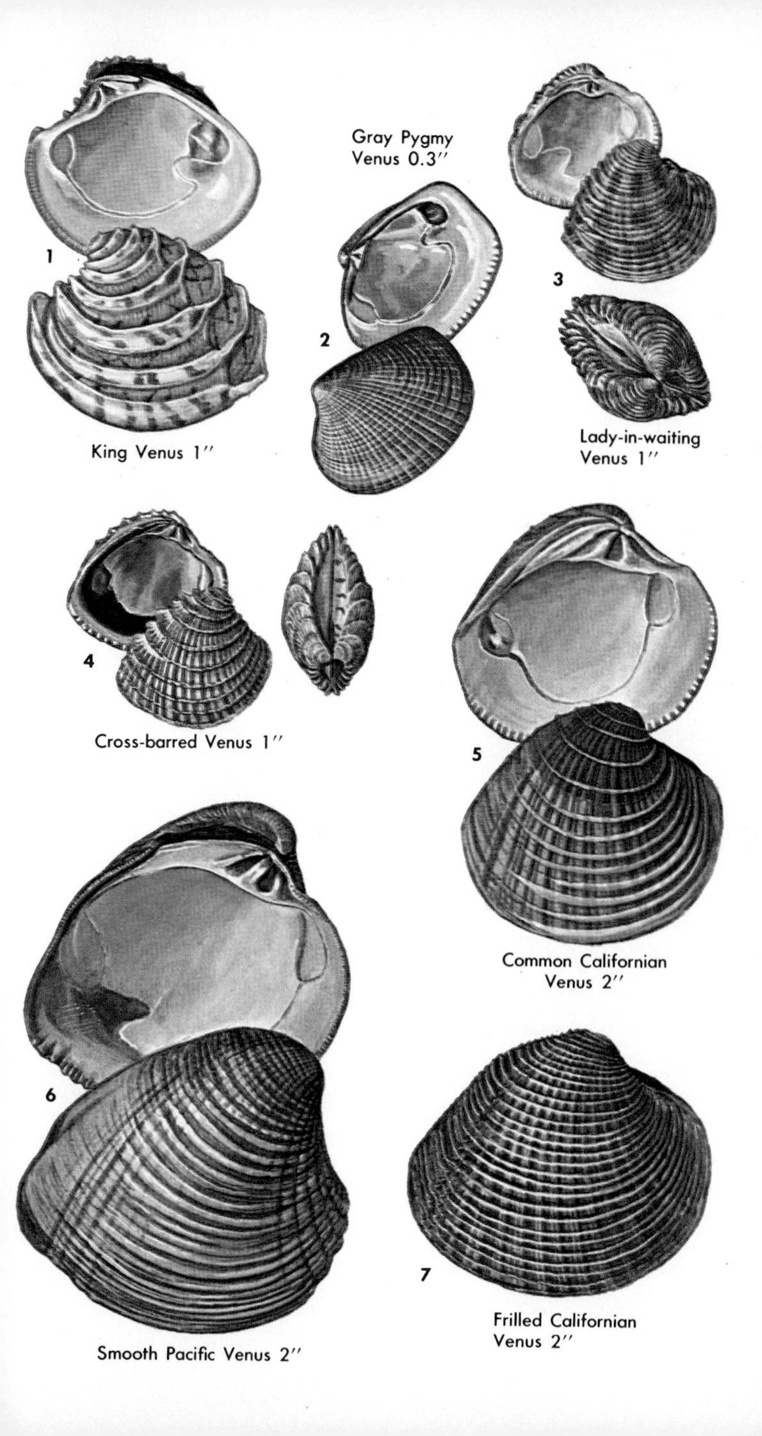

Gray Pygmy
Venus 0.3″
1
King Venus 1″
2
3
Lady-in-waiting
Venus 1″
4
Cross-barred Venus 1″
5
Common Californian
Venus 2″
6
Smooth Pacific Venus 2″
7
Frilled Californian
Venus 2″

VENUS CLAMS—continued

The once abundant and popular Pismo Clam of southern California is gradually being reduced in numbers, despite a bag limit of ten clams per person per day, a minimum size of 5 inches, and no summer or commercial collecting. It takes nearly seven years for a clam to grow 5 inches. The largest clam recorded was about 26 years old, weighed nearly 4 pounds, and was approximately 7½ inches long. The dark rings in the shell denote annual growth stoppages in August during the spawning season, at which time a 5-inch female produced an estimated 30 million eggs. In one year, a 5-inch clam filters about 10,000 gallons of sea water through its gills to extract algal food.

Among the highly esteemed food clams of the Pacific Coast are the Pacific and Japanese Littlenecks. The latter was probably accidentally introduced into the Northwest at the turn of the century when Japanese oysters were being imported.

1. **PACIFIC LITTLENECK.** *Protothaca staminea* (Conrad). Alaska–Baja Calif. Shell has numerous concentric and radial ribs. Radial ribs stronger on central portion. Beaks quite smooth. Color variable, ranging from pure white to gray, rusty brown to dark chocolate-brown; sometimes has a mottled pattern (fig. 1a) or reddish brown chevrons. Sold commercially as Hard-shell Clams in some areas of the Pacific Coast.

The large, unmarked forms from north of the Columbia River have coarse, distinct, somewhat ridged concentric ribs. Exterior of colorful S. Calif. form crisscrossed and beaded. Very common; mainly intertidal.

2. **JAPANESE LITTLENECK.** *Tapes philippinarum* (Adams and Reeve). Introduced from Japan; Wash.–Calif. Similar to Pacific Littleneck but longer and more compressed. The lunule and escutcheon are fairly distinct and quite smooth. Color tan or brown but quite variable and often variegated. Common; in intertidal zone.

3. **PISMO CLAM.** *Tivela stultorum* (Mawe). Central–Baja Calif. Shell heavy and rather smooth. Periostracum thin, glossy. Ligament strong and conspicuous. Hinge teeth well developed. Mauve stripes vary in size. Common; burrowing just under sand in the intertidal zone.

4. **MORRHUA VENUS.** *Pitar morrhuana* (Linsley). E. Canada–N.C. Shell gray-brown, rather thin, moderately inflated, with numerous growth lines. Lunule large and long. Color variable, grayish white to rust-brown. Interior white. Fairly common; dredged from sand at 20–600 ft.

5. **LIGHTNING VENUS.** *Pitar fulminata* (Menke). N.C.–Fla. and Caribbean. Shell plump, with crowded, prominent growth lines. Exterior whitish with brown markings. Lunule very large and outlined by an impressed line. Moderately common; in offshore sands, 12–200 ft.

6. **GLORY-OF-THE-SEAS VENUS.** *Callista eucymata* Dall. N.C.–Texas; Caribbean. Shell somewhat thin, oval, with about 50 rather flattened concentric ribs. Color ranges from rather glossy white to waxy pale brown, with reddish-brown zigzag markings and splashes. No escutcheon. Margins rounded. Rare; dredged, 40–600 ft.

7. **ATLANTIC CYCLINELLA.** *Cyclinella tenuis* (Récluz). S. Fla.–Caribbean. Shell thin, dull white, almost circular, sculptured with numerous irregular growth lines. This species is frequently mistaken for a small *Dosinia*, but the latter has concentric ridges and a more dorsally located muscle scar. Uncommon; from 6–36 ft.

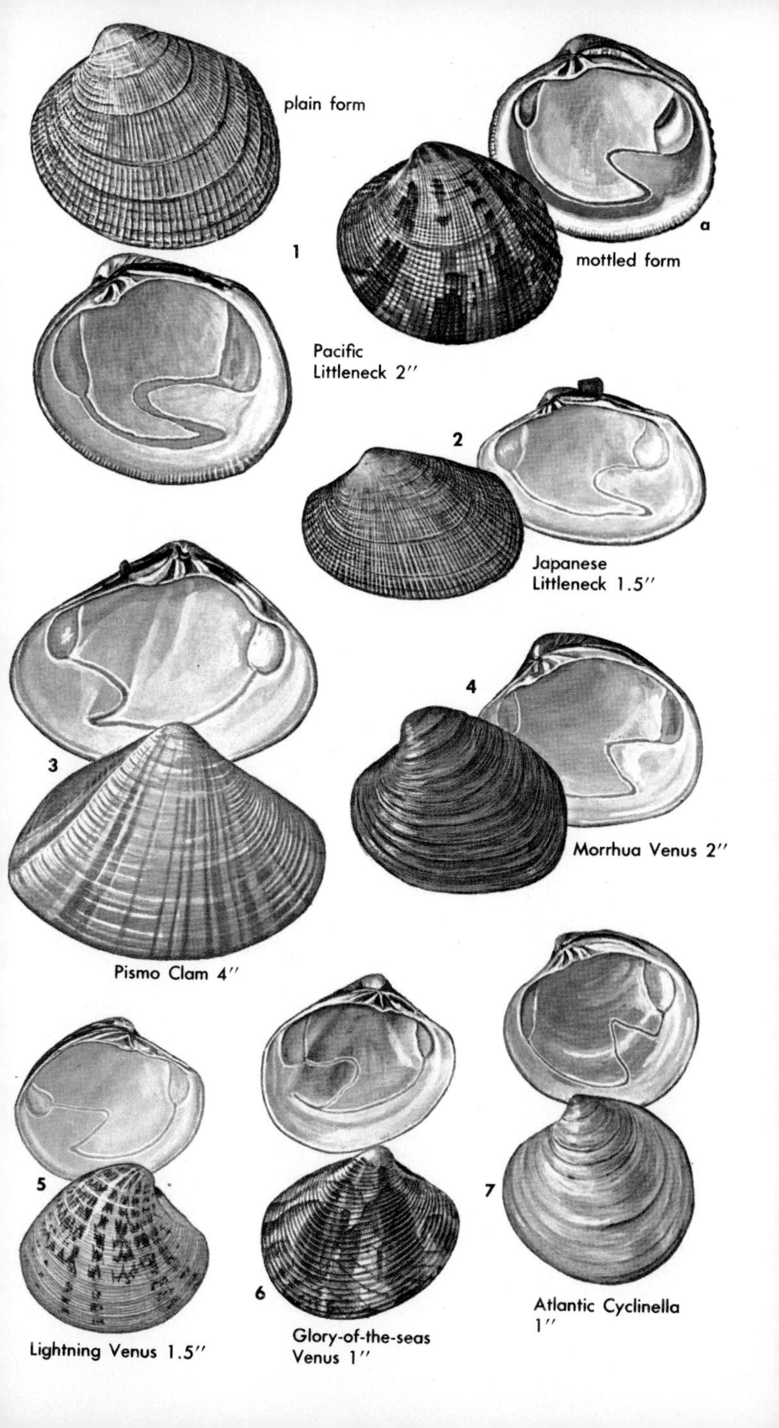
plain form
a
1
mottled form
Pacific
Littleneck 2″
2
Japanese
Littleneck 1.5″
4
Morrhua Venus 2″
3
Pismo Clam 4″
5
7
6
Atlantic Cyclinella
1″
Lightning Venus 1.5″
Glory-of-the-seas
Venus 1″

The smooth, colorful shells of *Macrocallista* have their beaks close to the anterior end. The teeth are concentrated under the beaks, with an anterior lateral tooth in the left valve. The flesh is pink and quite delicious. The *Saxidomus* clams are limited to the Pacific coast and are a favorite food. All along the Atlantic coast the tiny Amethyst Gem Clam abounds in fine sand in water just offshore. It was introduced to the Pacific coast about 1890. Females brood about 200 young within the gill chambers. The life span is less than two years. Their enemies include young moon snails, crabs, ducks, fish, and, curiously, a burrowing sea anemone *(Paranthus)*. The great numbers of gem clams serve as a "buffer" against these predators for commercially important Venus clams which live in the immediate area.

1. **SUNRAY VENUS.** *Macrocallista nimbosa* (Lightfoot). N.C.–Gulf of Mexico. Shell elongated, compressed, and smooth and glossy with a thin, shiny periostracum. Color lavender-gray with darker radial markings. Sunbaked specimens are pinkish. Fairly common; in sandy mud, low tide to 12 ft.

2. **CALICO CLAM.** *Macrocallista maculata* (Linné). N.C.–Caribbean. Shell sturdy, smooth, egg-shaped; exterior cream or tan with brown checkerboard markings. Pallial sinus quite large, flushed with very light pink. Occasional albino and melanistic specimens are found. Fairly common; in sand, 6–60 ft. This is a delicious-tasting species. It was introduced to Bermuda about 1962.

3. **PACIFIC WHITE VENUS.** *Amiantis callosa* (Conrad). S. Calif.–Mexico. Shell strong and heavy. Exterior a glossy ivory, with numerous neat, concentric ridges. Lunule small, heart-shaped, somewhat indented under the beak. Common; on sandy bottoms, low-tide mark to 30 ft. Often washed ashore.

4. **SMOOTH WASHINGTON CLAM.** *Saxidomus giganteus* (Deshayes). Alaska–Calif. Shell solid and heavy, rather chalky, with few concentric ridges. Interior pure white, somewhat glossy. Fished commercially in Alaska. Extremely common; in gravel sand; intertidal zone to 12 ft.

5. **COMMON WASHINGTON CLAM.** *Saxidomus nuttalli* Conrad. Calif.–Baja Calif. Also called Butter Clam. Very similar to the preceding species, but with more concentric ridges. Exterior dull brown-gray; interior usually purple on upper margin. Very common; in mud-sand; intertidal to 30 ft.

6. **AMETHYST GEM CLAM.** *Gemma gemma* (Totten). Nova Scotia–Texas. Introduced to Wash.–Calif. Shell very small. Color whitish to tan with purplish tints. Pallial sinus short and narrow and points toward the beak. Exterior has microscopic concentric lines. Common; in sand, 1–20 ft.

7. **ELEGANT DOSINIA.** *Dosinia elegans* Conrad. Fla.–Texas; Caribbean. Shell white, strong, circular; moderately compressed, with numerous evenly spaced concentric ridges (20–25 per in. in adult specimens). Periostracum extremely thin, glossy. Both valves, still attached, are quite often washed ashore. Common; in sand, 6–36 ft.

8. **DISK DOSINIA.** *Dosinia discus* (Reeve). Va.–Gulf of Mexico. Exterior white and shining, very similar to Elegant Dosinia but relatively smooth to the touch, having about twice as many, much finer concentric ridges (about 50 per in.). Periostracum very thin, shiny. Frequently washed ashore after storms. Common; 6–40 ft.

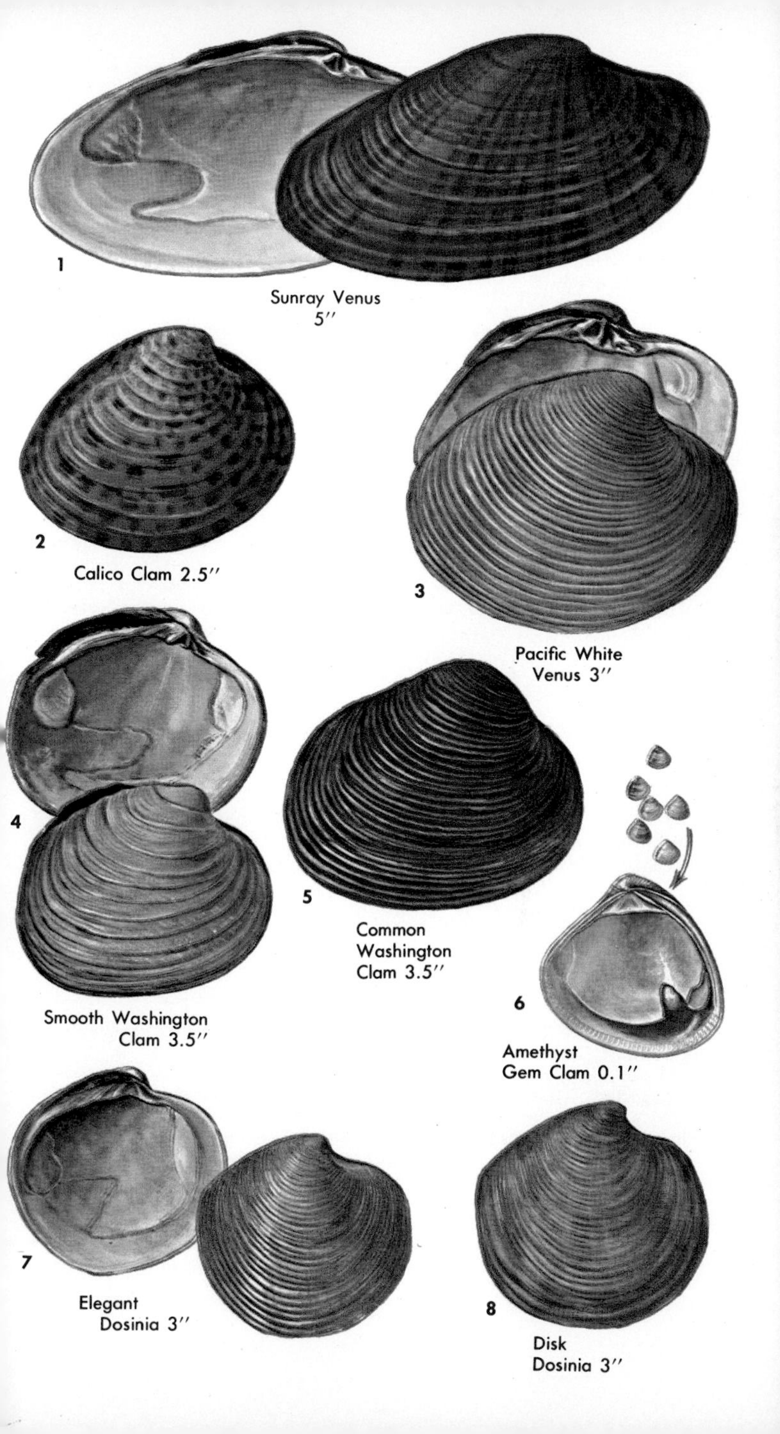
1
Sunray Venus
5″
2
Calico Clam 2.5″
3
Pacific White
Venus 3″
4
5
Common
Washington
Clam 3.5″
Smooth Washington
Clam 3.5″
6
Amethyst
Gem Clam 0.1″
7
Elegant
Dosinia 3″
8
Disk
Dosinia 3″

FALSE ANGEL WINGS

Most of the dozen members of the family Petricolidae bore into rock, coral, clay, shale, and peat, and so large populations may contribute to serious erosion of the shore line. Some species merely settle in any available crevice. As a result, the shells usually become considerably deformed. The Petricolidae lack the lunule, escutcheon, and lateral teeth of the Venus clams (p. 230), although the cardinal teeth are quite similar.

1. **FALSE ANGEL WING.** *Petricola pholadiformis* Lamarck. E. Canada–Gulf of Mexico and Caribbean. Introduced to Calif. and Wash. Shell rather thin, elongate, chalky white, with numerous strong radiating ribs. Ligament external and posterior to beaks. Common; in stiff clay or peat, intertidal zone.

Boring Petricola, *P. lapicida* (Gmelin), is a similar but smaller species (not illus.) from S. Fla. and Caribbean. It is chalky white, egg-shaped, with crisscrossed threaded sculpturing. Beaks close together with a long furrow between. Fairly common; in coral rock.

2. **HEARTY RUPELLARIA.** *Rupellaria carditoides* (Conrad). B.C.–Baja Calif. Shell variable in shape but usually oblong, chalky white to grayish, with coarse, irregular concentric growth lines. Radially sculptured, with very fine, scratched lines. Fairly common; boring in hard rock, intertidal zone to 6 ft.

3. **ATLANTIC RUPELLARIA.** *Rupellaria typica* (Jonas). N.C.–Fla. and Caribbean. Shell whitish gray, strong, quite variable but usually oblong, with numerous coarse ribs that are narrower anteriorly. Growth lines uneven. Common; in coral to 12 ft.

TELLINS

The very large, worldwide family Tellinidae includes many genera and several hundred species. Most are marine, burrowing in sand. Tellins, with long, slender siphons, are detritus feeders. The shells vary in shape and size, but the slight twist at the posterior end and the two cardinal teeth in the hinge help to place the species in the Tellinidae. One major group of tellins has lateral teeth, but the *Macoma* clams do not.

Shells in the genus *Tellina* are generally shiny, often quite colorful, and rounded to elongate in shape. The pallial sinus scar on the inside of the shell is wide and deep. The horny, brown ligament is external.

4. **SUNRISE TELLIN.** *Tellina radiata* Linné. S.C.–Caribbean. Shell elongate, thin but strong, moderately inflated, smooth and very highly polished. Interior flushed with yellow. Most specimens have wide rays of pink. Beaks may have red stripe. Exterior rarely all white (form *unimaculata* Lamarck). Uncommon except in West Indies; in sand, 3–48 ft.

5. **SMOOTH TELLIN.** *Tellina laevigata* Linné. N.C.–Caribbean. Shell oval, glossy, except for microscopic radial lines. Uncommon; in sand.

6. **GREAT TELLIN.** *Tellina magna* Spengler. N.C.–S. Fla. and Caribbean. Right valve has fine concentric lines. Left valve smoother and less colorful. Uncommon; in sand, 2–10 ft.

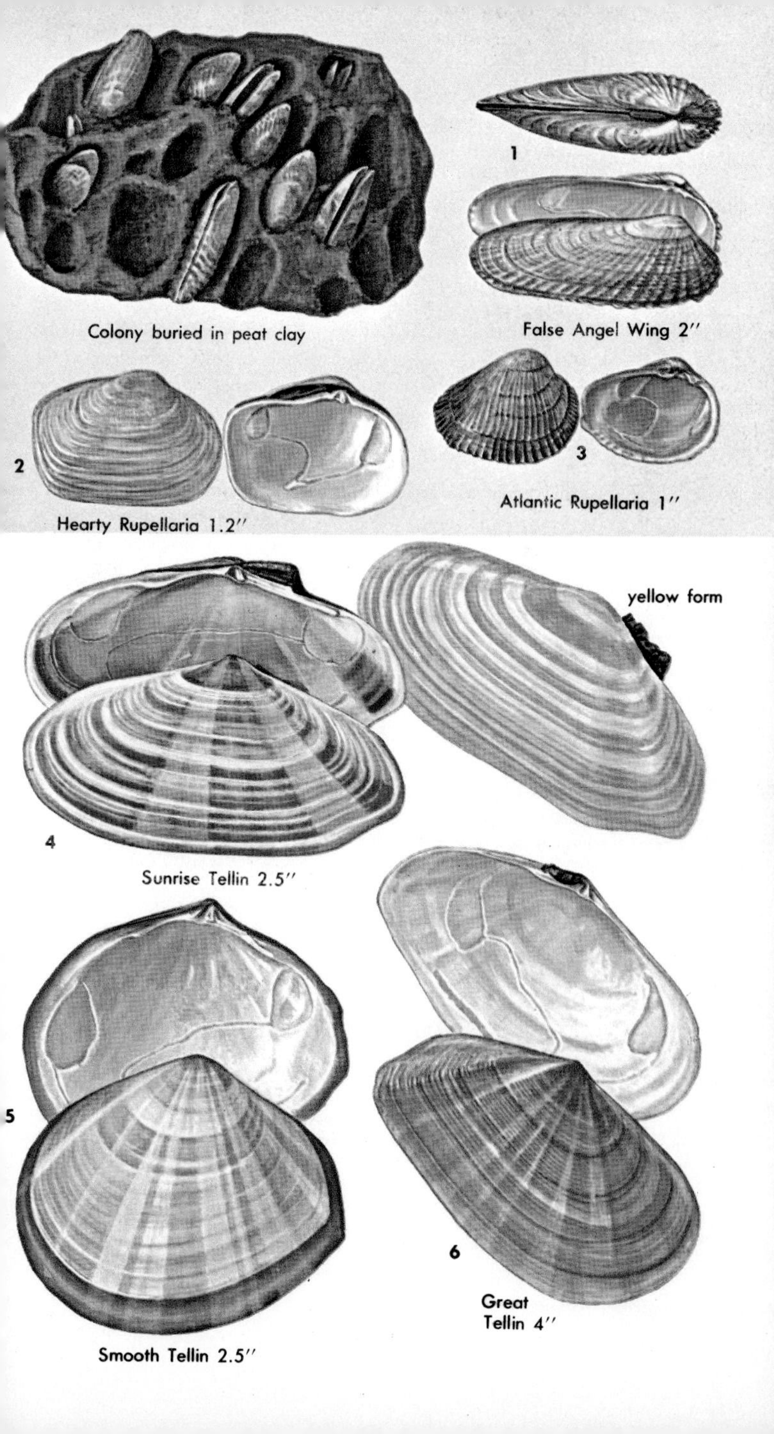

Colony buried in peat clay

False Angel Wing 2″

Hearty Rupellaria 1.2″

Atlantic Rupellaria 1″

Sunrise Tellin 2.5″

Smooth Tellin 2.5″

Great Tellin 4″

Tellin clams feed by means of two long, slender siphons, which can be extended several times the length of the shell, thus permitting the clam to live well below the surface of the sand. In some large, oval species, the shell is buried in a horizontal position, rather than in the conventional hinge-up position. The inhalant siphon sweeps an area of the bottom clean of algal detritus. The foot of tellins is large and suitable for rapid and deep burrowing. The clams travel extensively under the sand, both vertically to escape enemies and horizontally in search of edible bottom deposits.

The morphological characters of the shells of tellins are very useful in distinguishing genera and species. The most important are the placement of the tiny teeth of the hinge, the configuration of the pallial sinus in relation to the muscle scars, and the details of the external sculpturing. Heavily shelled species are found in coarse, sandy gravel, while the delicate, thinly shelled species live in soft mud. The mud species may have oblique sculpturing in addition to the concentric ridges, thus giving a crisscross, or "scissulate," pattern.

1. **SPECKLED TELLIN.** *Tellina listeri* Röding. N.C.–Brazil. Shell elongate, not polished, with many evenly spaced, concentric growth lines. Color speckled, rarely all white or all purple. Shape varies, the young being more elongate. Common; on coarse sand, from 6–300 ft. Formerly called *T. interrupta* Wood.

2. **NORTHERN DWARF TELLIN.** *Tellina agilis* Stimpson. E. Canada–Ga. Shell fragile, glossy, and iridescent. Color varies from white to rose. Sculpture of fine, concentric lines, but sometimes has coarse growth lines. Common; in sandy mud, 3–150 ft.

3. **MERA TELLIN.** *Tellina mera* Say. S. Fla. and Caribbean. Shell moderately inflated, thin but strong, opaque white in color, rarely pinkish. Exterior smooth, with fine, irregular, concentric growth lines. Common; to 12 ft., usually in sandy mud.

4. **DEKAY'S DWARF TELLIN.** *Tellina versicolor* DeKay. Mass.–Texas; Caribbean. Very similar to Northern Dwarf Tellin, but more iridescent, longer, and much fatter. Ventral margin nearly straight. Rarely, has pink rays. Moderately common; in sand, 6–300 ft.

5. **TAMPA TELLIN.** *Tellina tampaensis* Conrad. S. Fla.–Texas. Shell similar to Mera Tellin, but posterior end is more pointed. Color slightly pinkish to orange. Pallial sinus almost reaches anterior muscle scar. Common; in sand, 3–50 ft.

The False Tampa Tellin, *T. paramera* Boss (not illus.), Fla.–Carib., is similar, ½ in., white. Pallial line touches anterior muscle scar. Uncommon; 8 ft.

6. **SALMON TELLIN.** *Tellina salmonea* Carpenter. Alaska–Calif. Shell white; interior delicate pink. Exterior has several dark brown-stained growth lines. Ligament prominent. Common; in sand, 1–200 ft.

7. **ALTERNATE TELLIN.** *Tellina alternata* Say. N.C.–S. Texas. Shell solid, compressed, shining, with numerous fine, evenly spaced, concentric growth lines. Color cream or pink. Ligament prominent. Common; 6–120 ft.

8. **ROSE-PETAL TELLIN.** *Tellina lineata* Turton. Fla.–Caribbean. Shell smooth, glossy, slightly opalescent. White to watermelon red, with fine concentric lines. Locally common; 3 in. or deeper, sand; intertidal to 60 ft.

Speckled Tellin 2.5″

Northern Dwarf Tellin 0.5″

Mera Tellin 0.6″

DeKay's Dwarf Tellin 0.5″

Tampa Tellin 0.7″

Salmon Tellin 0.5″

Rose-Petal Tellin 1″

Alternate Tellin 3″

1. **GREAT ALASKAN TELLIN.** *Tellina lutea* Wood. Arctic Ocean–Alaska; Japan. Shell long, quite compressed, white, rarely pinkish. Periostracum glossy, greenish yellow in young; dark brown in adults. Ligament prominent. Common; intertidal zone to 130 ft. There is an ecologic form that has brownish cracks in the shell.

2. **CANDY-STICK TELLIN.** *Tellina similis* Sowerby. S. Fla.–Caribbean. Shell moderately long and compressed, thin but quite strong, with fine concentric growth lines, crossed at an oblique angle by fine concentric threads. Color opaque white, somewhat yellow-tinged, with 6–12 short, reddish radial rays. Common; in sand, intertidal to 40 ft.

The Iris Tellin, *T. iris* Say (not illus.), N.C.–Gulf of Mexico, is similar but very thin-shelled, translucent, and longer. Common; intertidal zone to 40 ft.

Candé's Tellin, *T. candeana* Orbigny (not illus.), lower Fla. Keys and Caribbean, is more wedge-shaped and usually yellowish white. The oblique sculpturing is microscopic. Uncommon; 6–50 ft.

3. **FAUST TELLIN.** *Tellina fausta* Pulteney. N.C.–Brazil. Shell moderately inflated, solid, smooth, with a tan periostracum, fine concentric scratches, and occasional heavy growth lines. Ligament prominent, almost black. Inside opaque white or yellow; muscle scars glossy. Common; in sand, intertidal zone to 12 ft.; buries 1 ft. in sand, in eelgrass beds, 3–90 ft.

4. **ROMBERG'S STRIGILLA.** *Strigilla rombergi* Mörch. S.E. Fla.–Caribbean. Shell magenta pink with fine growth lines, crossed in center and posteriorly by numerous oblique radial lines. Anteriorly the oblique lines are wavy and run in the opposite direction. Pallial sinus does not reach the anterior muscle scar. Uncommon; 10–40 ft.

Large Strigilla, *S. carnaria* (Linné), is rare in Fla., common in Caribbean. This species (not illus.) closely resembles Romberg's Strigilla, but the markings are somewhat coarser, and the upper line of the pallial sinus runs directly to the anterior muscle scar. In sand, 6–36 ft.

5. **WHITE STRIGILLA.** *Strigilla mirabilis* (Philippi). N.C.–Texas; Caribbean. Shell white, small, oval, inflated, and shining, with an oblique crisscross sculpturing that resembles that of *Divaricella* (p. 220). Inner margins of shell smooth. Common; 3–24 ft.

Pea Strigilla, *S. pisiformis* (Linné), Fla. Keys–Caribbean, is similar but smaller (not illus.). Interior pink with white margins. Abundant; in sand, 3–30 ft. This and the previous species are locally very abundant, especially in the Bahamas, and are used extensively in the shellcraft business.

6. **LINTEA TELLIN.** *Tellina aequistriata* Say. N.C.–Texas; to Brazil. Shell always white, slightly inflated, fairly strong; has numerous sharp, slightly raised, concentric lines and 2 posterior radial ribs in right valve, 1 in the left. Posterior twist pronounced. Uncommon; 6–100 ft. Formerly called *T. lintea* Conrad.

7. **CRENULATE TELLIN.** *Phyllodina squamifera* (Deshayes). N.C.–Gulf of Mexico. Shell elongate, with fine, evenly spaced, concentric ridges. The strong crenulations on the posterior dorsal margin serve to separate this species from all other American tellins. Fairly common; 30–360 ft., in clear sand.

8. **WHITE-CRESTED TELLIN.** *Tellidora cristata* Récluz. N.C.–Texas. Shell somewhat triangular, compressed, always pure white. The large sawteeth along the entire dorsal margin make this very unusual species easy to identify. Uncommon; from low-tide line to 40 ft., in mud or sand where it burrows an inch deep.

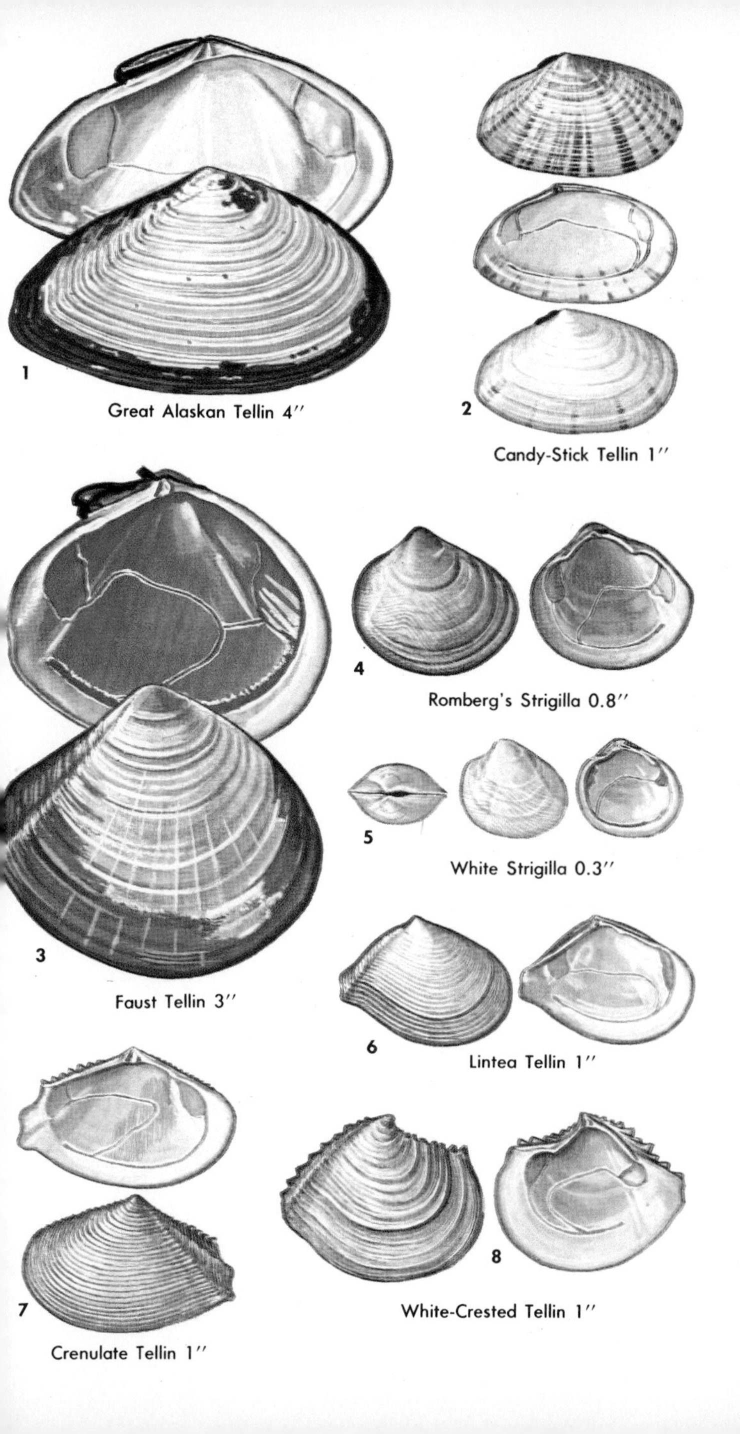

Great Alaskan Tellin 4″

Candy-Stick Tellin 1″

Romberg's Strigilla 0.8″

White Strigilla 0.3″

Faust Tellin 3″

Lintea Tellin 1″

White-Crested Tellin 1″

Crenulate Tellin 1″

MACOMA CLAMS

Macomas, also members of the Tellinidae, are most prolific in cold or temperate seas. They differ from the rest of the tellins in not having lateral teeth, even when very young. *Macoma* shells are usually rather dull white or chalky, never shining. The posterior twist is more pronounced than that in *Tellina*. Curiously, the pallial sinus in one valve of a specimen is often considerably larger than the other. Many of the macomas are intertidal dwellers and are to be found burrowing in the muddy bottoms of quiet bays and inlets, where they are harvested as an important source of food. Some species live over 1,000 feet deep, and serve as a major fish food. Although there are several rather large, edible species along the Pacific Coast, these delicious clams are not often used as food, because they are usually clogged with mud. They may be cleaned by keeping them alive in baskets in changing sea water or hanging them overboard in mesh bags. The clams live in 1 to 2 feet of mud and, because they keep changing the positions of their long siphons, they are difficult to locate.

In the genus *Psammotreta* the shell is strongly twisted. The hinge, only partially external, is rather weak. The Florida species is rarely washed ashore intact, though single valves are quite common.

1. **WHITE SAND MACOMA.** *Macoma secta* (Conrad). W. Canada–Baja Calif. Shell thin, glossy white or cream color. Left valve almost flat; right valve somewhat inflated. Ligament short, wide, partially internal. Common; from intertidal zone to 150 ft.

Bent-nose Macoma, *M. nasuta* (Conrad) (not illus.), Alaska–Baja Calif., is like the White Sand Macoma. However, the pallial sinus in the left valve reaches the anterior muscle scar. Very common; in mud, in quiet waters, intertidal to 130 ft.

2. **BALTHIC MACOMA.** *Macoma balthica* (Linné). Arctic Seas–Ga. and Calif. Shell variable, often oval; moderately compressed; dull white, occasionally flushed with pink. Periostracum thin, flaky. Common; 3–60 ft.

3. **CHALKY MACOMA.** *Macoma calcarea* (Gmelin). Arctic Seas–N.Y. and Calif. Distinguished from the Balthic Macoma by larger size and longer shape. The upper line of the pallial sinus runs nearly to the anterior muscle scar in the left valve. Common; 3–100 ft.

4. **CONSTRICTED MACOMA.** *Macoma constricta* (Bruguière). N.C.–Texas and Caribbean. Shell white, moderately inflated, twisted to right posteriorly and narrowing to a blunt point. Periostracum light yellow-brown. Common; from low tide to 30 ft.

5. **ATLANTIC GROOVED MACOMA.** *Psammotreta intastriata* (Say). S. Fla.–Caribbean. Shell thin but strong, strongly twisted. Right valve has strong posterior radial rib; left valve with radial groove. Common; 6–60 ft. Formerly in *Apolymetis*.

6. **TENTA MACOMA.** *Macoma tenta* Say. Mass.–Fla. and Caribbean. Shell tellin-like, fragile, smooth, white or yellowish, and slightly iridescent. Posterior end slightly twisted to the left. Common; in sand or mud, 6–100 ft.

Yoldia-shaped Macoma, *Macoma yoldiformis* Carpenter (not illus.), ranges from Alaska–Calif. Quite similar to Tenta Macoma, but is distinctly twisted to the left. Color uniformly white, rarely translucent or opalescent. Common; intertidal zone to 200 ft.

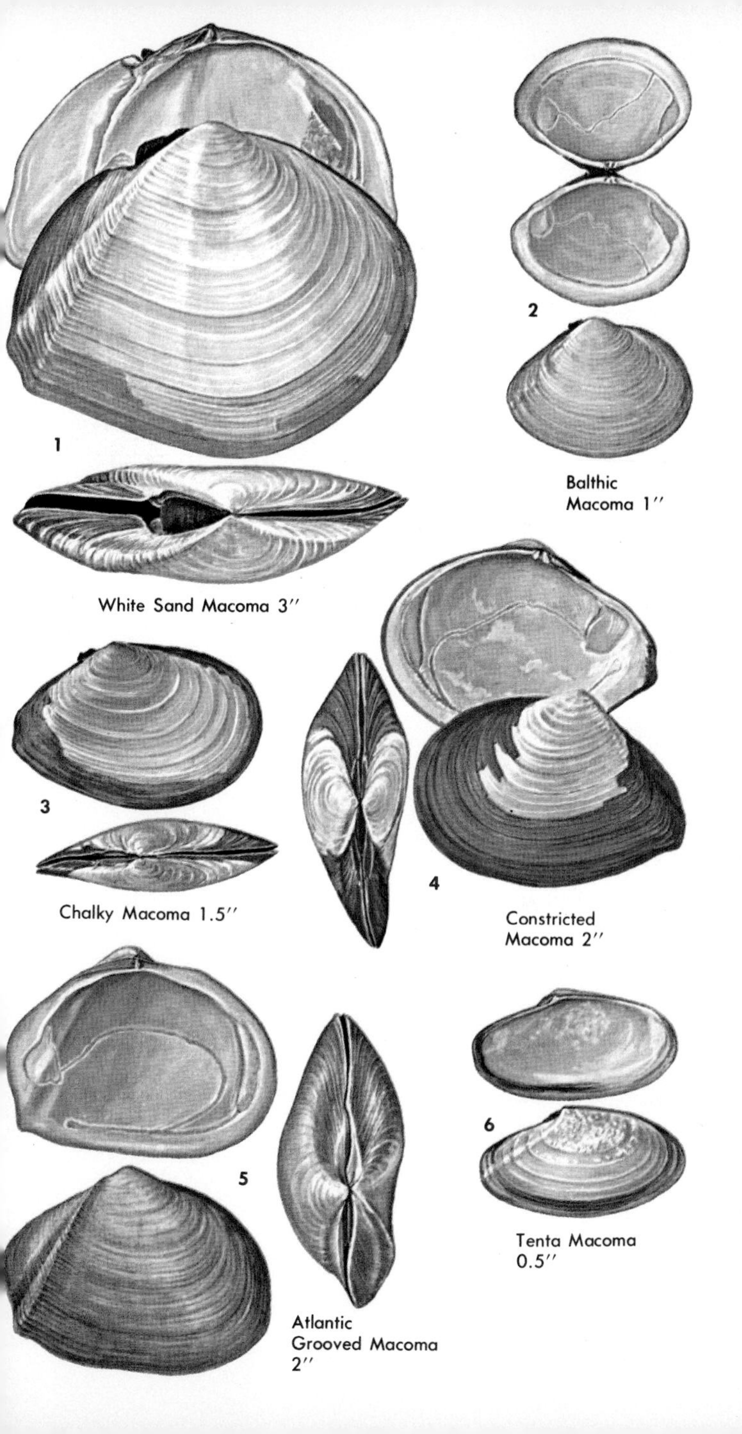

Balthic Macoma 1″

White Sand Macoma 3″

Chalky Macoma 1.5″

Constricted Macoma 2″

Tenta Macoma 0.5″

Atlantic Grooved Macoma 2″

SEMELE CLAMS

The colorful semele clams (family Semelidae) contain a great number of warm-water species, many of which resemble tellin clams. The pallial sinus is well rounded. The ligament is external, but below there is a chitinous resilium embedded in a diagonal groove parallel to the hinge line. In *Cumingia,* a genus of rather small, white species, the resilium is round and set inside a spoon-shaped chondrophore not unlike that of the *Mactra* surf clams.

Semeles live in soft sand or mud, usually in shallow waters, but a few are deep-sea dwellers. The siphons are separate and quite long. The foot is large and blunt and without a byssus. Some *Cumingia* live in crevices in rocks or pilings or grow embedded in sponges.

The largest American species occur along the Pacific Coast, and although delicious to eat, they are not abundant enough to be fished commercially. The more colorful, thin-shelled species from southern Florida and the Caribbean are used extensively in the shellcraft industry. The heavy, barklike periostracum in some species protects the shell from erosion.

1. **WHITE ATLANTIC SEMELE.** *Semele proficua* (Pulteney). N.C.–Caribbean. Shell quite round, usually dull white, with beaks almost central. Interior glossy, yellowish, or tinted and speckled with pink or purple. Lunule small, depressed. Common; in sand, 1–30 ft.

2. **PURPLISH SEMELE.** *Semele purpurascens* (Gmelin). N.C.–Uruguay. Shell thin, strong; growth lines fine, evenly spaced, crossed obliquely by fine concentric lines. Exterior colors variable and dull; interior glossy purple, brownish, or orange. Common; 1–50 ft.

3. **CANCELLATE SEMELE.** *Semele bellastriata* (Conrad). N.C.–Caribbean. Surface strongly cancellate; yellowish to tan with reddish flecks, or a purplish gray with brownish rays. Interior white, cream, or reddish. Fairly common; offshore to 100 ft.

4. **ROCK–DWELLING SEMELE.** *Semele rupicola* Dall. Calif.–Baja Calif. Shell yellowish, often irregular in shape. Surface has numerous concentric wrinkles. Interior glossy, white, purple-tinged at margins, or orange-stained. Common; in rocky areas, 6–60 ft.

5. **BARK SEMELE.** *Semele decisa* (Conrad). S. Calif.–Baja Calif. Shell heavy, brownish, coarsely wrinkled. Interior white, polished, tinged with purple, particularly on the hinge and margins. Common; on rocky bottoms, 6–200 ft. Dead valves without periostracum are pinkish brown.

6. **TELLIN–LIKE CUMINGIA.** *Cumingia tellinoides* (Conrad). Nova Scotia–N.E. Fla. Shell small, white, quite thin, with numerous small, sharp, raised concentric lines; closely resembles a tellin externally. Posterior end gaping. Fairly common; in mud, 6–60 ft.

7. **SOUTHERN CUMINGIA.** *Cumingia coarctata* Sowerby. S. Fla.–Caribbean. Similar to Tellin-like Cumingia, but with stronger, more widely separated concentric ridges. Color white or yellow. Moderately common; in clear sand to 200 ft.

8. **CALIFORNIAN CUMINGIA.** *Cumingia californica* (Conrad). Calif.–Baja Calif. Shell white, oval to triangular, usually distorted, with numerous raised concentric lines. Pallial sinus very long. Common; in rock crevices and on pilings, from shore to 200 ft.

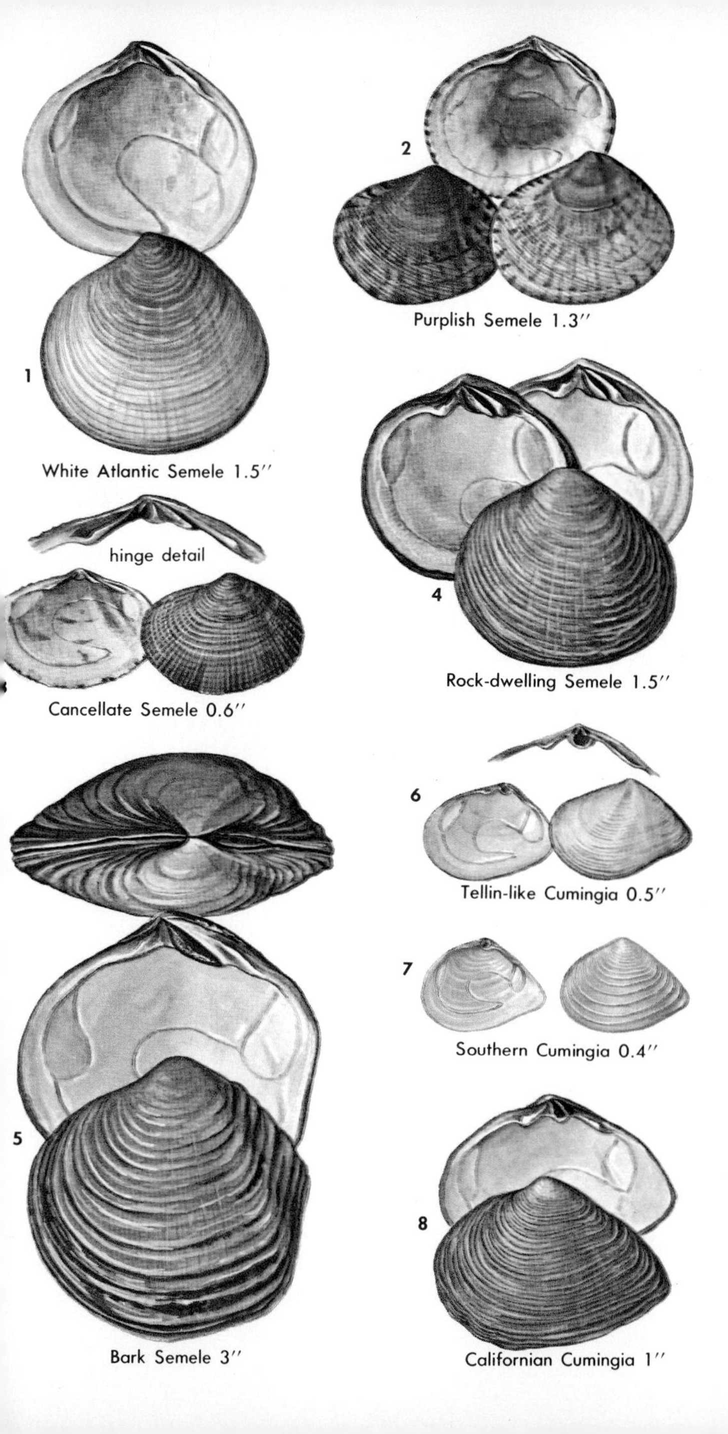

White Atlantic Semele 1.5″

Purplish Semele 1.3″

Cancellate Semele 0.6″

Rock-dwelling Semele 1.5″

Tellin-like Cumingia 0.5″

Southern Cumingia 0.4″

Bark Semele 3″

Californian Cumingia 1″

DONAX CLAMS

The bright colors of the solid, little, wedge-shaped donax, or wedge clams (family Donacidae), make them very noticeable on Florida's sandy beaches. Several other, less colorful species live in American waters. There are about 50 in the world. Many of them are used to make broth, but most are not large enough to be of commercial value.

Donax clams live in large colonies just below the surface of the sand on the slopes of ocean beaches. With purposeful use of the foot and siphons, they can control their migration up and down the beach slope so as to remain in the ideal wash zone of the intertidal area. Some species jump out of the sand as soon as they feel the acoustic shock of a breaking wave. The rushing water then carries them farther up the beach. Each species has its preference for the type of beach, distance from water, size of the sand grains, and slope of the beach. The density of populations varies at different localities and may fluctuate from one year to another. Overcrowding may change the shape of the shells. Crabs, carnivorous snails, and wading birds take a heavy toll. *Donax* are capable of surviving in dry sand for three days.

The large False Donax, genus *Iphigenia,* has a similar shell, but the inner margins of the valves are smooth. They live in quiet waters near the shore and are seldom found on beaches.

1. **FLORIDA COQUINA.** *Donax variabilis* Say. Va.–Fla. and Texas. Shell small, sturdy, with a strong external ligament. Surface striations very fine, becoming quite strong posteriorly. Color extremely variable, ranging from white to yellow, purple, and even deep red; commonly with dark rays or with a plaid pattern.

The subspecies, or ecologic form, *rɔemeri* Philippi, very common along the Texas coast, is shorter, the posterior end is blunter, and the ventral margin has a pronounced downward dip. Both common.

2. **FOSSOR DONAX.** *Donax fossor* Say. N.Y.–N.J. Shell similar to Florida Coquina (and may be only a form of it), but lacks the strong surface striations. It is yellowish white, with or without purplish rays. A study showed that over 1,500 specimens may live in 1 sq. ft. of beach.

3. **GOULD'S DONAX or BEAN CLAM.** *Donax gouldi* Dall. S. Calif.–Mexico. Small, obese, truncate posteriorly. Surface smooth, glossy, with fine axial threads anteriorly. Margins often purple-tinged. Locally common; intertidally.

4. **CALIFORNIAN DONAX.** *Donax californicus* Conrad. Calif.–Baja Calif. Shell quite thin, longer, and somewhat larger than Gould's Donax. Exterior yellowish white, sometimes faintly rayed. Interior white, usually with a purple tinge. Periostracum tan to greenish. Common; in intertidal zone of bays and coves.

5. **FAT GULF DONAX.** *Donax tumidus* Philippi. N.W. Fla.–Texas. Shell small, obese, quite triangular. Color whitish, rarely rayed. Blunt posterior end is minutely beaded. Uncommon; at 2 or 3 ft.

6. **GIANT FALSE DONAX.** *Iphigenia brasiliana* (Lamarck). S. Fla.–Brazil. Shell quite large, heavy, moderately inflated. Beaks prominent, centrally located. Interior creamy, purplish near beak. Periostracum tan. Fairly common; in sand to 6 ft. Formerly *I. brasiliensis.*

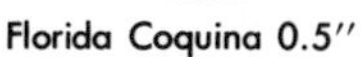

Florida Coquina 0.5″

Florida Coquina *(Donax)* burrowing back into the sand

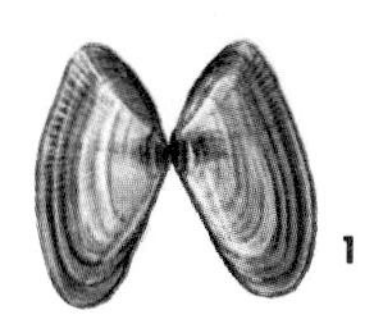

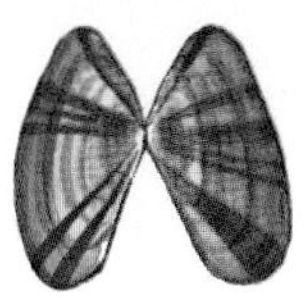

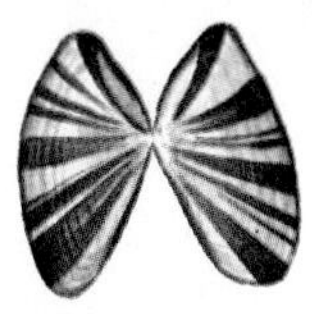

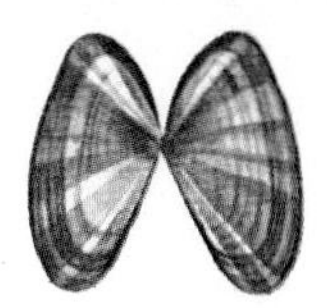

color variations of the Florida Coquina

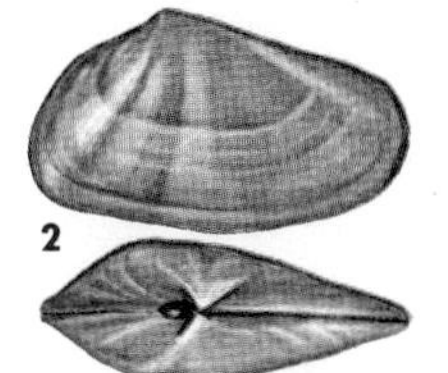

Fossor Donax 0.3″

Gould's Donax 0.6″

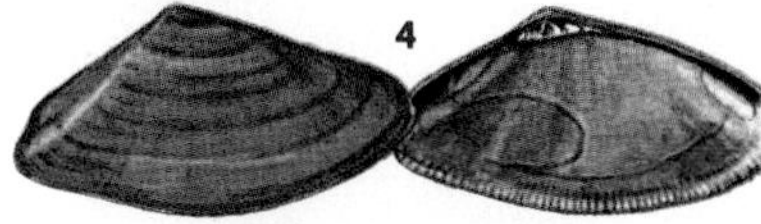

Californian Donax 1″

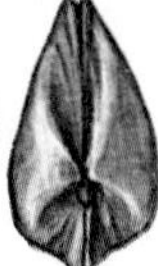

Fat Gulf Donax 0.3″

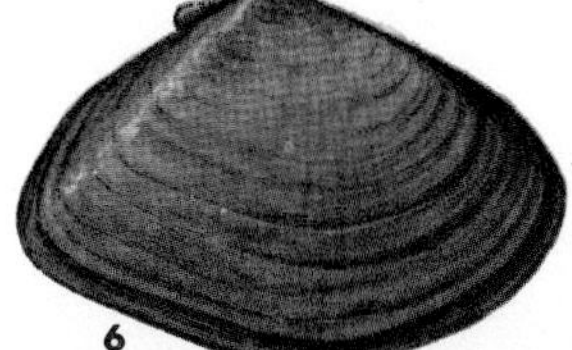

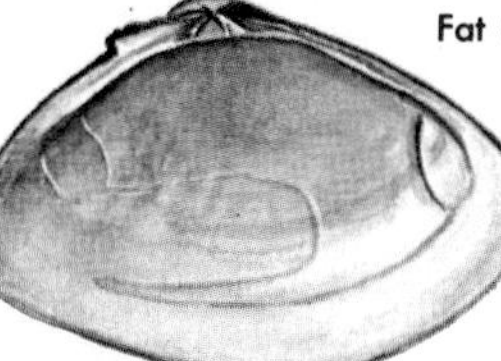

Giant False Donax 2.5″

SANGUIN CLAMS

The tropical family of sanguin clams (Sanguinolariidae, or Gariidae) contains a diverse assemblage of shallow-water genera, from the colorful *Asaphis* to the drab *Tagelus*. The siphons are separate and long, hence the pallial sinus is large. The hinge teeth are usually small, and, to compensate for this, there is a large, strong, external ligament. Although some species resemble the tellins in outline, there is never a posterior twist. The valves gape at the siphonal end. In *Sanguinolaria*, one valve may be considerably flatter than the other.

Members of this family are rapid burrowers and live in muddy areas, usually near mangroves or brackish water. Many of the species are fished commercially, especially in Asia.

Asaphis is represented by only one species in North America. It has colorful and roughly sculptured valves. It lives in warm waters from Florida to Brazil. Curiously, this species is also present in the Indian and Southwest Pacific oceans. It is believed that their ranges were continuous many eras ago, when the ancient Tethys Sea connected the Caribbean and Mediterranean with the Indian Ocean.

In the genus *Gari* of the Pacific Coast, there is a moderately wide opening at the posterior end. The large, brown ligament is mounted above a large shell bar on the top of the hinge.

1. **ATLANTIC SANGUIN.** *Sanguinolaria sanguinolenta* (Gmelin). S. Fla. (rare)—Caribbean (common). Shell glossy, quite smooth, moderately compressed, with slight posterior gape. Left valve very slightly flattened. Lives in sandy mud, 6–20 ft. Erroneously called *S. cruenta*.

2. **NUTTALL'S MAHOGANY CLAM.** *Sanguinolaria nuttalli* Conrad. S. Calif.–Baja Calif. Shell whitish with purple rays, covered with a strong, glossy, rich brown periostracum. Right valve almost flat. Common; in bays and estuaries.

3. **GAUDY ASAPHIS.** *Asaphis deflorata* (Linné). S.E. Fla.–Brazil. Variably colored, interior glossy. Shell strong, moderately inflated, with numerous radiating, coarse threads. Moderately common; in gravel in intertidal zone.

4. **PURPLISH TAGELUS.** *Tagelus divisus* Spengler. Mass.–Texas and Caribbean. Shell fragile, smooth, with a weak radial rib internally in each valve. Periostracum thin, shining. Common; in sandy mud, intertidally.

5. **STOUT TAGELUS.** *Tagelus plebeius* (Lightfoot). Mass.–Fla. and Texas. Shell quite strong, somewhat cylindrical, gaping. Exterior smooth, with fine concentric wrinkles. Abundant; in sandy mud, intertidally to 25 ft.

6. **CALIFORNIAN TAGELUS.** *Tagelus californianus* (Conrad). S. Calif.–Panama. The largest Pacific species of *Tagelus*. Yellowish tan shell with brown periostracum. Pallial sinus quite short, never extending beyond the beaks. Common; on mud flats.

Affinis Tagelus, *T. affinis* (C. B. Adams), is found over much the same range as the Californian Tagelus. It is similar (not illus.) but is 1 or 2 inches long, chubbier, and the pallial sinus extends beyond the beaks. Commonly washed ashore. Lives in muddy sand, 10–120 ft.

7. **CALIFORNIAN SUNSET CLAM.** *Gari californica* (Conrad). Alaska–S. Calif.; Japan. Shell quite heavy, with strong, irregular growth lines. Whitish, with reddish or purple rays or tints. Common; washed ashore after storms. Lives from 1–150 ft.

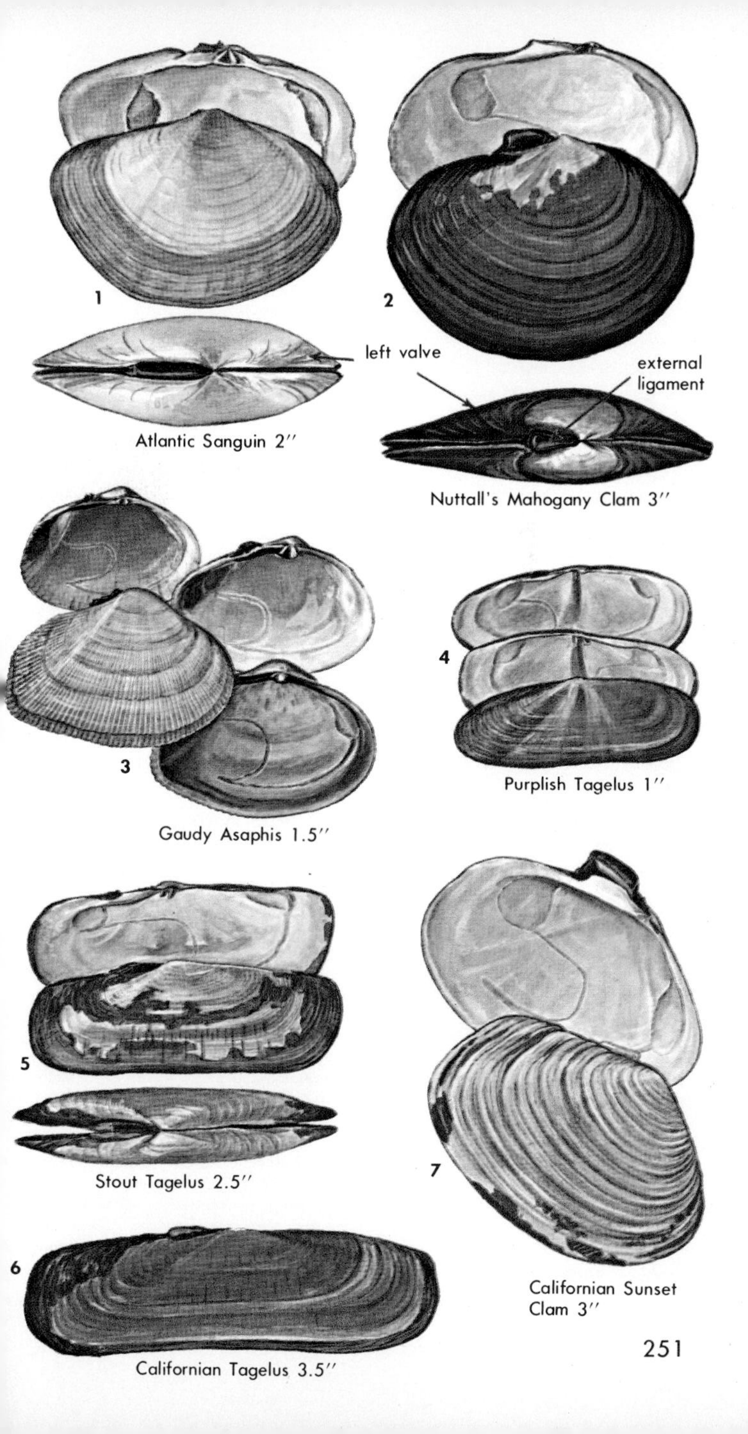

Atlantic Sanguin 2″

Nuttall's Mahogany Clam 3″

Purplish Tagelus 1″

Gaudy Asaphis 1.5″

Stout Tagelus 2.5″

Californian Tagelus 3.5″

Californian Sunset Clam 3″

JACKKNIFE AND RAZOR CLAMS

The family Solenidae, a familiar, worldwide group of long, greenish brown shells, is well represented in American waters. Usually the name "jackknife" refers to the genera *Ensis* and *Solen,* and the name "razor" to the broader forms of *Siliqua* and *Solecurtus.* Jackknife clams are dwellers of soft, muddy, intertidal flats in temperate seas. They are extremely agile, burrowing to safety very rapidly. They are also capable swimmers, darting about erratically with considerable speed. Care should be taken when digging them with bare hands because of the razor-sharp edges of the valves. All species are edible, and some *Siliqua* on the Pacific Coast are commercially fished. The Corrugated Razor clam (fig. 5) superficially resembles the true razor clams but belongs to the family Solecurtidae.

The shells of *Ensis* and *Solen* are remarkably alike, but in *Solen* there is only one cardinal tooth on each valve, while in *Ensis* there are two cardinal teeth in the left valve. *Ensis* in America are usually quite large and curved, often reaching a length of 7 or 8 inches. Table salt poured down the hole of a jackknife clam will sometimes stimulate the clam to jump out. Rapid propulsion is accomplished by the contraction of the shell valves and foot, which creates a powerful jet of water.

1. **ATLANTIC JACKKNIFE CLAM.** *Ensis directus* Conrad. Canada–S.C. Shell thin, gaping, moderately curved, with sharp edges; about 6 times longer than high and covered with a thin, glossy, olive to brownish periostracum. Very common; found in colonies in sandy mud near the low-water mark. A popular food.

Small Jackknife Clam, *E. minor* Dall, (not illus.), Fla.–Texas, is much smaller (shell 2.5 in.) and more fragile, about 9 times longer than high. Interior white with purple stains.

2. **CALIFORNIAN JACKKNIFE CLAM.** *Ensis myrae* S. S. Berry. S. Calif. This, the only *Ensis* in California, closely resembles the Atlantic Razor Clam, but is smaller. Uncommon; in sand, below low-tide mark. Formerly miscalled *californicus* Dall.

3. **GREEN JACKKNIFE CLAM.** *Solen viridis* Say. Rhode Island–N. Fla. and Texas. Shell thin, gaping. Ventral margin slightly curved. Periostracum thin, glossy, usually greenish. Fairly common; on intertidal sand bars.

4. **BLUNT JACKKNIFE CLAM.** *Solen sicarius* Gould. W. Canada–Baja Calif. Shell thin, moderately inflated, slightly curved. Periostracum thin, glossy, yellowish to olive green. Locally common; mud flats to 150 ft.

5. **CORRUGATED RAZOR CLAM.** *Solecurtus cumingianus* Dunker. N.C.–Fla. and Texas. Shell rectangular, gaping, with coarse growth lines and sharp, oblique threads. Uncommon; in sand or mud, 12–80 ft. (Solecurtidae).

6. **ATLANTIC RAZOR CLAM.** *Siliqua costata* Say. E. Canada–N.J. Shell somewhat oval, fragile, and smooth. Periostracum greenish, shining. Interior has strong, raised rib from hinge. Common; in sand, below low-tide mark.

7. **PACIFIC RAZOR CLAM.** *Siliqua patula* Dixon. Alaska–Calif. Shell fairly thin, compressed, with glossy, olive-green periostracum. Interior white, flushed with purple. Common; intertidally. A favorite food.

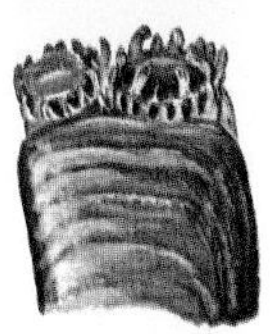

siphon detail

hinge detail

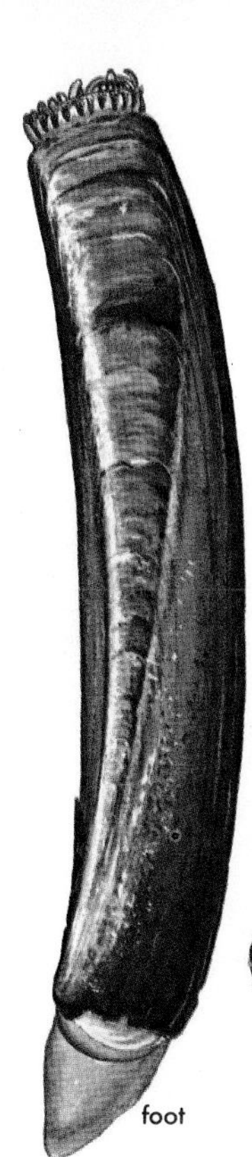

foot

1

Atlantic Jackknife Clam 6″

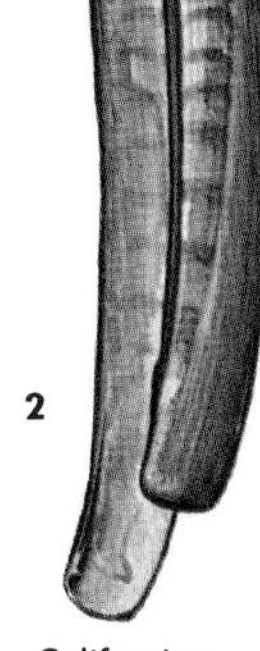

2

Californian Jackknife Clam 2″

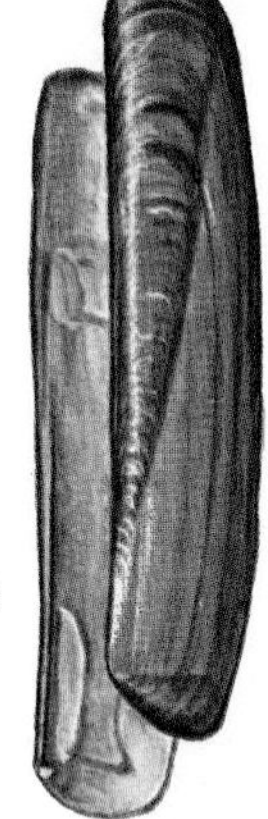

3

Green Jackknife Clam 2″

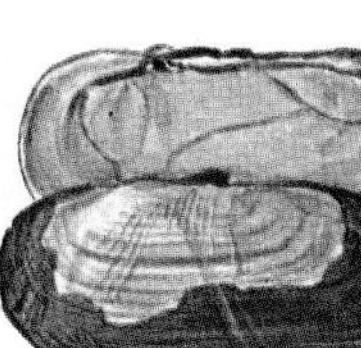

5

Corrugated Razor Clam 2″

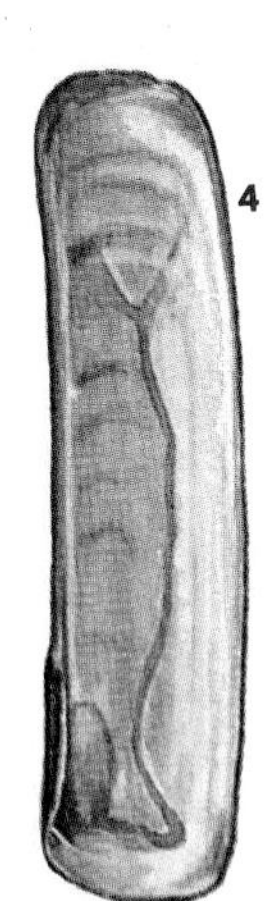

4

Blunt Jackknife Clam 3″

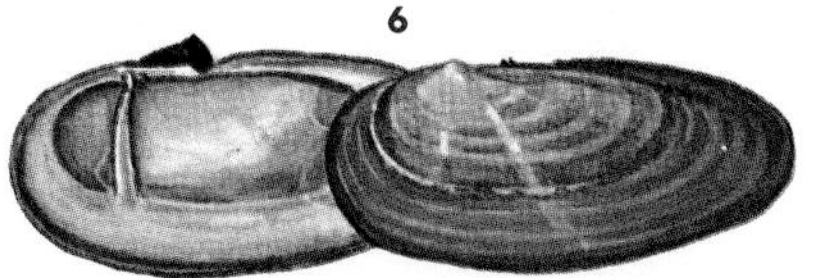

6

Atlantic Razor Clam 2″

7

Pacific Razor Clam 5″

MACTRA SURF CLAMS

Members of the family Mactridae are characterized by large, ovate shells, with a strong hinge bearing a large spoon-shaped depression, the chondrophore. This oval pocket holds a thick, horny pad, the resilium, which keeps the valves slightly ajar. The family is worldwide in distribution, with about two dozen American species. Many of the larger species are commercially important. Over 70 percent of all clams harvested in North America are the Atlantic Surf Clam from the New Jersey coast. Millions are dredged annually at depths from 30 to 80 feet. Gaper Clams, *Tresus,* of the Pacific Coast are often used as food. The heavy-shelled Rangia Clam of eastern brackish waters is used as a major source of road-bed material.

1. **ATLANTIC SURF CLAM.** *Spisula solidissima* Dillwyn. Nova Scotia–S.C. Shell cream-tan, strong, oval, quite smooth, with fine growth lines. Periostracum thin, yellowish tan. Pallial sinus slopes slightly upward. Abundant; in sand, from shore to 100 ft.

A southern subspecies, *raveneli* Conrad, is smaller but more elongate, and with a flatter anterior slope. The pallial sinus does not slope upward. Common. Misidentified as *similis* (Say).

2. **FRAGILE ATLANTIC MACTRA.** *Mactra fragilis* Gmelin. N.C.–Texas and Caribbean. Thin but strong. Periostracum silky gray. The back slope has 2 small ridges, one of which is very close to the dorsal margin of the valve.

3. **CALIFORNIAN MACTRA.** *Mactra californica* Conrad. Wash.–Panama. Shell has concentric undulations on both beaks. Locally common; intertidal in bays and lagoons.

4. **COMMON RANGIA.** *Rangia cuneata* (Gray). Va.–Texas. Shell very thick, heavy; exterior rough, white, with gray-brown periostracum. Interior smooth, off-white. Common; in brackish-water marshes.

5. **PACIFIC GAPER.** *Tresus nuttalli* (Conrad). Calif.–Baja Calif. Rather oblong. Common; buried deeply in intertidal mud. The subspecies *T. n. capax* (Gould), Calif.–Alaska, is much more oval and obese. Formerly in *Schizothaerus.*

6. **CHANNELED DUCK CLAM.** *Anatina plicatella* (Lamarck). N.C.–Texas; Caribbean. Shell has concentric ribs on the exterior. White, thin, fairly strong. Commonly washed ashore; 6–40 ft. Formerly in *Labiosa.*

7. **SMOOTH DUCK CLAM.** *Anatina anatina* (Spengler). N.C.–Texas. Shell cream, thin, strong, gaping, with distinct rib on back slope. Uncommon; 6–40 ft. Formerly *Labiosa lineata* Say.

Outlines of Pacific Coast Surf Clams *(Spisula)*

Hemphill's Surf Clam. S. Calif.–Pan. *Spisula hemphilli* (Dall). 4″

Hooked Surf Clam. Wash.–Calif. *Spisula falcata* (Gould). 3″

Catilliform Surf Clam. Wash.–Calif. *Spisula catilliformis* (Conrad). 5″

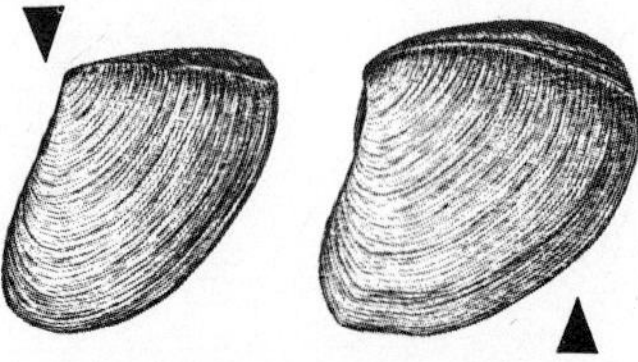

Hatchet Surf Clam. Calif.–Mexico. *Spisula dolabriformis* (Conrad). 5″

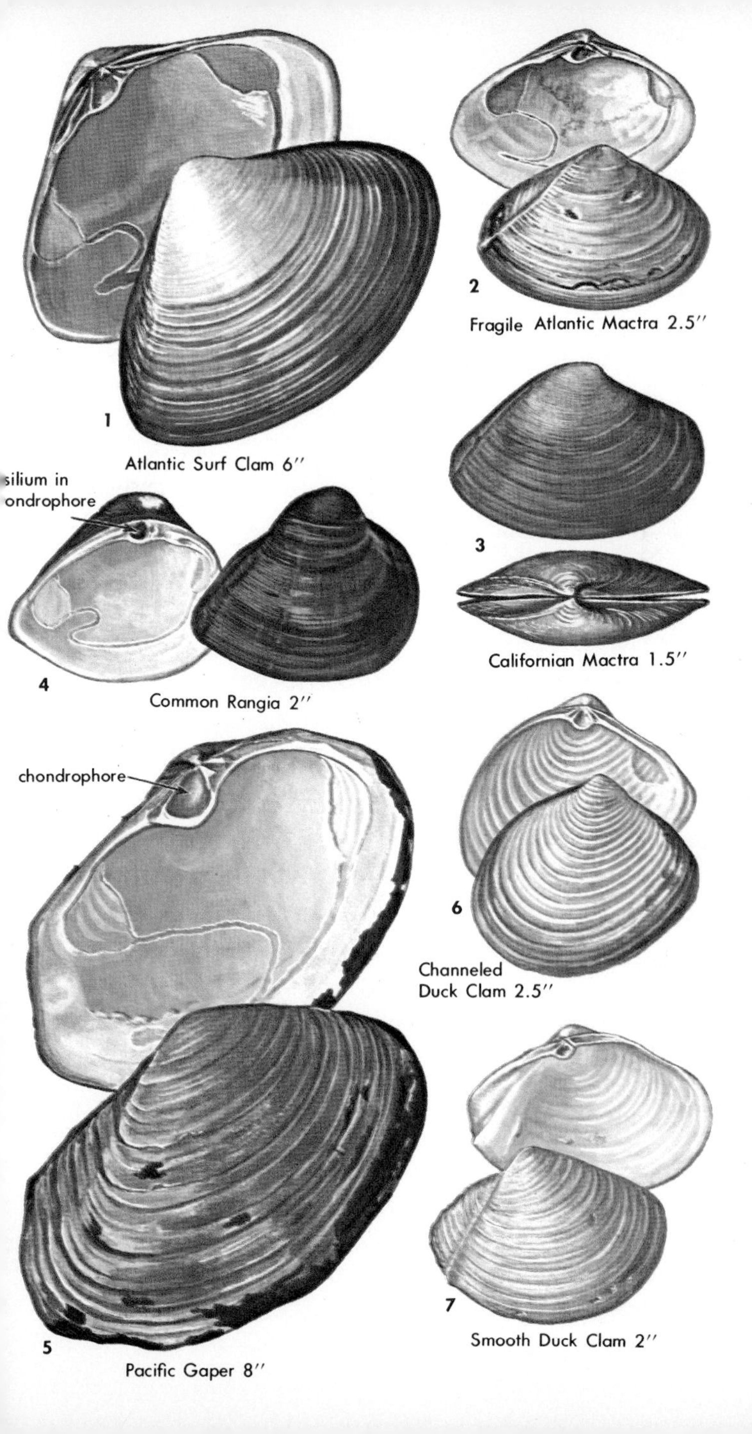
1
Atlantic Surf Clam 6″
2
Fragile Atlantic Mactra 2.5″
3
Californian Mactra 1.5″
silium in
ondrophore
4
Common Rangia 2″
chondrophore
5
Pacific Gaper 8″
6
Channeled
Duck Clam 2.5″
7
Smooth Duck Clam 2″

SOFT-SHELL CLAMS AND THEIR ALLIES

Several minor families of bivalves are represented in American waters by a few very common, and in some cases economically important, species. The family Myacidae includes the edible soft-shell clam of the eastern United States and the Chubby Mya of the Pacific Coast. The long retractible siphons are encased in a single, long tube. A host of small *Corbula* clams (family Corbulidae) are found mainly offshore. The Arctic Wedge Clam represents the family Mesodesmatidae and may be said to be the "coquina clam" of New England. It is common on some Atlantic northern beaches.

The diverse family Hiatellidae includes the small saxicave clams of cold waters and the huge *Panopea* geoducks (pronounced "gooee-duck"). The Pacific species is the largest American clam, with a 9-inch shell and siphons that may stretch to 2 feet. To conserve these giant, excellent eating clams, some states have a daily bag limit of three specimens.

1. **SOFT-SHELL CLAM.** *Mya arenaria* Linné. Also called "long-neck," "gaper," and "steamer" clam. Labrador–S.C. Introduced to Pacific Coast. Shell elliptical. Left valve flatter and has spoon-shaped chondrophore. Pallial sinus V-shaped. Common; in sandy mud, intertidally to 30 ft. Commercially fished; many are canned.

Truncate Soft-shell Clam, *Mya truncata* Linné (not illus.), Arctic Seas–Maine and also to Wash. State, has shell 1–3 in., widely gaping at truncate posterior end. Pallial sinus U-shaped. Common; clay and mud, 12–100 ft.

2. **DIETZ'S CORBULA.** *Corbula dietziana* C. B. Adams. N.C.–Caribbean. Right valve thicker and slightly larger than left. Resilium and ligament internal. Blushed with pink and rose, with brown rays. Has coarse concentric ridges. Common; in sand, 20–100 ft.

Contracted Corbula, *Corbula contracta* Say (not illus.), Mass.–Caribbean, is similar to Dietz's Corbula, but gray to yellow. Valves (0.3 in.) are same size and have many weak concentric ridges. Common; in sand, 3–50 ft.

3. **CHUBBY MYA.** *Platyodon cancellatus* (Conrad). W. Canada–Calif. Shell white or grayish, fairly thick, chalky, with well-defined, concentric growth lines; occasionally has weak radial grooves. Beak of left valve slightly overlaps that of right. Periostracum thin, yellowish to rusty brown. Fairly common; 12–200 ft.

4. **ARCTIC SAXICAVE.** *Hiatella arctica* (Linné). Arctic Seas–Caribbean; also Pacific Coast–off Panama. Adult shells variable, generally rectangular, chalky white, with irregular growth lines. Periostracum thin, gray, flaky. Common; intertidal to 1,200 ft., in crevices of rock or in sponges.

5. **ARCTIC WEDGE CLAM.** *Mesodesma arctatum* (Conrad). E. Canada–Va. Shell quite thick, strong, wedge-shaped, not unlike a *Donax*. Chondrophore prominent. Lateral teeth are strong and bear fine denticles on each side. Periostracum smooth. Common; in sand, 1–300 ft.

6. **PACIFIC GEODUCK.** *Panopea generosa* Gould. Alaska–Baja Calif. Shell dirty white to cream. Concentric sculpturing coarse and wavy. The fused siphons are large and cannot be withdrawn completely. Common; in 2 or 3 ft. of mud.

Atlantic Geoduck, *P. bitruncata* Conrad (not illus.), N.C.–Texas. 3–4 in.; similar to Pacific Geoduck. Uncommon; 1–50 ft.

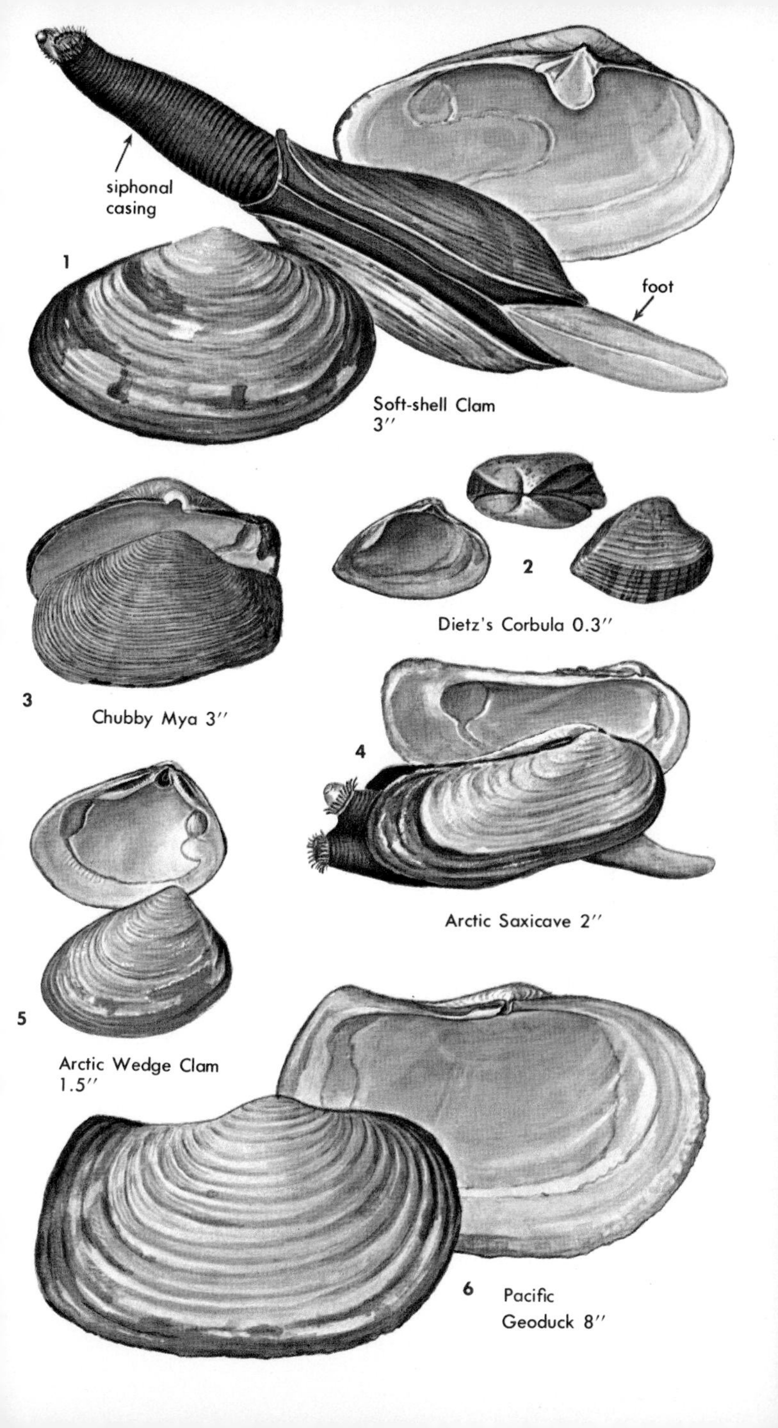
siphonal
casing
1
foot
Soft-shell Clam
3″
2
Dietz's Corbula 0.3″
3
Chubby Mya 3″
4
Arctic Saxicave 2″
5
Arctic Wedge Clam
1.5″
6
Pacific
Geoduck 8″

ANGEL WINGS AND PIDDOCKS

The boring clams of the family Pholadidae dig into soft mud, stiff clay, hard shells, and even shale and gneiss rocks. Many species are very destructive to cables, wood, and cement-covered pilings. Larger species are used as food. There are three dozen American species, most of them living along the Pacific Coast.

Boring is done by the rough anterior end of the shell, which is rocked back and forth by the adductor muscles and foot. Jets of water from the mantle cavity flush debris from the burrow. Angel wings and piddocks can move up and down their mud burrows, and feeding is done through the siphon, which protrudes from the burrow. In some species the sexes are separate, and the expelled eggs are fertilized by free-swimming sperm. Some species are hermaphrodites; some brood the early veliger stages within the mantle cavity.

Pholad clams have small shelly or chitinous plates, usually attached to the dorsal line or surrounding the siphon. Within the shell cavity, and tucked in under the beaks, are a pair of shelly projections, or apophyses, to which the muscles of the foot are attached. Of the thirteen genera found in American waters, *Cyrtopleura, Zirfaea,* and *Barnea* contain the largest and better-known species.

The popular Angel Wing is abundant in many muddy areas near mangroves along the west coast of Florida. They may live as deep as 3 feet in the mud. Digging should be done along the side, not directly on top, of the siphonal holes in order not to break the shell.

1. **ANGEL WING.** *Cyrtopleura costata* (Linné). N.J.–Brazil. Shell usually pure white, thin, moderately fragile, covered with a thin, gray, periostracum. Has 26 or more well-developed radial ribs, with blunt scales where they cross the concentric ridges. The apophyses, one in each valve, are short, wide, and spoon-shaped. Interior may be marked with chalky pink, from feeding on certain algae. Moderately common; in colonies in mud and clay, living about a foot below the surface. Edible. Over-collecting is making them scarce. A pink-stained ecologic form is rare.

2. **CAMPECHE ANGEL WING.** *Pholas campechiensis* (Gmelin). N.C.–Brazil. Also West Africa. Notable for the rolled-over plate, supported by about 12 braces, which covers the beaks. Do not confuse with False Angel Wing (p. 238). Uncommon; offshore. Rarely collected alive.

3. **FALLEN ANGEL WING.** *Barnea truncata* Say. Eastern U.S.–Brazil. Shell fragile, widely gaping at both ends, truncate posteriorly. Common; in intertidal clay.

Pacific Mud Piddock, *B. subtruncata* Sowerby (not illus.), Ore.–Chile. Shell is almost identical.

4. **GREAT PIDDOCK.** *Zirfaea crispata* (Linné). E. Canada–N.J.; Europe. A strong, indented radial line divides each valve into 2 distinct areas, the anterior portion with radial rows of raised scales. Fairly common; in mud and sand, from intertidal zone to 250 ft.

5. **FLAT-TIPPED PIDDOCK.** *Penitella penita* (Conrad). Arctic Seas–Baja Calif. Adults have two chitinous flaps that protect the siphon, and a shelly callus plate that closes the large anterior gape. Common; boring in hard clay, sandstone, and even cement.

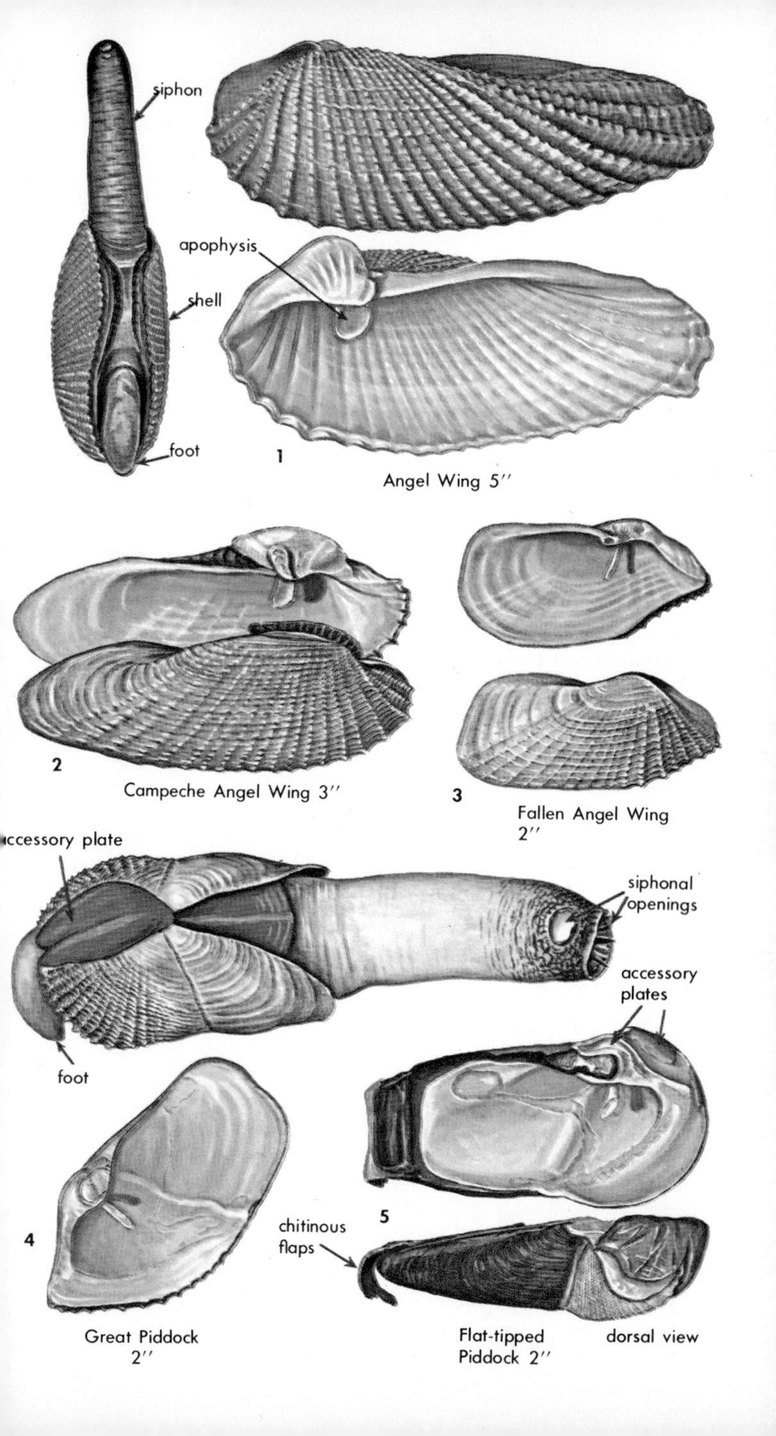
siphon
shell
foot
apophysis
1
Angel Wing 5″
2
Campeche Angel Wing 3″
3
Fallen Angel Wing
2″
ccessory plate
siphonal
openings
accessory
plates
foot
4
5
chitinous
flaps
Great Piddock
2″
Flat-tipped
Piddock 2″
dorsal view

MARTESIA WOOD BORERS

The small, chubby wood borers of the genus *Martesia* are the most destructive of the family Pholadidae. They occur throughout the world in warm waters and are frequently found in wood and large floating seeds. The Striate Martesia has found its way to most oceans of the world. Members of this genus have an oval accessory plate sitting on top of the beaks of the valve. Just posterior to this is a long, narrow accessory plate, and on the ventral side is still a third plate. Note the deep, funnel-shaped pit just in front of the rolled-in umbo region. Many species have been described.

1. **STRIATE MARTESIA.** *Martesia striata* (Linné). N.C.–Brazil; Baja Calif.–Peru. Shell variable, usually long and pear-shaped, with a shelly callus covering the front end. Juveniles are nearly circular and have no callus. Very common; in wood or soft stone. A very destructive borer.

2. **WEDGE-SHAPED MARTESIA.** *Martesia cuneiformis* Say. N.C.–Texas–Brazil. Shell smaller, fatter, and shorter than the Striate Martesia. Has a thick, oval accessory plate on top of the umbones which is centrally grooved. Very common; in wood. Rarely comes north in driftwood.

SHIPWORMS

Every year, throughout the world, the boring clams, of the family Teredinidae, take a heavy toll drilling rope, boat bottoms, and wooden wharf pilings. It has been estimated that the annual amount of damage caused by the various species of shipworms greatly exceeds the total income received from all mollusks taken for food and other purposes.

Both *Bankia* (fig. 3) and *Teredo* (fig. 4) are known as shipworms or pileworms. Identification by species requires special techniques. Scientists use the pallets found at the end of the worm-shaped clam. These pallets, which close off the end of the burrow, are made up of two parts, blade and stalk. In *Bankia* the blade is composed of several somewhat conical cups; in *Teredo* it consists of a single, paddle-shaped club.

There are about 14 genera of shipworms, each having special adaptations and habits. Most require normal ocean conditions in order to spawn, but some live in brackish waters, others at great depths, and most can withstand exposure to ice or fresh water for short periods. Some feed by filtering for algae, while others are also capable of "digesting" some wood. One group of shipworms sheds eggs freely into the ocean, while others keep the larvae in brood pouches. Shipworms settling on new wood can burrow in, grow to several inches in length, and completely riddle the interior within a few months.

Even the hardest of woods succumb to the borings of shipworms. When too many larval clams invade a piece of wood, over-population creates extremely stunted, or stenomorphic, individuals. Soft woods may be bored at the rate of about 4 inches in a month's time.

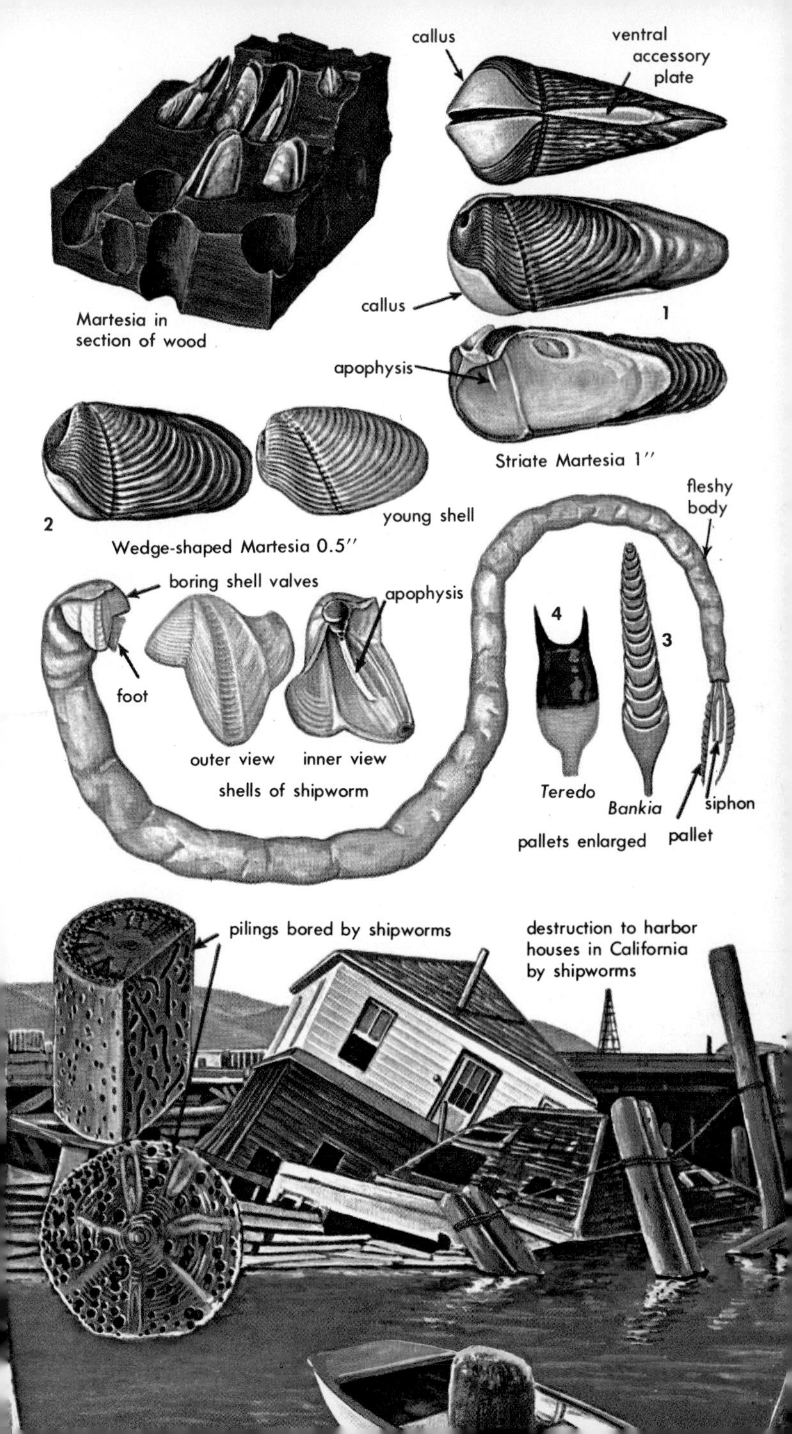
callus
ventral
accessory
plate
Martesia in
section of wood
callus
1
apophysis
Striate Martesia 1″
2
young shell
Wedge-shaped Martesia 0.5″
fleshy
body
boring shell valves
apophysis
4
3
foot
outer view
inner view
shells of shipworm
Teredo
Bankia
siphon
pallets enlarged
pallet
pilings bored by shipworms
destruction to harbor
houses in California
by shipworms

PANDORA AND SPOON CLAMS

The family Lyonsiidae has many American species of *Lyonsia*. The shell is thin, gaping, pearly, with a toothless hinge and a small, shelly ossicle, or loose "bone," up under the beaks. *Entodesma,* from the Pacific Coast, has a heavy periostracum.

The family Pandoridae, with flat, pearly shells, has over 20 American species of *Pandora*. Most live in cold waters.

The family Thraciidae has a dozen American cold-water species. One of the beaks is usually punctured by the other.

The family Periplomatidae has many pearly, thin-shelled species with an oblique, spoon-shaped chondrophore and a tiny, free ossicle.

1. **GLASSY LYONSIA.** *Lyonsia hyalina* Conrad. E. Canada–S.C. Shell very thin, fragile. Periostracum coated with sand grains. Common; in sandy mud, 1–100 ft.

2. **NORTHWEST UGLY CLAM.** *Entodesma saxicola* Baird. Alaska–Baja Calif. Oblong, misshapen, and fragile. Periostracum thick. Common; intertidally to 100 ft.

3. **GOULD'S PANDORA.** *Pandora gouldiana* Dall. Quebec–N.J. Shell whitish, chalky, often worn down to pearly underlayers. Common; in sand in rocky areas and in oyster beds.

4. **WAVY PACIFIC THRACIA.** *Cyathodonta undulata* (Conrad). Central Calif.–Mexico. Shell thin, fragile, with oblique undulations on anterior end of shell and with minutely granular surface. Uncommon; down to 240 ft.

5. **CONRAD'S THRACIA.** *Thracia conradi* Couthouy. E. Canada–N.Y. Hinge without teeth but thickened posterior to the beaks. Right valve has hole punctured by left beak. Fairly common; in mud, 6–200 ft.

Common Pacific Thracia, *Thracia trapezoides* Conrad (not illus.), Alaska–Calif. 2 in. long and somewhat resembles Conrad's Thracia, but much more elongate. Has an oblique furrow and ridge from beak to hind end. Hole in right beak. Grayish. Common; in sand, 12–100 ft.

6. **COMMON WESTERN SPOON CLAM.** *Periploma planiusculum* Sowerby. S. Calif.–Peru. Shell thin, with weak, concentric growth lines. Right valve fatter than left. Chondrophore directed forward. Common; 6–120 ft.

Unequal Spoon Clam, *Periploma inequale* C. B. Adams (not illus.), S.C.–Texas. Almost identical to the Common Western Spoon Clam, but slightly more elongate. Left valve fatter. Beaks have minute slit. Common; in sand, 6–50 ft.

7. **ROUND SPOON CLAM.** *Periploma discus* Stearns. Calif.–Baja Calif. Shell thin, fragile, almost circular; translucent white. Chondrophore quite prominent. Uncommon; in mud, 6–36 ft., but sometimes washed ashore.

DIPPER CLAMS

The strange family of Cuspidariidae, in the order Septibranchia, has over 50 American, mainly deep-water, species. The small, fragile shells lack hinge teeth, and the front end is produced into a "dipper handle."

8. **GLACIAL DIPPER CLAM.** *Cuspidaria glacialis* Sars. E. Canada–Md.; Alaska. Valves fat, rounded, and with a rather long, compressed "snout" at front end. Concentric growth lines coarse. Shell white, but covered with thin, gray periostracum. Common; in sand, 50–8,000 ft.

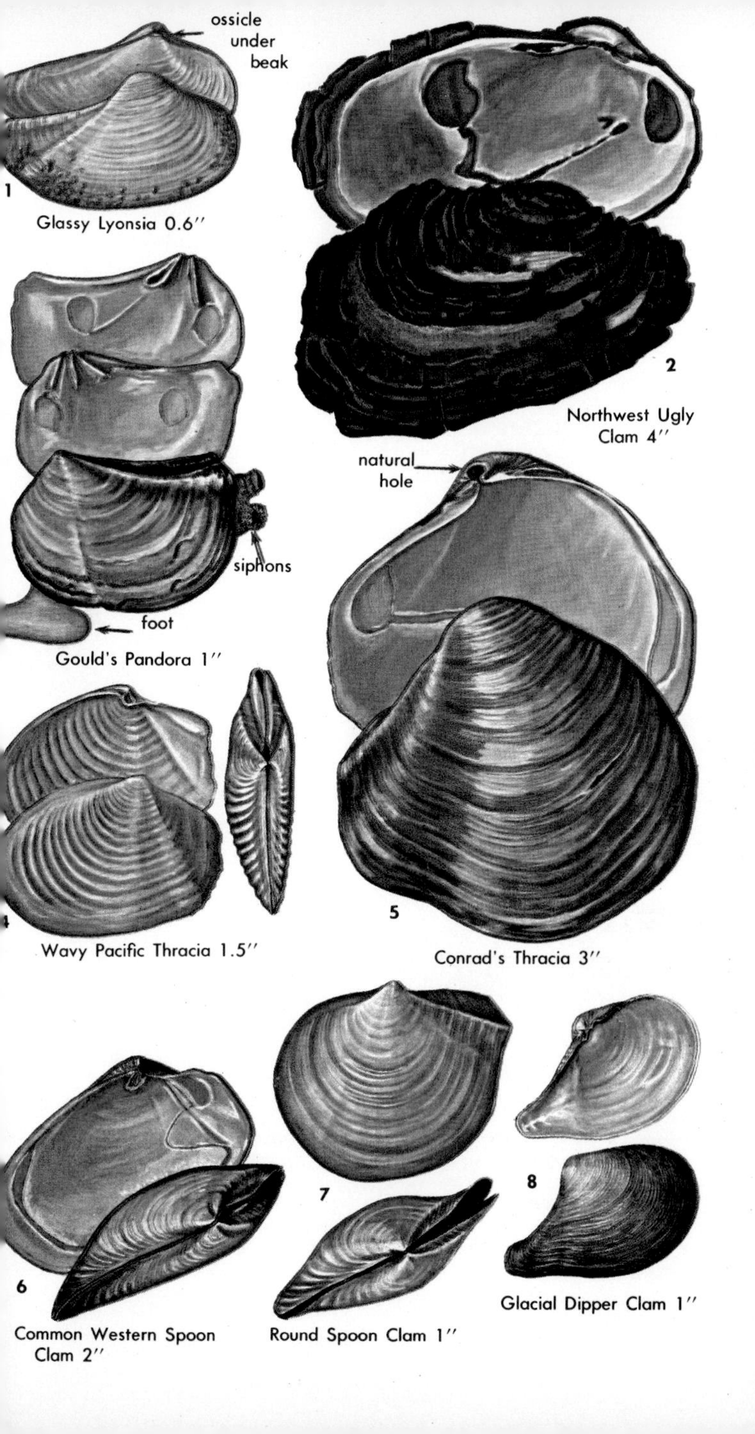
ossicle
under
beak
1
Glassy Lyonsia 0.6″
2
Northwest Ugly
Clam 4″
siphons
foot
Gould's Pandora 1″
natural
hole
5
Wavy Pacific Thracia 1.5″
Conrad's Thracia 3″
7
8
6
Glacial Dipper Clam 1″
Common Western Spoon
Clam 2″
Round Spoon Clam 1″

CLASS CEPHALOPODA (Octopus, squids)

This class includes the most active and highly evolved of the mollusks. They are characterized by having eight or more arms surrounding the mouth. The cephalopods were once a dominant group during the Mesozoic era, some 200 million years ago. Today there are only about 650 species, including the squids, the nautiluses of the southwest Pacific, and the octopuses. Except for the pearly nautilus, all have two gills, two kidneys, and three hearts—a main one to pump the blood through the body and two smaller branchial ones to force the blood through the gill capillaries. The blood is blue in color because of the oxygen-carrying haemocyanin. All species are believed to be carnivorous and most have a strong set of horny beaks and a ribbon bearing the strong radular teeth. Cephalopods squirt ink into the water to confuse attacking fish.

There are about 150 species of *Octopus,* with a dozen or so species in American waters. All have eight arms, a beak, a radula, but no internal shell. Largest is the Common Pacific Octopus, which is found along the entire Pacific Coast. In Alaska its arms may have a radial spread of 32 feet. Most octopuses are relatively harmless, although deaths from bites and from encounters have been authenticated in Australia and Europe. American species also have a painful bite. The Common Atlantic Octopus reaches a weight of about 20, rarely, 40, pounds. They live in lairs in shallow water and feed mainly upon crabs, snails, and clams. The sexes are separate. The elongate eggs are ⅛ inch and are laid in long clusters. The female guards the eggs, keeping them clean and aerated with gentle caressing, until they hatch in about ten weeks. The octopus can change its color at will.

The arms of the octopus vary somewhat in size, depending upon the species. They are used for crawling and for grasping objects. In the males, one of them, the hectocotylus arm, has a small, characteristic pad, or *ligula,* at the tip, which is used in transferring sperm to the female.

1. **UMBRELLA OCTOPUS.** *Tremoctopus violaceus* (Delle Chiaje). Worldwide; pelagic. Arms 3–6 ft. in length, with skin webs between 4 of the arms. Skin cape has a short to deep embayment between the 2 middle arms. Color variable, ranging from light grayish silver to dark purplish. This species moves about in schools and feeds upon the Portuguese Man-O-War "jellyfish," using the stinging strands at a later time to assist in killing its fish prey. Young brooded at base of arms. Uncommonly seen in the beach surf of Florida.

2. **COMMON ATLANTIC OCTOPUS.** *Octopus vulgaris* Lamarck. Conn.–Caribbean; Europe. Color changeable. Skin smoothish to warty and rough. The arms, reaching a length of 3 feet, are 4 times as long as the mantle. Common; 3–30 ft., among large rocks.

Common Pacific Octopus, *Octopus hongkongensis* Hoyle (not illus.), Alaska–Baja Calif.; also east Asia. Color variable, but generally a deep reddish brown. Arm length usually 2–3 ft., but in Alaska may be 14 ft. Common; shore to 600 ft.

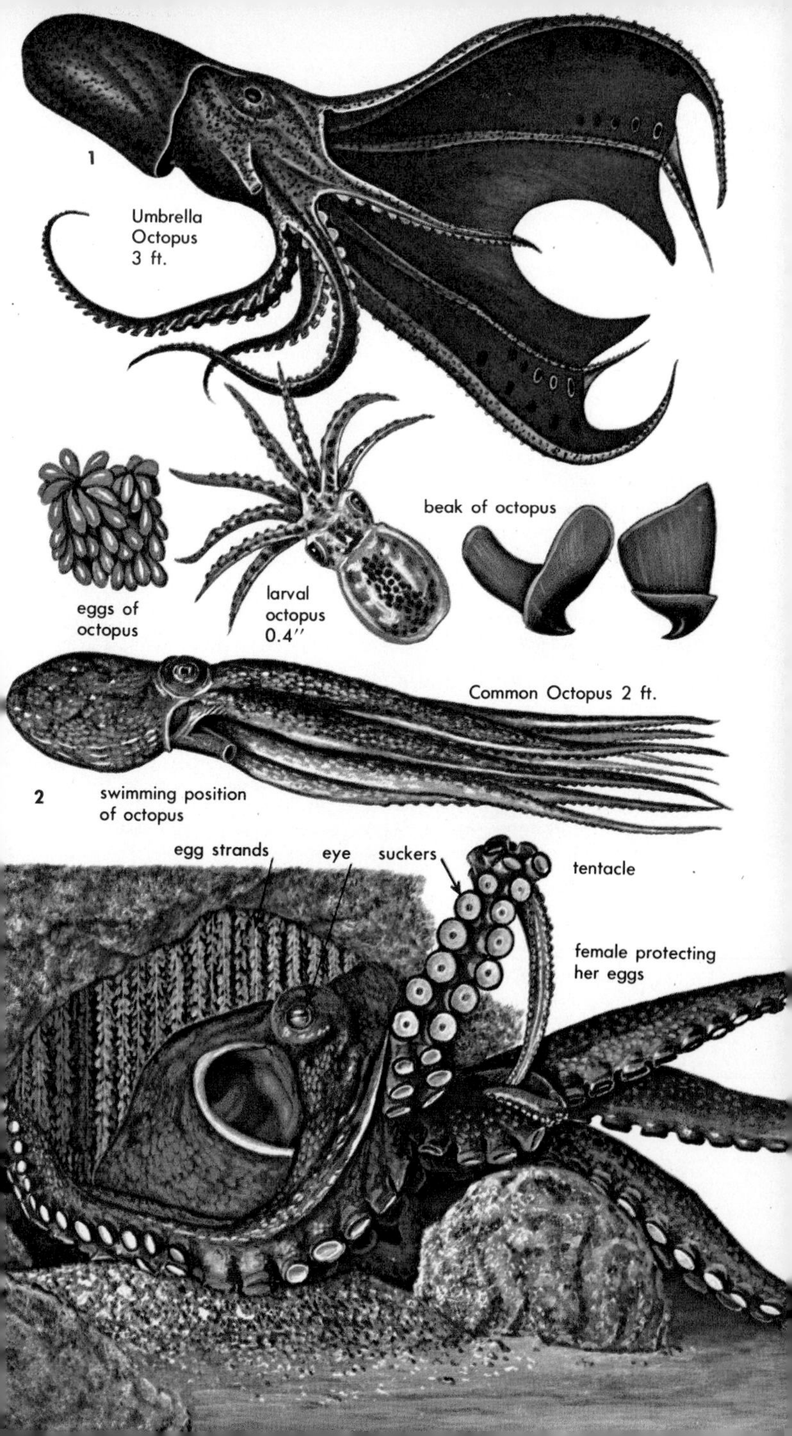
1
Umbrella
Octopus
3 ft.
eggs of
octopus
larval
octopus
0.4″
beak of octopus
Common Octopus 2 ft.
2
swimming position
of octopus
egg strands
eye
suckers
tentacle
female protecting
her eggs

ARGONAUTS (Paper Nautilus)

There are six worldwide, pelagic species of Paper Nautilus, the commonest being *Argonauta argo* Linné, which is found in most tropical and temperate seas. The family Argonautidae is very closely related to the octopuses. The female argonaut has two highly modified arms that secrete a pearl-white, parchment-like egg case into which she places hundreds of tiny eggs. While the female may grow to a length of 2 feet and produce a "shell" 14 inches long, the male is rarely over half an inch in size. The argonauts swim near the surface of the water, and during storms they are occasionally washed ashore. They eat pteropod mollusks and small pelagic fish. Large sailfish prey upon them.

The thimble-sized males live in the vicinity of the larger female and at times rest inside the "shell." One of the arms, the hectocotylus, is contained in a sac that later ruptures and permits the sperm-charged arm to wiggle forth. It is usually ten times as long as the male itself, has numerous small suckers, and a long, threadlike male organ. The arm breaks off and continues to wiggle and use its suckers independently for many hours, until it has entered the mantle cavity of the female. Until 1853, this tiny arm found inside the female was thought to be a parasitic worm, and, indeed, was given the name *Hectocotylus*.

1. **COMMON PAPER NAUTILUS.** *Argonauta argo* Linné. Warm worldwide seas. "Shell" laterally compressed, the keel being very narrow and bearing many sharp nodules. Parchment white in color, with a purplish stain in the early part. Empty "shells" washed ashore from Texas to N.J., but most frequently in Fla.

2. **BROWN PAPER NAUTILUS.** *Argonauta hians* Lightfoot. Warm worldwide seas. "Shell" fragile, tan to light brown. Keel broad and bounded by 2 rows of rather large, dark brown nodules, usually larger at the end of the "shell." Infrequently cast ashore after storms, especially Fla. and S. Calif.

SPIRULA

3. Although the inch-sized, white, spiral shell of *Spirula spirula* (Linné) is commonly found on the beach in many parts of the world, especially the West Indies, the squid animal producing it is very rarely captured. Normally, this 1½-inch-long squid lives at depths of from 600 to 3,000 feet. When disturbed, the *Spirula* withdraws its head, eight arms, and two tentacles within the cylinder-shaped mantle cavity. The animal usually hovers in a vertical position with the tentacles hanging downward. At the top end there is a small luminescent disc that continuously gives off a yellowish green light, evidently to enable a school to keep together. The shell has 25 to 37 gas- and air-filled chambers, with a delicate tube or siphuncle running through them. When the squid animal dies, the body rots away and the shell floats to the surface of the ocean. The Common Spirula belongs to the order Decapoda.

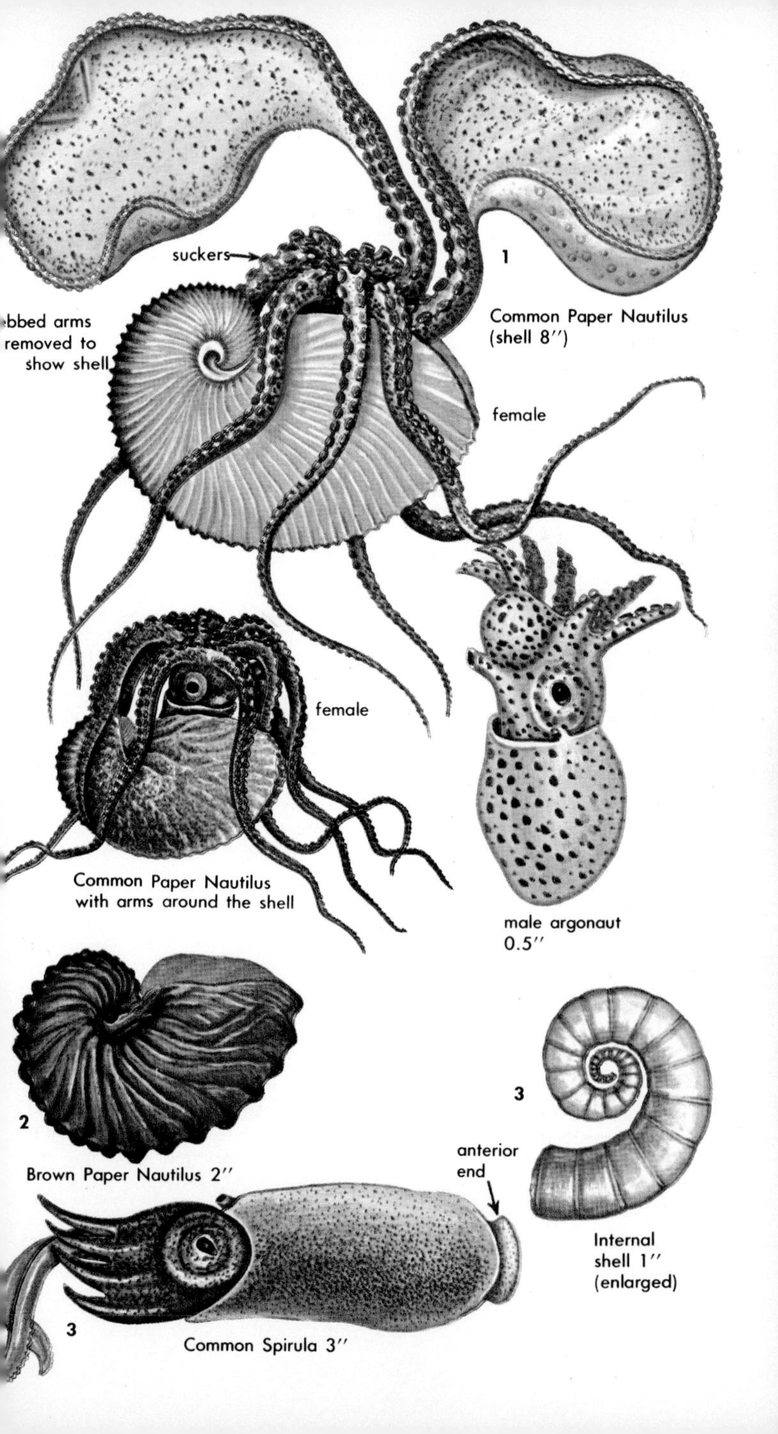
suckers
1
Common Paper Nautilus
(shell 8″)
ebbed arms
removed to
show shell
female
female
Common Paper Nautilus
with arms around the shell
male argonaut
0.5″
2
Brown Paper Nautilus 2″
3
anterior
end
Internal
shell 1″
(enlarged)
3
Common Spirula 3″

SQUIDS

The order Decapoda includes the approximately 350 species of squids, most of which have a cylindrical body, eight arms, and two long, armlike tentacles. In size, they vary from the inch-long Sandalops Squid to the 60-foot *Architeuthis* of the North Atlantic. Although many live in shallow coastal water or as pelagic species on the open seas, a number of them live at great depths and have highly colored luminescent organs. The American squids, of which there are about 60 species, have a thin, transparent, stiff, internal pen. The Cuttlefish, *Sepia,* of the Old World and Asia has a chalky internal "bone." Brown ink was obtained from the *Sepia* squid by the ancients of the Mediterranean world.

Squids are very active, probably reaching a swimming speed of 30 miles an hour. They feed on small fish and smaller squids, using their strong, parrot-like beaks to tear chunks of flesh from their prey. Some species will hydroplane across the surface of the water much in the manner of flying fish. On rare occasions they land on the decks of ships.

Females are fertilized by sperm cells contained in tubular, saclike spermatophores, which are transferred to the inside of the female by the hectocotylus arm of the male. The eggs of *Loligo* are only 1/10 inch long and are laid in jelly strands that are attached to some hard object on the sea bottom. At times, many acres of sea bottom may be covered with squid egg masses.

Of the many families of squid in America, two contain the common species that are used extensively for food and fish bait. The family Loliginidae contains the Pacific and Atlantic Long-finned squids, as well as the Brief Squid, *Lolliguncula.* They have large eyes. The family Ommastrephidae, with small eyes, contains the Common Short-finned Squid, *Illex.* These relatively small squid travel in schools up and down the coast in search of small fish. They are fished commercially.

1. **ATLANTIC LONG–FINNED SQUID.** *Loligo pealei* Lesueur. E. Canada–Caribbean. Adults are 2–3 ft. overall. The long triangular fins at front end, about half the length of the body, are characteristic. Color variable, speckled with purplish red and pink. Eyes large. Abundant; in shallow waters. Used as fish bait.

Common Pacific Squid, *Loligo opalescens* Berry (not illus.), Wash.–S. Calif. 1.5–2 ft. long. Very similar to the Atlantic Long-finned Squid, but fins are slightly shorter. Color variable, often with an iridescent sheen. Abundant. Fished commercially for foreign export.

2. **BRIEF SQUID.** *Lolliguncula brevis* (Blainville). N.J.–Brazil. Rarely exceeds 9 in. overall. Fins short and oval. Eyes large. Upper arms short. Brown-spotted, with underside of fins white. Common; shallow waters, especially in warm waters.

3. **COMMON SHORT–FINNED SQUID.** *Illex illecebrosus* (Lesueur). Greenland–Gulf of Mexico. 1–15 ft. long. Eye openings small and with notch at front edge. Fins only ⅓ the length of body. Color very variable. Abundant; especially in summer in New England in shallow water. Used as fish bait.

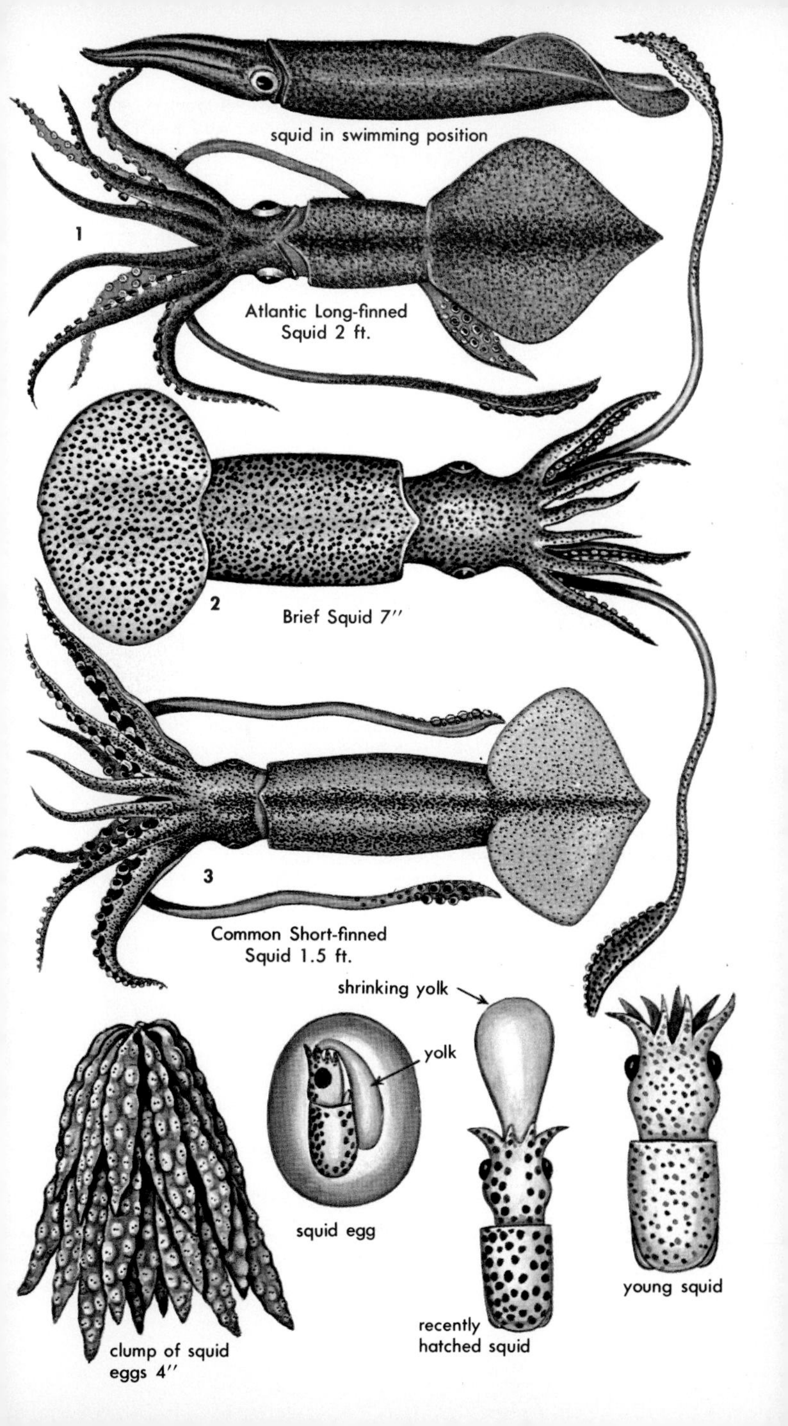

squid in swimming position
1
Atlantic Long-finned
Squid 2 ft.
2
Brief Squid 7″
3
Common Short-finned
Squid 1.5 ft.
shrinking yolk
yolk
squid egg
young squid
recently
hatched squid
clump of squid
eggs 4″

BIBLIOGRAPHY

Many thousands of interesting research articles on mollusks are listed by subject matter and by author in BIOLOGICAL ABSTRACTS and the ZOOLOGICAL RECORD (section on Mollusca). A list of worldwide popular books may be obtained by writing to the Division of Mollusks, U.S. National Museum, Washington, D.C. 20560. Below is a selected list of books and periodicals. A recording of the correct pronunciations of the scientific names is this book (12-inch, 33⅓ RPM, high fidelity) by R. T. Abbott, 1969, is available from the Delaware Museum of Natural History, Greenville, Delaware 19807.

GENERAL BOOKS ON MOLLUSKS

Abbott, R. Tucker. *Introducing Seashells.* D. Van Nostrand, Princeton, N.J. 1955.

Abbott, R. Tucker. *Sea Shells of the World,* a Golden Nature Guide. Golden Press, N.Y. 1962.

American Malacological Union. *How to Collect Shells.* A symposium by many experts. A.M.U., Box 318, Route 2, Marinette, Wisconsin. 1966.

Arnold, Winifred H. *A Glossary of a Thousand-and-one Terms Used in Conchology.* Veliger, supplement to vol. 7, 1965.

Dance, S. Peter. *Shell Collecting, an illustrated history.* Univ. Calif. Press, Berkeley. 1966.

Fretter, V., and A. Graham. *British Prosobranch Mollusks.* Ray Society, London. 1962.

Hyman, Libbie H. *The Invertebrates,* vol. 6, *Mollusca.* McGraw-Hill, N.Y. 1967.

Johnstone, Kathleen Y. *Sea Treasure, a Guide to Shell Collecting.* Houghton Mifflin, Boston. 1957.

Lane, Frank W. *Kingdom of the Octopus.* Sheridan House, N.Y. 1960.

Moore, R. C. (editor). *Treatise on Invertebrate Paleontology.* Part I, Mollusca, vol. 1, pp. 1–351. Geol. Soc. America, N.Y. 1960.

Morton, J. E. *Molluscs, an Introduction to Their Form and Functions.* Harper Torchbooks, no. TB 529, N.Y. 1960.

Pelseneer, Paul. *Mollusca. A Treatise on Zoology,* vol. 5, Asher and Co., Amsterdam (reprint 1964). 1906.

Petit, R. E. *Directory of Conchologists.* P.O. Box 133, Ocean Drive Beach, South Carolina. 1968.

Wagner, R. J. L. and **R. T. Abbott** (editors). *Van Nostrand's Standard Catalog of Shells* (2nd edition). D. Van Nostrand, Princeton, N.J. 1967.

Wilbur, K. M. and **C. M. Yonge.** *Physiology of Mollusca. Academic Press, N.Y. Vol. 1, 1964; Vol. 2, 1966.*

BOOKS ON ATLANTIC COAST MOLLUSKS

Abbott, R. Tucker. *American Seashells.* D. Van Nostrand, Princeton, N.J. 1955.

Abbott, R. Tucker. *How to Know the American Marine Shells.* Signet Key paperback, N.Y. 1961.

Bousfield, E. L. *Canadian Atlantic Sea Shells.* National Museum of Canada, Ottawa. 1960.

Jacobson, M. K. and **W. K. Emerson.** *Shells of the New York City Area.* Argonaut, Rye, N.Y. 1961.

Johnson, C. W. *List of Marine Mollusca of the Atlantic Coast from Labrador to Texas.* Proc. Boston Soc. Nat. Hist. 1934.
La Rocque, A. *Catalogue of the Recent Mollusca of Canada.* Bull. 129, National Museum of Canada, Ottawa. 1953.
McLean, James H. *Marine Shells of Southern California.* Los Angeles County Museum of Natural History. 1969.
Morris, Percy A. *A Field Guide to the Shells of Our Atlantic and Gulf Coasts.* Houghton Mifflin, Boston. 1951.
Olsson, A. A. and **Anne Harbison.** *Pliocene Mollusca of Southern Florida.* Academy Natural Sciences, Phila., Pa. 1953.
Perry, L. M. and **J. S. Schwengel.** *Marine Shells of the Western Coast of Florida.* Paleontological Research Institution, Ithaca, N.Y., 1955.
Warmke, G. L. and **R. T. Abbott.** *Caribbean Seashells.* 1961, Livingston, Narberth, Pa. 1961.

BOOKS ON PACIFIC COAST MOLLUSKS

Abbott, R. Tucker. *American Seashells.* D. Van Nostrand, Princeton, N.J. 1955.
Keen, A. Myra. *An Abridged Check List and Bibliography of West North American Marine Mollusca.* Stanford Univ. Press, Calif. 1937.
Keen, A. Myra and **E. Coan.** *Marine Molluscan Genera of Western North America.* Stanford Univ. Press, Calif. 1963.
Keen, A. Myra. *Sea Shells of Tropical West America.* Stanford Univ. Press, Calif. 1958.
Keep, J. and **J. L. Baily, Jr.** *West Coast Shells.* Stanford Univ. Press, Calif. 1935.
La Rocque, A. *Catalogue of the Recent Mollusca of Canada.* Bull. 129, National Museum of Canada, Ottawa. 1953.
Morris, Percy A. *A Field Guide to Shells of the Pacific Coast and Hawaii.* Houghton Mifflin, Boston. 1952.
Ricketts, E. F. and **J. Calvin.** *Between Pacific Tides* (3rd edition). Stanford Univ. Press, Calif. 1952.

PERIODICALS

In addition to the leading journals listed below, there are many interesting and useful shell-club publications, such as the New York Shell News. An up-to-date list of shell clubs appears in the latest edition of *Van Nostrand's Standard Catalog of Shells.*

Hawaiian Shell News. A popular monthly devoted to Pacific Ocean shells. Write: 2777 Kalakaua Ave., Honolulu, Hawaii.
Johnsonia. Monographs of the Marine Mollusca of the Western Atlantic. Many groups monographed: Volutidae, Epitoniidae, Conidae, Melongena, Littorina, Cassis, Pinna, Pholadidae, Cardiidae, Cymatiidae, Murex, Calliostoma, etc. Write: Dept. Mollusks, Museum of Comparative Zoology, Cambridge, Mass.
Malacologia. An international journal of malacology (technical). Write: Museum of Zoology, Univ. Michigan, Ann Arbor, Mich.
The Nautilus. A quarterly devoted to the interests of conchologists (technical). Write: 11 Chelten Road, Havertown, Pa. 19083.
The Veliger. A quarterly devoted to all aspects of mollusks from the Pacific region. Write: 12719 San Vicente Boulevard, Los Angeles, Calif. 90049.

INDEX

Individual species names, both common and scientific, are indicated with the text page (even). Most species are illustrated on the facing page (odd).

MEASURING SCALE (IN MILLIMETERS AND CENTIMETERS)

1 2 3 4 5 6 7 8 9 10 11 12 13 14 15

MEASURING SCALE (IN 10THS OF AN INCH)